City Competitiveness and Improving Urban Subsystems:

Technologies and Applications

Melih Bulu
Istanbul Sehir University, Turkey

Information Science
REFERENCE

Managing Director:	Lindsay Johnston
Book Production Manager:	Sean Woznicki
Development Manager:	Joel Gamon
Development Editor:	Hannah Abelbeck
Acquisitions Editor:	Erika Carter
Typesetter:	Milan Vracarich, Jr.
Print Coordinator:	Jamie Snavely
Cover Design:	Nick Newcomer

Published in the United States of America by
Information Science Reference (an imprint of IGI Global)
701 E. Chocolate Avenue
Hershey PA 17033
Tel: 717-533-8845
Fax: 717-533-8661
E-mail: cust@igi-global.com
Web site: http://www.igi-global.com

Library of Congress Cataloging-in-Publication Data

City competitiveness and improving urban subsystems: technologies and applications / Melih Bulu, editor.
p. cm.
Includes bibliographical references and index.
Summary: "This book uses information science perspectives to improve working subsystems in transportation, sewage, electricity, water, communication, education, health, governance, and infrastructure since their efficient and synchronized operation is vital for a competitive city"--Provided by publisher.
ISBN 978-1-61350-174-0 (hbk.) -- ISBN 978-1-61350-175-7 (ebook) 1. Municipal services--Information technology. 2. Infrastructure (Economics) 3. Information technology--Economic aspects. 4. Urban economics. 5. City planning. I. Bulu, Melih, 1970-
HD4431.C48 2012
363.60285--dc22

2011013919

British Cataloguing in Publication Data
A Cataloguing in Publication record for this book is available from the British Library.

Editorial Advisory Board

List of Reviewers

Table of Contents

Foreword

THE ROLE OF SUSTAINABLE URBAN AND INFRASTRUCTURE DEVELOPMENT IN AUGMENTING CITY COMPETITIVENESS

This book, entitled "*City Competitiveness and Improving Urban Subsystems: Technologies and Applications,*" explores a range of interests and activities concerning city competitiveness and urban systems from distinct disciplinary perspectives, including research from experts focusing on technology, planning, governance, and engineering. The primary focus of the book is one of the most important topics of our time, city competitiveness and urban systems. This book more specifically concentrates on the interconnectedness of social, economic, environmental and governing issues.

The book makes an important contribution to city competitiveness literature, highlighting the role of urban systems and bringing together the recent research findings from many disciplines in a single edited collection. This book is an invaluable resource as it highlights alternatives and approaches in urban systems technology and applications for not only academics, but also city administrators and the urban systems development industry. The book also provides useful directions for policy-makers to set sensible strategies for building prosperous and competitive cities for their citizens.

I was honoured when asked to write the foreword of this book and here below are some thoughts and insights from my academic, public sector, and professional practice experiences on the very topic of the book—urban competitiveness and urban systems.

CITIES AND COMPETITIVENESS

Cities in the course of history have always been the centres of civilisation and major economic activities. Throughout history, knowledge and creativity have been vital resources for building and sustaining a strong economy, society, and culture (Van Doren, 1992). Cities originally emerged as places of exchange of goods, nevertheless, production of these goods and the establishment of these cities relied heavily on knowledge, and today many of our modern cities are specialised havens for the production and exchange of knowledge, networks, and material goods (Yigitcanlar, 2010c). Furthermore, during the last few decades cities have become critical platforms for shaping and leveraging human capital into a collective intellectual capital, which is one of the main triggers of knowledge production, innovation, and competitiveness (Edvinsson, 2010).

The emergence of the first major cities and civilisation occurred in six widely separated places around the world dating from around 6,000 years ago: Mesopotamia, India, Egypt, China, Central America and Peru, and cities have been in competition with each other since then. The agricultural revolution turned the ancient cities of Sumerian's Babylon, Egyptian's Cairo, Hittite's Hattusa, Greek's Athens and many others into melting pots of cultures and centres of technical and scientific innovation, creativity, trade, and finance. The industrial revolution accelerated the quantity of mass production and the speed of the physical and economic development of our cities in the course of the era of industrialisation (Mumford, 1968).

During the last three decades the shift from muscle-power based production to brain-power based production, along with rapid advancements in the information, communication, and transportation technologies, have initiated the revolution of the early 21st century, the so called information or knowledge revolution. This era of knowledge has brought numerous and tremendous changes in our lives and in the mechanisms that our cities function (Van Winden et al., 2007). The broad areas that have been impacted from the knowledge revolution are grouped under four major areas – economy, society, environment, and governance (Yigitcanlar, 2010c).

In history, the idea of globalisation-that was first put forward in action by Alexander the Great, the Macedonian Emperor, and then by Roman, British, and American imperialism, respectively-has been finally realised at the end of the 20th century. As a result of a rapidly globalising world and the move from an industrial to a knowledge era has impacted the economy and resulted in a shift from neo-traditional economic views and practices to knowledge-based economic ones (Metaxiotis et al., 2010).

The change in the economy did not only occur at the national level, but also had reflections in the local, regional, supra-national, and global economic levels. The new economy, the knowledge economy removed boundaries and provided opportunities for distribution and movement of the talent, investment, and finance across the globe, opportunities which have significantly changed the lifestyles of many residents of the world's global cities. Societies in these cities have become more diverse in terms of cultural and habitual perspectives and preferences (Durmaz et al., 2010). Although at times important social conflicts rose, tolerance to others has become the motto of these global cities in order to establish and maintain a cohesive society.

In this era of knowledge, major metropolises continue to attract more and more talented and lay populations; for the first time, in early 21st century the urban population figure exceeded the 50 percent mark. Although this rapid urbanisation helped local economies to flourish, due to increased local consumption and the service sector's rising role in the economy, both developed and developing countries have suffered from the environmental implications of this not so well planned sudden urban growth (Yigitcanlar and Dur, 2010).

Especially following the fear of peak-oil, global warming, climate change, and scarcity of agricultural lands to feed the world, local, regional, national, supra-national, and international organisations gave priority to properly deal with the negative impacts of rapid urbanisation and to target a development that is sustainable. This required a change of mind regarding the ill practices of the era of industrialisation, a restructuring of some of the institutions that oversee the governance of socio-economic and enviro-urban development, and the achievement of a triple-bottom-line sustainability (Teriman et al., 2009).

This need for restructuring gave birth to a new city concept of knowledge cities, and a new urban development paradigm of knowledge-based urban development (Yigitcanlar, 2010d). In the era of knowledge, that is already upon us, knowledge-based urban development is seen as a new form of development to bring not only economic prosperity, but also environmental sustainability with a just socio-spatial order in cities, and aims to produce a knowledge city purposefully designed to encourage the production, exchange, and commercialisation of knowledge in an environmentally conserved, economically secure, and socially just human setting (Yigitcanlar, 2010c).

Knowledge cities are globally highly competitive cities as city competitiveness is basically determined by the ease with which value-creating activities or knowledge-based activities can take place (Yigitcanlar, 2009). Successful knowledge cities attract talented young highly-skilled knowledge workers, are centres of innovation and entrepreneurship, and are competitive locations for global and regional headquarters (Yigitcanlar and Velibeyoglu, 2008). However, building prosperous knowledge cities is a big, expensive, long-term, and challenging task as they require attracting and retaining both endogenous and exogenous talent and investment (Yigitcanlar et al., 2008a; 2008b).

Today talented knowledge workers, or in other words, the creative class of knowledge workers, have become perhaps the most important actors of knowledge production and city competitiveness, as investment is now following talent (Baum et al., 2009). Knowledge workers are highly mobile and are expected to change jobs, if not occupations, more frequently. Thus, they favour urban environments that contain thick labour markets offering opportunity to advance their careers by moving between employers (Yigitcanlar et al., 2007). Additionally, beyond a thick labour market, attracting and retaining knowledge workers involves building world-class and sustainable urban infrastructure and amenities and improving the quality of life and place in cities (Yigitcanlar et al., 2008c; 2008d).

SUSTAINABLE URBAN AND INFRASTRUCTURE DEVELOPMENT

A great deal of world population is now living in cities and the urbanisation trend in both developed and developing countries is following an upward trajectory. Many of the rapidly growing cities are developing strategies in becoming more competitive not only through their urban economies that offer employment, but also through lifestyle options that provide quality of life and place for their residents (endogenous), and strategies to become magnets for global talent (exogenous) by investing in their tangible and intangible assets (Velibeyoglu and Yigitcanlar, 2010).

Urban settings have substantial impacts on people's lifestyles, behaviours, and consumption patterns. Therefore, a development that is sustainable, as well as knowledge-based, is crucial in increasing the competitive edge of cities (Teriman et al., 2010). Much like knowledge-based urban development, sustainable urban development also aims for a balanced development of economy, society and environment (Yigitcanlar, 2010b). The incorporation of sustainability in an urban context is centred upon density, urban form, urban design, and infrastructure, and amenities. (Yigitcanlar and Dur, 2010).

As O'Connor (2010) states, the common current policy perspective is based on the idea that we can shape a city's development and vitality principally by containing and constraining the density and location of its residential development. However, the historical perspective, on the other hand, alerts us to the need to give much more attention to the way our investment in infrastructure is central to our capacity to invent, innovate, adapt, and become more competitive.

The changing circumstances of our cities today, such as coping with sprawl and rapid urbanisation, globalisation, climate change, the impacts of knowledge economy on urban space and place (e.g. knowledge city transformation), and a wider comprehension of sustainable urban development, call for different types and models of infrastructure provision. In recent years a vast amount of attention has been directed to this area (Yigitcanlar, 2010a; 2010b).

For example, O'Connor (2010) suggested that smaller units in urban sub-regional networks are potentially more effective at meeting the new pattern of infrastructure demand, as diseconomies are felt when the original, large systems attempt to reach out into very much larger service areas. These smaller, regionally focused units have greater potential to enhance sustainability as they can mould and change to meet the specific needs of particular parts of cities. This is also a valid view for the new form of urban food production such as urban farming, which is the Holy Grail of sustainable cities (Yigitcanlar, 2010a).

Searle (2010) pointed to the cost issues surrounding the infrastructure and noted that the provision of an infrastructure that is sustainable is a central challenge of contemporary urban governance and city competitiveness. Despite the availability of the necessary technology, progress continues to be slow. There are a number of reasons for this. A key reason is that the externality costs of unsustainable infrastructure are not priced properly or otherwise taken into account. More fundamentally, this arises to a significant extent because governments lack sufficient popular legitimacy to price or regulate such costs; this lack of legitimacy, in turn, largely stems from a lack of public awareness of the full costs of unsustainability.

Probably the most popular infrastructure type that sits in the middle of the sustainability debate is transport infrastructure. This is mainly because a more sustainable transport system combined with good urban design (e.g. transit oriented development) is proven to minimise transport disadvantages and the amount of private motor vehicle trips, increase the sustainability of urban environments, and contribute to the quality of life and place in cities (Duvarci and Yigitcanlar, 2007; Yigitcanlar et al., 2007; 2007).

However, beyond typical hard infrastructure (e.g. transport, water, and power) sustainability of the soft infrastructure also plays considerable roles in building sustainable urban environments. Soft infrastructures; such as emergency, health, employment, and education services, offer necessary community services and assist in forming a cohesive society and contribute immensely to the liveability of cities. In the era of knowledge, additionally knowledge infrastructures such as history, culture, and identity of the place are also considered as essential infrastructures contributing to cities standing out in the global competition.

In the development of a sustainable urban infrastructure, the biggest challenge, beyond the cost, has been the integration of the entire, both hard and soft infrastructure systems and their effective and efficient delivery and management (Yigitcanlar, 2010a; 2010b). In recent years, provision, monitoring, integration, management of infrastructure, service, and amenities have started being supported with advance information and communication technologies (Lee et al., 2008; Yigitcanlar and Han, 2010). Wired and wireless networks, ubiquitous technologies, urban and infrastructure management systems, geographic Information Systems, global positioning systems, intelligent transport technology, and e-government are among some of these rapidly developing and converging technologies (Yigitcanlar et al., 2003; Baum et al., 2004; Yigitcanlar, 2005; Yigitcanlar, 2006; Baum et al., 2007).

CITY COMPETITIVENESS AND SUSTAINABLE INFRASTRUCTURE

Previous research has shown how dramatic growth in the size, health and prosperity of cities over the thousands of years of civilisation has depended on a city's ability to invent and install appropriate urban infrastructures (Yigitcanlar, 2010a). Starting with access to water, roads, and waste disposal, through provisions for health and education, to the more recent additions of power and electronic communications, the extraordinary growth of cities to their current role of accommodating more than half the world's six billion plus people on a tiny fraction of the earth's surface has been made possible by successive waves of new urban infrastructure (UN, 2001).

It is widely accepted that an appropriate urban infrastructure of our time is a sustainable urban infrastructure, which refers to anything built or used in a way that contributes to the overall sustainability of our economy, society, and natural resources; in most cases, this is related to energy consumption and water use, two of the highest profile conservation areas. The availability of a sustainable urban infrastructure helps cities maintain their competitive edge in the highly competitive global environment, the evolving knowledge economy, and the knowledge society (Corey, 2004).

The book on competitive cities in the global economy, by OECD (2006), emphasises the importance of sustainable urban and infrastructure development, and suggests government action in cities to combine 'remedial' actions aimed at correcting the usual negative effects of urbanisation, such as sprawl and social deprivation, with 'proactive' measures for competitiveness, such as enhancing local social capital networks, developing links between higher education institutions and industry, and strengthening infrastructures, from roads and airports to information highways. OECD also recommends (2006):

- A flexible strategic vision is necessary to foster competitiveness, ensure a diversified range interdependent ventures, and information and transportation links between universities, researchers, technicians, and manufacturers.
- Liveable cities with high-quality infrastructure, green spaces, and inner city residential areas and public projects can contribute to economic success, attracting foreign investors as well as highly qualified professionals and tourists.
- Effective governance of cities depends on leadership from the national government to encourage reform, a formal government at the metro-regional level, and lower level local networks that include non-governmental actors, associations and businesses, which can deal with social tensions and understand market realities.
- To balance the financial needs of cities with those of the rest of the country, cities can diversify tax revenues with 'smart taxes' such as congestion charges and use public-private partnerships to raise money for public projects. Equalisation payments between metropolitan regions can be effective, but national equalisation schemes to redistribute resources from richer to poorer regions sometimes disregard the higher spending needs of cities and act as a disincentive to poorer regions to increase tax revenues.

CONCLUDING REMARKS

Building cities is about building competitiveness and attractiveness, as well as being drivers of local, regional, and global economies. In other words, cities in the era of knowledge need to be competitive in their own leagues. Global cities, such as London, New York, Paris, Sydney, and Tokyo, have been in the global race for centuries competing head on with each other. Established knowledge cities, such as Boston, Barcelona, Melbourne, Singapore, and Stockholm, have been in a tough global competition for quite some time in order to attract and retain talent and investment that is now highly mobile. Global and knowledge cities are considered well-established cities in terms of their economies, societies, environments and governing institutions. Furthermore, in the international city rankings these cities receive high scores in terms of competitiveness, quality of life, knowledge city status, eco-city status, city branding and so on.

Global competition in attracting talent, investment and finance is much harder for the newly emerging knowledge cities, such as Bangalore, Brisbane, Istanbul, Shenzhen, and Valencia. These cities have the advantage to learn from other prosperous global and knowledge city experiences and develop their own development pattern by not copying others, but being more creative and innovative and putting citizens in the heart of the development that is sustainable and knowledge-based. Moreover, for these cities investing on sustainable urban and infrastructure development will help in their progress and in augmenting their city competitiveness globally.

Balancing economic, socio-cultural, enviro-urban and institutional development, and providing sustainable infrastructure to support development are the cruxes of the challenge facing cities in today's globalised world. While dealing with this challenge, it should be always kept in mind that, as Shakespeare put it, "what is the city but the people?" and a human and environment oriented policy framework should be put into action.

Tan Yigitcanlar
Queensland University of Technology, Australia

***Tan Yigitcanlar** has a multi-disciplinary background and over two decades of work experience in private consulting, government and academia. Dr Yigitcanlar currently holds a senior academic position at the School of Urban Development, Queensland University of Technology, Brisbane, Australia, and a managerial position as the Events Program Executive Director at the World Capital Institute, which is an international Think Tank. The main foci of his research are advocating and promoting sustainable knowledge-based urban development, and infrastructure and transport development in city-regions. These two broad research areas are clustered around several themes: knowledge economy, knowledge-intensive activities, knowledge city, ubiquitous city, local governance, civic engagement, sustainable urban and infrastructure development, urban and transportation modelling, accessibility planning, online planning support systems, and understanding urban structures, processes and driving forces in the knowledge era. Dr Yigitcanlar is one of the leading researchers in the field of 'knowledge-based and sustainable urban development', and he has been responsible for research, teaching, training and capacity building programmes on urban and regional planning, transport modelling, environmental science, policy analysis and Information and Communication Technologies in respected Turkish, Japanese and Australian universities. He is the Editor of five books entitled 'Knowledge-based urban development', 'Creative urban regions', 'Knowledge-based development for cities and societies', 'Sustainable urban and regional infrastructure development' and 'Rethinking sustainable development'. He is the founding Editor-in-Chief of the 'International Journal of Knowledge Based Development', Special Issue Co-Editor of the 'Journal of Knowledge Management', and the Chairman of the 'Knowledge Cities World Summits'.*

REFERENCES

Baum, S., O'Connor, K., & Yigitcanlar, T. (2009). The implications of creative industries for regional outcomes. *International Journal of Foresight and Innovation Policy*, *5*(1-3), 44–64. doi:10.1504/IJFIP.2009.022098

Baum, S., Van Gellecum, Y., & Yigitcanlar, T. (2004). Wired communities in the city: Sydney, Australia. *Australian Geographical Studies*, *42*(2), 175–192. doi:10.1111/j.1467-8470.2004.00274.x

Baum, S., Yigitcanlar, T., Mahizhnan, A., & Andiappan, N. (2007). Singapore government online: A consideration of e-government outcomes. *Journal of E-Government*, *3*(4), 65–84. doi:10.1300/J399v03n04_04

Corey, K. (2004). Moving people, goods, and information in Singapore: Intelligent corridors . In Hanley, R. E. (Ed.), *Moving people, goods, and information in the 21st century: The cutting-edge infrastructure of networked cities*. New York, NY: Routledge.

Durmaz, B., Platt, S., & Yigitcanlar, T. (2010). Creativity, culture, tourism and place-making: Istanbul and London Film Industries. *International Journal of Culture, Tourism and Hospitality Research*, *4*(3), 198–213.

Duvarci, Y., & Yigitcanlar, T. (2007). Integrated modeling approach for the transportation disadvantaged. *Journal of Urban Planning and Development*, *133*(3), 188–200. doi:10.1061/(ASCE)0733-9488(2007)133:3(188)

Edvinsson, L. (2010). Understanding and cultivating social knowledge . In Metaxiotis, K., Carrillo, J., & Yigitcanlar, T. (Eds.), *Knowledge-based development for cities and societies: An integrated multi-level approach* (pp. 10–15). Hershey, PA: Information Science Reference.

Lee, S., Yigitcanlar, T., Hoon, H., & Taik, L. (2008). Ubiquitous urban infrastructure: Infrastructure planning and development in Korea. *Innovation: Management . Policy & Practice*, *10*(2/3), 282–292. doi:10.5172/impp.453.10.2-3.282

Metaxiotis, K., Carrillo, J., & Yigitcanlar, T. (Eds.). (2010). *Knowledge-based development for cities and societies: An integrated multi-level approach*. Hershey, PA: Information Science Reference. doi:10.4018/978-1-61520-721-3

Mumford, L. (1968). *The city in history: Its origins, its transformations, and its prospects*. New York, NY: Mariner Books.

O'Connor, K. (2010). Infrastructure and urban sustainability: Looking back and looking forward . In Yigitcanlar, T. (Ed.), *Sustainable urban and regional infrastructure development: Technologies, applications and management*. Hershey, PA: Information Science Reference.

OECD. (2006). *Competitive cities in the global economy. OECD Territorial Reviews*. Paris, France: OECD Press.

Searle, G. (2010). Sustainability of infrastructure: Achieving sustainable cities with sustainable infrastructures . In Yigitcanlar, T. (Ed.), *Sustainable urban and regional infrastructure development: Technologies, applications and management*. Hershey, PA: Information Science Reference.

Teriman, S., Yigitcanlar, T., & Mayere, S. (2009). Urban sustainability and growth management in South-East Asian city-regions: The case of Kuala Lumpur and Hong Kong. *Planning Malaysia*, *7*(1), 47–68.

UN. (2001). *World urbanization prospects*. New York, NY: United Nations Press.

Van Doren, C. (1992). *A history of knowledge: Past, present and future*. Toronto, Canada: Random House Publishing.

Van Winden, W., Berg, W., van Den, L., & Peter, P. (2007). European cities in the knowledge economy. *Urban Studies (Edinburgh, Scotland)*, *44*(3), 525–549. doi:10.1080/00420980601131886

Velibeyoglu, K., & Yigitcanlar, T. (2010). An evaluation methodology for tangible and intangible assets of city-regions: The 6K1C framework. *International Journal of Services Technology and Management*, *14*(4), 343–359.

Yigitcanlar, T. (2005). Is Australia ready to move planning to an online mode? *Australian Planner*, *42*(2), 42–51. doi:10.1080/07293682.2005.9982417

Yigitcanlar, T. (2006). Australian local governments practice and prospects with online planning. *URISA Journal*, *18*(2), 7–17.

Yigitcanlar, T. (2009). Planning for knowledge-based development: Global perspectives . *Journal of Knowledge Management*, *13*(5), 228–242. doi:10.1108/13673270910988079

Yigitcanlar, T. (Ed.). (2010a). *Sustainable urban and regional infrastructure development: Technologies, applications and management*. Hershey, PA: Information Science Reference. doi:10.4018/978-1-61520-775-6

Yigitcanlar, T. (Ed.). (2010b). *Rethinking sustainable development: Urban management, engineering and design*. Hershey, PA: IGI Global. doi:10.4018/978-1-61692-022-7

Yigitcanlar, T. (2010c). Making space and place for the knowledge economy: Knowledge-based development of Australian cities . *European Planning Studies*, *18*(11), 1769–1786. doi:10.1080/09654313.2010.512163

Yigitcanlar, T. (2010d). Spatial restructuring of cities in the age of knowledge economy: & from Australian cities . *Urban Insight*, *4*(1), 14–26.

Yigitcanlar, T., Baum, S., & Horton, S. (2007). Attracting and retaining knowledge workers in knowledge cities. *Journal of Knowledge Management*, *11*(5), 6–17. doi:10.1108/13673270710819762

Yigitcanlar, T., Baum, S., & Stimson, R. (2003). Analysing the patterns of ICT utilization for online public participatory planning in Queensland, Australia. *Assessment Journal*, *10*(2), 5–21.

Yigitcanlar, T., Dodson, J., Gleeson, B., & Sipe, N. (2007). Travel self containment in master planned estates: Analysis of recent Australian trends. *Urban Policy and Research*, *25*(1), 133–153. doi:10.1080/08111140701255823

Yigitcanlar, T., & Dur, F. (2010). Developing a sustainability assessment model: The sustainable infrastructure land-use environment and transport model . *Sustainability*, *2*(1), 321–340. doi:10.3390/su2010321

Yigitcanlar, T., & Han, H. (2010). Ubiquitous eco cities: Telecommunication infrastructure, technology convergence and urban management. *International Journal of Advanced Pervasive and Ubiquitous Computing*, *2*(1), 1–17. doi:10.4018/japuc.2010010101

Yigitcanlar, T., O'Connor, K., & Westerman, C. (2008). The making of knowledge cities: Melbourne's knowledge-based urban development experience. *Cities (London, England)*, *25*(2), 63–72. doi:10.1016/j.cities.2008.01.001

Yigitcanlar, T., Sipe, N., Evans, R., & Pitot, M. (2007). A GIS-based land use and public transport accessibility indexing model. *Australian Planner*, *44*(3), 30–37. doi:10.1080/07293682.2007.9982586

Yigitcanlar, T., & Velibeyoglu, K. (2008). Knowledge-based urban development: Local economic development path of Brisbane, Australia. *Local Economy*, *23*(3), 197–209. doi:10.1080/02690940802197358

Yigitcanlar, T., Velibeyoglu, K., & Baum, S. (Eds.). (2008a). *Knowledge-based urban development: Planning and applications in the information era*. Hershey, PA: Information Science Reference. doi:10.4018/978-1-59904-720-1

Yigitcanlar, T., Velibeyoglu, K., & Baum, S. (Eds.). (2008b). *Creative urban regions: Harnessing urban technologies to support knowledge city initiatives*. Hershey, PA: Information Science Reference. doi:10.4018/978-1-59904-838-3

Yigitcanlar, T., Velibeyoglu, K., & Martinez-Fernandez, C. (2008). Rising knowledge cities: The role of knowledge precincts. *Journal of Knowledge Management*, *12*(5), 8–20. doi:10.1108/13673270810902902

Preface

Cities attract people by offering a job and a livable environment. A job provides a kind of status in the society as well as an income to meet the needs of the person. A city means a home to live, schools for education, hospitals for health problems, police for safety, and a market for the daily food requirements. Besides, the city provides opportunities for leisure time activities. Cinemas, exhibitions, and sport arenas draw the attention of people. Obviously all these opportunities are available if people have jobs, which will means that people will have incomes.

The balance between jobs and opportunities of a city worked quite well after cities became centers of industrial production; cities attracted people from rural areas for more than a century. As a result of this, people migrated from rural areas to cities. This migration finished in developed nations, however, it continues in developing nations. According to UN forecasts, people are living in cities more than rural areas for the first time in the world history since 2007.

On one hand people are migrating from rural areas to cities, on the other hand inhabitants of cities migrate to other cities as well. People look for jobs to have income together with a better environment to live. Therefore an ideal city for people should establish a balance between creating job opportunities and providing a livable environment. Cities that have this balance become more attractive to people so that people prefer to migrate to these types of cities. A migration wave generally brings a new momentum to a city, because immigrants are more motivated to contribute to their new surroundings. Besides, the market size of a city enlarges. Therefore at the beginning, a migration fuels the city economy; however, as a migration continues, the size of the city increases and problems due to over-population may occur. An infrastructure designed for less people may not meet the requirements of an increasing population. If a city could solve its problems due to over-population, its development period will continue, otherwise, a decline period is unavoidable.

Today, cities have become the wealth producing centers of the national economies. Factories produce their products for the people of the city as well as other customers living in the neighbor cities. Service companies aim to meet the needs of people and other companies in the city. At the same time, cities are attractive markets for companies, because the denser spatial distribution in cities as compared to rural areas means that the people there with have more purchasing power. Companies want to be close to cities so that they are closer to their markets. Besides, companies close to skilled manpower, that is important for their success, mainly live in cities. As a result, we see a close interdependency between the company and the city.

As the capabilities of transportation and communication increase, production and service companies of the city offer their outputs to other cities. Thus, production and service companies target not only their domestic market but also the people living in other cities. Thus, companies enlarge their market size so that they can use the advantage of economies of scale. This development causes a natural rivalry among the companies of different cities, because the companies of other cities also want to sell their products in cities other than their own. Competition among the companies of different cities pushes city decision makers to support their own companies. Some companies even relocate their facilities to another city, one where better conditions are available. Sometimes, a city will offer companies special opportunities in order to convince them to migrate to its own borders. Thus cities compete to attract companies in today's world.

People who are looking for a good job will look for their ideal city to live. Skilled people are especially important for the future of a city, because those people are ones to be the main initiators of wealth creation. Another important player is the entrepreneur. The entrepreneur is the initiator of the job creation as well as the wealth. The appropriate city to invest in is an interesting topic for entrepreneurs. Richard Florida finds a positive relationship between the number of people who have a higher education and the development of high-tech sectors in a specific region in his bohemian index study. The diversity index is another index used by Florida which compares the metropolitan areas of US. Florida says overall diversity is a strong indicator of a metropolitan area's high-technology success. Attracting skilled people and entrepreneurs are vital for city leaders. These people are one of the main inputs of a wealthy city. In order to attract skilled people and entrepreneurs, city leaders benchmark their cities with other cities. By analyzing the benchmark data, they see the potential areas to be improved. An important way of doing this is to use city competitiveness indexes.

Harvey analyses the changing pattern of capitalism and its effects on urban governance which caused a shift from managerialism to urban entrepreneurship since 1980s. He says "neoliberal policies have advocated the rising competitiveness between cities and regions to attract hyper mobile foreign capital." Harvey considers alternative competitiveness strategies for cities such as acquisition of key control and command functions in finance, government, or information gathering and processing. He also states that a competitive edge with respect to redistributions of surpluses through central governments is still of tremendous importance.

According to Jiang and Shen, studies assume urban competitiveness to be equal with urban success, which is understood in narrow economic terms, such as economic output, income, and employment growth. However, given its great relevance for public policy formulation, competitiveness should be understood from a longer-term perspective. Such a perspective should acknowledge the close relationship between the economic and noneconomic facets of urban life, highlight balanced development, and serve the overall public interest.

In some cities, one or a few companies dominate the city so that these companies mainly direct leaders of the city. Thus, the elected or assigned governors become the facilitator of company life in their cities. They aim to provide better infrastructure to companies. They try to attract skilled people to come to their cities. They promise a peaceful environment, and efficient and effective working systems. As a result, competition among cities starts.

The effects of globalization decreased the barriers of sending the outputs of companies to other cities in different countries. Thus the people living in cities meet their needs cheaper even though the service/product has high quality. On the other hand, globalization increased the pressure on companies, because they had already been in competition mainly with the companies of the same city or same country, however, today, their competitors are the companies from all over the world.

Today's corporate world, the weight of companies in the decision-making process of developed cities is getting more influence. Therefore, the competition among the companies somehow has a reflection on their city. The leading companies of a city seek better facilities from their home city. By knowing this, city leaders aim to provide an effective and efficient working city system. They know that an efficient and effective working city system will naturally bring a competitive edge to whole system.

A city can be thought of as a whole system composed of subsystems. Thus a city has various subsystems such as transportation, sewage, water, health, etc… Each of these subsystems has an aim that serves the city. The better a subsystem works, the better it serves the inhabitants. For example, a good working health subsystem will provide trust to all citizens. On the other hand, if the road subsystem did not work properly, the fire service could not respond to a fire on time. An inefficient fire service is a threat to all other city subsystems. Therefore, for a good working city system, all the subsystems should be working properly and consistently with each other.

Many times city subsystems cannot jointly work with each other. This creates a significant inefficiency for the city. For example, water pipe renewal under a bottleneck road may prevent spectators of the city's football team to reach their championship match on time. Or the weakness of design education prevents a huge design company to come to the city even though all other conditions are sufficient. Therefore, compatible working subsystems are vital for the competitive city. If the education system is working effectively, this will provide an advantage to whole companies in the city against their rivals in other cities. If the communication infrastructure is not sufficient to daily information transformation, an entire city's pace will be slower. As a result, the efficient and effective working subsystems are not sufficient separately for a competitive city but the synchronization of each subsystem is vital.

Today, technology is one of the most powerful tools for upgrading a city subsystem. Developments in technology find application areas in city subsystems: GPS technologies created various ease in transportation of people and freight; database technologies provided opportunities for health tracking systems; pattern recognition software created new chances for security systems; mobile technologies became indispensable for individuals in order to follow city activities. Besides, technology provides very useful opportunities for the synchronization of city subsystems. New technological applications are transforming the city onto a new level. Technology upgrades city subsystems so that the whole city system is being upgraded.

Cities that understand the importance of technology and apply new technology in related subsystems gain a competitive advantage over their competitors. For example, an inefficient working transportation system can cost an average of two more hours for each person in the city. On the other hand, efficient working tax collecting system via Internet supported by effective database system provides advantages to both leaders of the city as well as entrepreneurs.

Older cities try to upgrade their subsystems in order to reach a more efficient working system, whereas new cities are being built by using latest developments in technology. Founding a new city has advantages compared to upgrading to a new city, because, while trying to upgrade a city system, leaders must deal with conflicting parties in the city. For example, a company wants to utilize its old factory building as a new office building whereas closer inhabitants oppose the new project saying they need a park for their living environment.

Today, cities want to upgrade themselves into being more competitive in the creative market for various technology and consultancy companies. IBM's "smart city", Siemens' "city of the future" and Cisco's "connected cities" are some examples of these companies and their products. Although these companies are very good at their main business, their city upgrading related products are at the beginning level. It seems that these companies should divert much more of their resources to these products. Also, academicians' interest from different disciplines is increasing on city competitiveness, especially, academicians from urban economy, and regional planning and strategy disciplines. The weight of technological application has increased on city subsystems, computer engineers, mechanical engineers, electronic engineers, and statisticians, even doctors have diverted their work of interest to city subsystems as well. Different pieces of city competition were under examination such as transportation, health, and utilities. The subsystems of cities are one of the topics that have not been elaborated as much. We know that efficient and effective working subsystems are crucial for a healthy system. Therefore, improving subsystems of cities is vital for a better working city system.

Various experts from different disciplines explain their views on how to reach a competitive city by improving the subsystems of cities in the book. Authors shared their chapters about increasing efficiency and effectiveness of city subsystems as well as the synchronization among them. While doing this, they concentrate on new technologies and their applications. Therefore, the book is composed of two parts. The first part includes chapters for technology related topics for city competitiveness. In this part, readers will find chapters that have more theoretical content. Authors' evaluations of various technologies that are related to urban subsystems will be in this part.

In the first chapter, Ahmet Bulut explains the I3 framework for the competitiveness of a city. I3 is composed of instrumentation, interconnection, and an intelligence framework. The first component of I3, instrumentation, is about data collection. Data is important because it provides measurability. This is based on the assumption that if a process can be measured, then it can be improved. The second component, interconnection, is used to discover the associations and relationships between seemingly independent subsystems of a city. Through interconnection, the effects of any policy action can be evaluated holistically rather than focusing on a single dimension. The "intelligence" in I3 comes from being able to test specific policy hypotheses and making informed decisions based on the results of various policies that are put in place. Bulut considers his findings about the I3 framework on city competitiveness in his chapter.

Diana Hamburger discusses the importance of spatial location on city competitiveness. She analyzes the meaning of distance for the city. The effect of Information Technologies on distance provides a new vision for the city planners. Evolution in the view of distance causes changes on the spatial position of cities and the urban settlements. Therefore the chapter has interesting discussion areas for the concepts like accessibility and centrality that are important parts of the transportation subsystem of the city. Moreover, utilities network with spatial reference is argued specially referring to the telecommunication networks in the chapter.

Rahaman and Lourenço examine the effects of the Geographic Information System on city competitiveness. The Geographic Information System is considered one of the most powerful tools of information and accessibility to information among different stakeholders and cities in the world by providing spatial data. In the third chapter, authors highlight the role of GIS technology to assess empirically the competition among cities or regions based on different criteria and values assembled by different individuals, businesses, institutions, etc… This chapter is structured with origins of urban competitiveness, dynamics and functions of competition, and GIS and its role as well as future research possibilities.

Maria Nadia Postorino's chapter is a nice evaluation of the airport as a subsystem of the city. Cities that have developed transport systems benefit from a high-quality accessibility, which makes them more attractive both to users and investors. These cities have a competitive advantage for economic growth. Particularly, an airport can play an important role in assuring the development of the city best linked to it, as it enlarges its external accessibility and attractiveness. Postorino's chapter describes the importance of improvements concerning the surface transport systems linking cities to their nearest airports as well as improvements assuring safer operations and capacity enhancement at the city airport airside.

Mark Deakin in the fifth chapter develops the notion of the intelligent city as the smart provider of electronically enhanced services. He identifies how this growing interest in the notion of intelligent cities has led universities to explore the possibilities of using communities of practice (CoPs) as a way of drawing upon the work-based learning that such knowledge-based organizations offer to be smart in developing integrated models of e-government (eGov) services. Deakin analyzes the attempts made by a consortium of leading European cities to use the intelligence CoPs offer as organizational means to be smart in developing models of eGov services capable of integrating the e-learning needs, knowledge transfer requirements, and capacity building commitments of their socially-inclusive and participatory urban regeneration programs.

Ori Gudes et al. investigate the healthy city as a fundamental prerequisite for a competitive city. The main focus of the chapter is on the role of e-health planning, by utilizing Web-based geographic decision support systems. The proposed decision support system provides a powerful and effective platform for stakeholders to access essential data for decision-making purposes. The chapter also highlights the need for a comprehensive information framework to guide the process of planning for healthy cities. Additionally, it discusses the prospects and constraints of such an approach. The chapter outlines the potential insights of using information based framework and suggests practical planning methods as part of a more broad e-health approach for improving the health characteristics of competitive cities.

In the seventh chapter, Erkan Özdemir examines the city competitiveness in online environment. He focuses on the role of city websites for building and communicating the city brand. City administrators are taking advantage of the branding techniques used by the private sector companies in order to differentiate their cities from the rest and build an image and identity for them in this atmosphere of intensive competition. Özdemir's chapter evaluates the role of the city websites in building city brands and promoting their brands. In order to fulfill the aim, the city branding, related concepts, and the role of the city websites in building city brands are introduced in the chapter. Randomly selected city websites from the City Brands Index, which are evaluated using content analysis, and the results of this analysis, are presented and discussed, respectively.

Evren Tok examines the technologies of neoliberal urbanization in reproducing new urban spaces that are oriented towards securing a competitive edge. His chapter explores the questions of: Are technologies instrumental to producing new city spaces; how do urban technologies influence the parameters of the reproduction of city spaces; and, which social representations, actors, and networks are involved in the process. He also contends that, in neoliberal times, new technologies of power enable the creation of more polycentric urban areas. This emerging role of the new technologies of power necessitates an analysis of the place of technologies in the mobilizing of low urban politics. In the analysis, it is important to be attuned to nuanced articulations of public interests and private initiatives that are enriched by representations of the spatial realities of cities. He says, "we are more likely to realize the importance of urban technologies and 'street level' everyday urban practices if, rather than treating them as divorced from each other, we ontologically recognize their marriage. Such an approach will help us to better diagnose the role that technologies play in moving us beyond dualities like public and private."

The second section of the book includes application related chapters of new technologies for gaining competitiveness. The chapters in this section include cases from various cities from around the world. Readers will find examples of applications of new technologies for improving the subsystems of the city.

In the ninth chapter, Brunello et al. analyze the possible impacts of a high speed railway (HSR) infrastructure in Northern Italy. The chapter considers small and medium agglomerations in the vicinity of HSR corridors, not always served by HSR stations. A methodology was developed in order to analyze the possible effects of an HSR for the northern part of the Italy. The results are significant for all countries in Europe and worldwide, not only for investing in an HSR infrastructure, but mostly in terms of building territorial cohesion, while seeking international recognition for developing successful new technologies and systems.

Luis Carvalho concerned with the recursive conflation between the concepts of city's competitiveness in ICT and different versions of the knowledge city concept based on ICT and digitalization, often responsible for ambiguous political discourses and unclear local economic development strategies. To overcome this problem, the chapter distinguishes both concepts, identifies links between them and illustrates a way through which national and urban policies can support ICT-related city competitiveness, knowledge city strategies, and the local development of innovation arenas. The chapter illustrates these notions with the case of the on-going development of the Songdo district in South Korea, its competences in ubiquitous computing and the connection between its technological prowess with its ambitions of creating one of the most advanced u-cities in the world.

Tanughichi and Duarte examines database integration from various municipal Information Systems as an essential tool for urban and strategic planning. The city of Curitiba, Brazil, has implemented several Information Systems to enhance its public services, from education to health care. Most of these systems use several non-integrated personal smart cards to provide access to services. In the chapter, authors analyze public Information Systems and their inputs and outputs in Curitiba. Authors advocate that the personal smart card already used in the metropolitan transportation system only as a fare collection card, is, among all others, the best option, from a technological and administrative standpoint, to integrate all municipal Information Systems, improving the effectiveness of public services and assisting for a comprehensive planning process.

In the twelfth chapter, Jacqueline Bueno Sousa provides an interesting field analysis from US. She discusses the effect of Google data centers on cities. She focuses in her study on three communities where Google built major data centers: Goose Creek, SC; Lenoir, NC; and Council Bluffs, Iowa. The author interviews economic development officials in each of the three communities. As the result of her analysis, Sousa concludes that the data centers haven't always lived up to its expectation.

Todaro and Cangel examine the urban geospatial maintenance system of Venice. The reader will see that Venice is not attractive only to tourists but also urban experts as well. The Venice geospatial maintenance system is based on five fundamental elements of the territory for its activities; canals, bridges, embankments, built architecture, and paving. The objective of the chapter is to highlight the fact that, in a city, it is important to start with operative processes and everyday work. It is important that the systems not only be containers for masters (even geometric or geo-referential masters). They must become a part of actual management systems based on geospatial objects.

As can be seen in the contents of the chapters, this book has a multidisciplinary approach due its scope. Since the subsystems of the city uses various branches of the technology in order to be efficient and effective, readers should be prepared to be faced with wide technological terminology. Besides, it is assumed that the reader has knowledge about subsystems of a city.

Since the chapters have a strategy component beside technology content, it was rare to meet such experts and academicians working on these two components of the city simultaneously. Therefore, I really appreciate the authors for their contributions. I believe that this book will be a pioneer in studies that bring technology and strategy components for the city. Therefore I am hoping that the number of books and articles will increase in the future.

Melih Bulu
Istanbul Sehir University, Turkey

Acknowledgment

I acknowledge Tan Yigitcanlar supported the book during the preparation time by giving his comments and wrote a visionary foreword for readers. My wife, Yasemin visited various cities throughout the world with me and provided a comfortable environment during the preparation of the book. Erika Carter, Jan Travers, Hannah Abelbeck, and Christine Bufton coordinated all of the technical details perfectly between authors and the publisher. Obviously, 25 authors from various parts of the world believed the importance of the topic and shared their knowledge in their chapters. I thank all the authors for their valuable chapters.

Melih Bulu
Istanbul Sehir University, Turkey

Section 1
Technologies for Improving Urban Subsytems

Chapter 1
City Competitiveness and Infrastructure

Ahmet Bulut
İstanbul Şehir University, Turkey

ABSTRACT

This chapter introduces a new framework called "I3: Instrument, interconnect, and cultivate intelligence framework." This framework can be used to drive the transformation of today's not so smart cities into the smart cities of tomorrow. In i3, instrumentation is used to collect data, which is important, because data provides measurability, and measurement can lead to improvement. Interconnection in i3 is used to discover associations and relationships between seemingly independent subsystems in a city. In i3, controlled experiments are easily setup and run to test each individual policy. The intelligence bit in i3 comes from being able to test specific policy hypotheses and conduct rigorous analysis and synthesis of the integrated data. The i3 framework helps city officials and researchers discover valuable knowledge, make informed decisions based on the results of various policies that are put in place, and facilitates the culture of experimentation at every policy decision level. The overarching goal in i3 is to discover routine and well-structured patterns in city management operations, turn them into best practices, and finally automate the execution of such practices so that the framework itself can take a major responsibility over city management. In this chapter, a roadmap is provided as a guideline for policymakers to successfully deploy i3 in their jurisdiction. By using i3 continually, a regular city can be transformed into a smart city faster.

DOI: 10.4018/978-1-61350-174-0.ch001

INTRODUCTION

The world is getting smarter. Over 1.3 billion Radio Frequency Identification (RFID) tags were produced in 2005. That figure was projected to soar to 33 billion by 2010, a 2500-fold increase, according to an In-Stat report (In-Stat, 2006). Sensors are being embedded in cars, home appliances, photo cameras, city roads, pipelines, medicine and livestock; and the list goes on. Instrumentation facilitates high-quality and real-time data acquisition. While being instrumented, the world is also becoming interconnected: soon, there will be two billion people on the Internet. All systems and objects will be able to "speak" to each other. A trillion of such interconnected devices can and will produce oceans of data for city officials to analyze. These front line interfaces are linked to powerful backend systems. These systems have advanced analytics processors that can process massive amounts of raw data to gather insight in real time. The challenge is to turn these massive amounts of data into actionable knowledge.

Mobile technologies and Internet enables the collection and analysis of data to enhance the safety, security, usability, and efficiency of city infrastructure. Siemens AG identifies three key design goals as the main drivers that should be present in a "successful" infrastructure solution that addresses city management (Siemens, 2010). These design goals are: (i) the solution has to be environment friendly, (ii) it has to increase economic competitiveness, and finally (iii) it has to improve the quality of life.

Recent advances in Micro-Electro-Mechanical Systems (MEMS) and RFID technology made it easier than ever to put computational capability into other entities that are not so obvious as computers. These entities can be an object, a person, a process, a service, or an organization. This practice results in transformation of the entity into a digitally aware, connected, and smart entity.

With so much technology and networking available at such low cost, it is hard to "not" connect and enhance social processes and physical systems. With all the information available to mine, it is hard to "not" provide a better service to a customer, a citizen, a student, or to a patient. Imagine a world where all resources in the world could be turned around and transformed to support non-military aid organizations when required. Smart and ubiquitous use of sensors and real time connectivity not only makes this vision possible, but also highly relevant.

The "instrument, interconnect, and cultivate intelligence (i3)" is the common theme that has been resonating in politics, business, and academic world in the last decade on how to use information science in creating the "smart" cities of tomorrow. In this chapter, the i3 methodology is presented in detail with its applications, and a roadmap is provided for policymakers to put the framework into use.

MOTIVATION

In order to concretize the ideas in the ensuing development, consider a prototype application design based on the i3 methodology. The application is an integrated surveillance framework for smart cities and their law enforcement units to monitor crime, and to counteract and prevent terrorist activities.

Millions of input channels at the edges of this intelligent network are used to gather real time state information and environment heartbeat. Interconnection through i3 is useful to identify organized crime involving more than one geographical region and that spans a time frame over a couple of days or weeks or months. On-demand reporting and registration of suspicious activity using wireless devices increase the continued use of the information channels at the outset. Resulting streams of audio, text and video data are cross-correlated and analyzed associatively to drive complex relationships that identify anomalous activities promptly. The framework is used to automatically monitor city "health" continuously, and can be generalized into a homeland security infrastructure within a county, or a state, or within a whole country.

An attractive advantage of building an integrated surveillance framework based on i3 is that online environment monitoring tools are going to empower citizens to hold city officials accountable for their policy actions. Furthermore, if the framework enables citizens to contribute their own assessment on how city policies work, then this practice will result in prompt and closer scrutiny over how city officials perform their duties in order to meet city governance expectations. Therefore within a successful system of operations, all citizens, city officials and all other major stakeholders have to take active roles in city management and governance.

With the reduced cost and size of such micro sensors, it has become feasible to deploy densely distributed i3-based Wireless Sensor Networks (WSNs) in a range of applications. These applications clearly demonstrate the long term economic value and competitive advantage that various business sectors can gain with the use of these frameworks. A typical WSN deployment contains hundreds or thousands of sensor nodes, each of which are equipped with sensing, computing, and communication abilities. A sensor node can monitor its own environment through sensing, perform simple computations, and communicate its local view with its peers (Karaki et al., 2006). Below for the interested reader, five example WSN deployment scenarios are provided:

1. **Environment monitoring and precision farming:** A United States based MEMS company called MEMSIC introduced a new generation of sensor integration and wireless technology called eKo series for precision agriculture, environmental research, and crop monitoring (Berry et al., 2008, Jardak et al., 2009). It is now possible both to save water and to increase farm-yield and harvest quality at the same time by data provided through a WSN. The system can be deployed on a farm to measure moisture levels at a granular level per farm section such that the sensory data can be used to control the amount of water applied to each section improving the overall harvest quality in fine grain.
2. **Structural health monitoring (SHM):** Bridges, highways, and civil buildings have been deteriorating due to age. Manual inspection of bridges costs millions of dollars. Furthermore, manual inspection process is relatively unreliable, and can only be carried out sporadically. Automated inspection with traditional methods is expensive (Elgamal et al., 2003). For example, the total system cost of the wired monitoring system (including installation) on the Bill Emerson Memorial Bridge in Cape Girardeau, Missouri, USA is approximately $1.3M for 86 accelerometers, which makes the average installed cost per sensor a little over $15,000. Tremendous attention is directed towards development of SHM strategies, with the primary purpose being reliability and reduced maintenance and inspection costs. MEMSIC's ISM400 sensor board provides the enabling technology that had been previously lacking to perform structural health monitoring using the Imote2 (Niu, 2010). The ISM400+Imote2 platform (referred to as" node") provides, for the first time, a commercial device with the ability to collect synchronized and high-fidelity vibration data wirelessly. The collected data can be used to understand the behavior of the structure in case of ambient forces, e.g., wind, traffic, as well as major natural forces, e.g., earthquakes, floods, and tornados. The cost of deploying a wireless monitoring infrastructure using these nodes is a fraction of the cost associated with a wired deployment: the cost per node is around $500. University Illinois Urbana-Champaign (UIUC) SHM group researchers performed

a full-scale monitoring of the Jindo Bridge in South Korea. With the installation of a sensor network consisting of 70 nodes, the deployment represents the largest of its kind for civil infrastructure to date (Rice et al., 2009).

3. **Habitat monitoring:** Contamination of edible water sources by manmade and/or natural causes is a serious threat to human life. Water quality in a water source is determined by presence of contaminants and the quantity of these contaminants, physical and chemical factors, e.g., pH and conductivity, quantity of salts present, and presence of nutrients. In order to understand larger ecological systems, the first step is to study smaller ecosystems, such as lakes, in a controlled fashion. Monitoring process in this case involves determining measurement location and periodicity, selection of quality metrics, and statistical analysis of measurements to determine the health of the ecosystem. Through continuous monitoring, hazards that undermine water quality can be detected on time and avoided if possible (Kang et al., 2008).
4. **Machine condition monitoring:** Machine condition monitoring is the process of monitoring the condition of a machine with the intent to predict mechanical wear and failure. Vibration, noise, and temperature measurements are often used as key indicators of the state of the machine. Trends in the data provide health information about the machine and help detect machine faults early, which prevent unexpected failure and costly repair (Djurdjaanovic et al., 2010). According to a conducted study, the estimated cost of maintenance for a selected group of companies increased from $200 billion in 1979 to $600 billion in 1989 (Wireman, 1990). Through continuous monitoring, machine maintenance schedules can be prepared more intelligently to save billions of dollars.

(5) Video surveillance: Sensors are extensively used in Mobile Electronic Systems Integration (MOBESE) projects. These projects are high-profile citywide surveillance projects. There is one such system currently being operated for all 81 provinces in Turkey. In a typical application of a MOBESE project, network cameras are mounted on the poles in critical areas. At the field under every pole there are field cabinets where network switches, video encoders, UPSs and some sensors are installed. The network cameras monitor and record live real-time activity video. These streams of video data captured in the field are stored at the command control and dispatch centers for detailed data analysis functions.

In the next two paragraphs, two successful i3-based framework deployments are presented. These deployments are in the area of gathering real time business intelligence based on real time field data. These examples are provided as to demonstrate to city officials how compelling and beneficial it is to put an i3-based system to work to their advantage.

Video Surveillance in Law Enforcement

Santa Barbara Sheriff's Department (SBSD) is using software and hardware combined solution by IBM Global Services to fight crime, protect its people, and manage its workforce more effectively (IBM, 2006). SBSD has invested in a video surveillance framework, which is being used to provide visual evidence supporting incident data reported by its field officers. Otherwise, the crucial workforce would spend a substantial amount of their limited work hours for appearing in court to defend themselves against lawsuits. Video data captured on the scene is digitally signed, watermarked, and transmitted to a base station using wireless communication. Two types of

Figure 1. Observe, Orient, Decide, Act (OODA) Loop. (Adapted from http://upload.wikimedia.org/wikipedia/commons/3/3a/OODA.Boyd.svg).

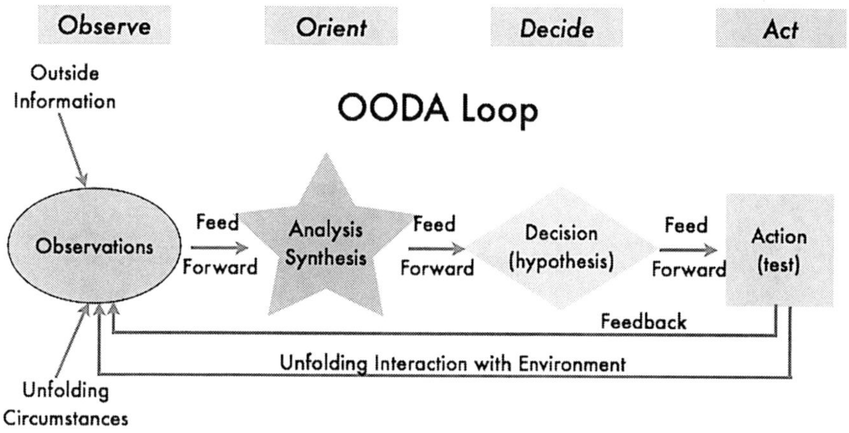

data channels are integrated in this application: (i) video data captured by on board sensors for visual evidence, and (ii) textual data that represents the on-the-scene interpretation and assessment of field officers. The complete system is implemented within six months.

RFID-Based Shipment Tracking

Driscoll's has started using an RFID enabled technology to track their shipment of berries from their cooling and distribution facilities in California (in the U.S.) and Baja California (in Mexico) to Florida based distribution centers across the U.S. The shipment is done via transportation trucks. Shipment tracking system consists of two 2.4 GHz RFID sensor tags installed within the container to measure inside temperature for spoilage assessment. A different type of sensor called harness is installed on the container door to track open and close activity for securing against theft. An RFID Reader with a GPS Unit and a GPRS Transmitter is mounted inside the driver cabin. Periodically, the sensors transmit their state information to the reader, which in turn collects measurements, piggybacks the latitude and longitude information, and sends them to a cell phone tower through a GPRS signal. The data is later sent to a backend server to be processed and visualized on an office dashboard.

Objective: The goal is to shorten the time it takes from the point of an observation being made to the point when an action is taken based up on this observation. This is where systems get smarter, organizations get more agile, and people and their countries win.

Observation, Orientation, Decision, and Action (OODA) Loop

The time it takes from the point of an observation being made to the point when an action is taken based up on this particular observation is the realization and one cycle in the Observation (O), Orientation (O), Decision (D), and Action (A) loop, i.e., OODA loop as shown in Figure 1 (Boyd, 1995). For most types of decision-making processes including city policy making, OODA presents a viable execution model.

The OODA loop in a smart city is orchestrated through rapid data collection, analysis of the collected data, visualization and sharing of outcomes, findings, and executables over a diverse set of communication channels. Repeated frame-

Figure 2. Steps of the i3 methodology

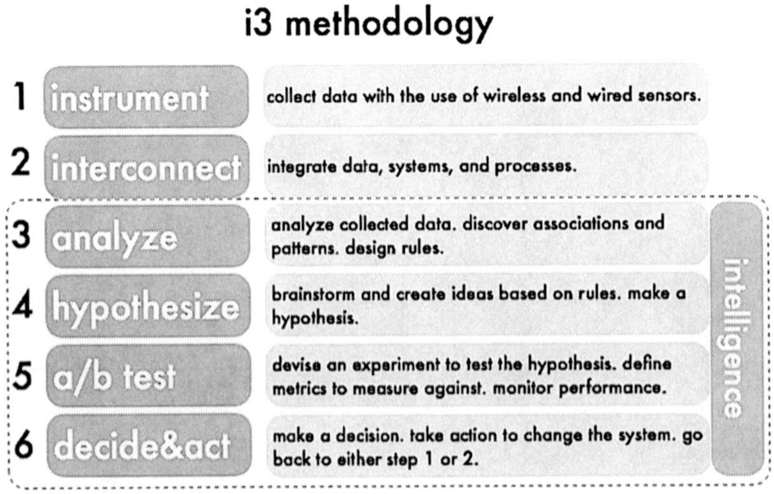

work use reveals frequently observed patterns, and the automation of functional components that correspond to such frequent patterns is vital in order to scale i3 from being used in citics to being used in countries.

i3 FRAMEWORK

In this section, the details of the i3 framework are presented. Three key attributes of i3 are measurability, comparability and being holistic. The i3 framework is:

a. measurable: the framework provides a platform to experiment with ideas and policies. This attribute enables rapid assessment of core subsystems and activities in a city. The measurability comes from the fact that the framework enforces and encourages determining key performance indicators to measure. The data collected is used to how specific policies affect the value of these performance indicators. These indicators are purely data centric, usually numeric values. The measurability feature is important to improve individual policies iteratively over time.

b. comparable: cities can be compared to each other in order to drive practical lessons. This way, lessons and best practices learned in one city can be applied in another city. Sharing of information and practices should be seamless. Rationality of the test methodology enables rapid deployment of policies and ideas at various regions across the city and other cities as well. Since performance indicators that are chosen to evaluate policy effectiveness are numeric, it is easy to test and compare results achieved in different cities and city locations on a common platform.

c. holistic: core subsystems and their constituent components are integrated via well-defined business processes. This makes it possible to carry out the performance evaluation of an idea on a broader scale. The effectiveness of a specific policy is evaluated thoroughly using a multitude of performance indicators that correlate data from multiple subsystems. This way, results of policy actions are assessed comprehensively. Otherwise, alleviating one problem in one part of the city may result in service or performance degradation in another part of the city that has not been accounted for. The holistic nature of the

Figure 3. Application of i3 framework to structural health monitoring. (Adapted from the model in Health Monitoring Framework for Bridges and Civil Infrastructure article by Elgamal et al., 2003).

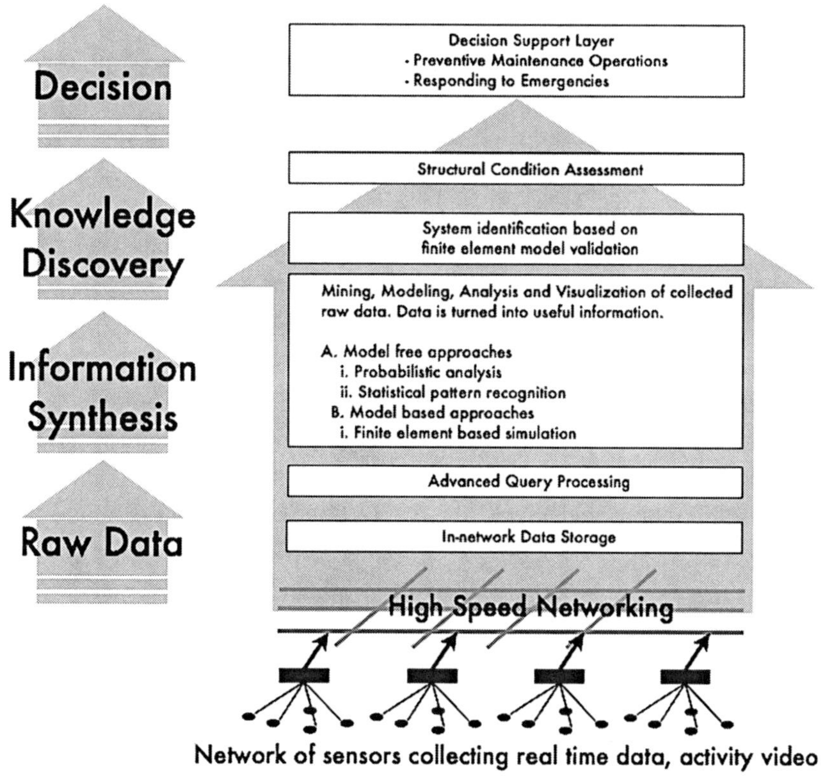

framework comes from collection and analysis of heterogeneous data.

Figure 2 shows the steps involved in a successful application of the i3 methodology within the i3 framework. Next, each of the three building blocks in i3 is explained with respect to how it contributes to creating a coherent framework.

Instrumentation

Figure 3 shows a health-monitoring network that can be deployed to monitor the health of civil infrastructures (Elgamal et al., 2003). At the outermost edges of the network, there is a network of sensors that collect real time data on the host system. The collected data corresponds to structural response data and activity video. These motes that form the sensor net are battery powered. They are either physically installed on or nearby a building, a highway, or a bridge. A sensor, e.g., IRIS, MICAz, Imote2 from MEMSIC Inc., is capable of transmitting a radio signal with a built-in 2.4 GHz antenna to a nearby base station. A base station offers the capability of a networked storage buffer for a short period of time until the field data is written into persistent storage. The collected field data is used in an information discovery pipeline that will be explained in "Intelligence" below.

Instrumentation is not necessarily physical as in the example above. Consider social networks and online activities: Facebook, Twitter, Google searches, and e-mails. A social network is a prime example of instrumenting human beings in order to trace their activities. On the go, people text message, update Facebook status, tweet, and search for directions, places, and news. There is a ton of textual data that represents what each person is doing or thinking at a given moment. The difference is that inhumane objects and systems are instrumented with sensors and their state information needs to be polled at periodic intervals; however, people push their state information into the data pool at their own will.

In essence, the instrumentation of people, processes, and systems give rise to a wealth of real time state and activity data. This data can provide invaluable insights into how each component of a large system works individually, how it affects the other components, and what kinds of associations can be drawn between these components. These insights are used in turn to improve societal processes in a city. For this purpose, heterogeneous data sources have to be interconnected and cross-analyzed.

Interconnection

In order to understand and detect the onset of an interesting phenomenon, a set of "homogeneous" readings does not provide adequate information for reliable assessment. Under the assumption that an interesting event means a non-trivial event, multiple and "heterogeneous" sensors need to collaborate in order to detect the occurrence of a complex phenomena. This assumption holds true in many real world scenarios. Consider the following three examples:

Example 1: A lower barometric reading, a sudden drop in temperature, an abrupt change in wind speed or direction, and a decrease in downwelling irradiance all combined indicate the onset of a thunderstorm.

Example 2: Automatic Assisted Living or Activities of Daily Living (ADL) refer to the automated health monitoring of people aged 65 or older in their own homes. A heterogeneous set of anonymous and binary sensors that include motion detectors, break-beam sensors, pressure mats, and contact switches are used to learn individual movement and activity patterns of home occupants (Wilson et al., 2005). If a pressure mat sensor is active, then the same occupant is not likely to be the one, who activates the break beam sensor monitoring room entrance activity. Rather than identification of human activities individually, a series of human activities over time indicates an inherent characteristic present in the home occupant being monitored. For this purpose, objects are instrumented with RFID tags. A person wearing an RFID reader as a bracelet uses these objects. A trace of object use over days, weeks, and months becomes indicative of the person leaving the trace itself, similar to a fingerprint. The next section, "Intelligence," shows how such insights can be used to build intelligence into the i3 framework. In order to develop ground truth for this model, textual descriptions of human activities that are available on the Internet can be used for the real-world performance of those activities (Wyatt et al., 2005).

Example 3: On a historical note, Chalk Ocean Airways failed to identify and properly repair fatigue cracks on their 1947 Grumman Turbo Mallard model airplane (Reuters, 2007). A few minutes after the takeoff for the Bahamas at 500ft, the plane lost its right wing and plunged into the shipping channel adjacent to the Port of Miami on December 19, 2005. On many occasions prior to the accident, the airplane in question had undergone maintenance operations. Most of these operations were shortsighted and fell short to address the actual structural problem on the plane that arose from operating on and off water. Instead of relying on man-made diagnostics that is error prone, the plane could have been turned into a digital network by proper instrumentation with structure sensors. Then, any type of structural problem that deviates from the structural norm of the plane would indicate a potential problem. In many instances similar to this unfortunate event, many small mechanical problems found in systems that are currently in use today usually magnify each other's seemingly negligible effect during operation and create a catastrophic chain of events that lead to disaster. Instead of interpreting a homogenous reading on its own, it is mandatory to interconnect subsystems and mine heterogeneous sensory data to draw associative relationships. In the airline case, "the fatigue crack on the airplane leads to the seepage in fuel tank over time" is an example of a relationship that can be drawn through association rule mining.

Intelligence

Once heterogeneous data is collected from multiple external interfaces, it is cross correlated and analyzed using advanced data mining tools. The objective is to draw actionable knowledge. In Example 3 above, "the fatigue crack on the airplane leads to the seepage in fuel tank over time" is a simple rule. Simple rules can be combined together to create more complex rules. Discovery of rules is only one side of the equation. Once initially "passive" rules are drawn, they are used in creating actionable knowledge. That is, these rules are a guideline in the process of building applicable actions and controls based on diagnosis of a situation. "If there is fatigue crack on the wing, then the fuel tank also needs to be checked along side the regular maintenance operation on the wing" is an applicable control. This is where the rules are used to identify a scenario, and hint on a specific action set.

The methodology outlined in Example 2 above can be applied to recognizing activities that span a diverse geography over a longer time frame. Individual sub-activities that comprise a complex activity can be detected by heterogeneous sensors, and can be registered into i3 at different locations. The challenge is to connect the pieces for detecting a higher-level activity, e.g., detecting traffic anomalies in a city. For this purpose, the social underlying system is abstracted as follows: the objects used to perform an activity represent that activity. Activities are characterized as distributions over the objects used. By comparing probability distributions using Kullback-Leibler divergence, activities can be discriminated and compared to each other. When an activity is represented as a sequence of events as in "crossing an intersection" means "(i) stopping at the red light, (ii) waiting for the green light, (iii) putting on the gas pedal, and (iv) driving through the intersection", then any violation indicates an anomaly that needs to be considered carefully. "A parking ticket" that is given on violation of parking rule may precede "crossing the intersection at red light barely safe", which may precede "an accident on 3rd street that involves a hit and run". This methodology empowers the law enforcement in a smart city to be preemptive rather than corrective. The difference between corrective action and preemptive action is that corrective action is not smart and detrimental for the city as a whole; whereas preemptive action is intelligent and is right for a smart city.

The i3 framework can detect anomalies using custom monitors, diagnose fault by comparing operational data with baselines, and finally apply control based on diagnosis (Han et al., 2006). This whole process describes a single iteration of the OODA loop. The actions that are made at the last step affect the underlying social and physical processes, and these give rise to a totally new set of observations. This in turn marks the beginning of the next iteration of the OODA loop.

IMPLEMENTATION

In this section, the implementation details of the i3 framework are presented from perspectives of technology, research, and prototyping & simulation. If a policymaker can test her hypothesis, measure its results, and compare these results with the status-quo, then she can make informed executive decisions for making her city smarter simply by improving processes that govern how each subsystem in a city work in accordance with each other. This is key to leaving guesswork behind. This is possible by organization from bottom up in the hierarchy.

Technology Infrastructure

Better information management drives informed decision making, which increases agility. An infrastructure that organizes information from multiple input channels is vital in a smart city. In the business world, IBM has been advocating its Cognos product line addressing Business Intelligence (BI) and Performance Management (PM) as viable solutions to organize business data at all levels for informed decision-making (IBM, 2010c). IBM Cognos provides dashboarding and scorecarding capability for best practice Key Performance Indicators (KPIs).

Decision Making and Action Patterns

Decisions broadly fall into two categories: tactical decisions and strategic decisions:

Tactical decisions are made by operations. Line of business workers need short term data to make informed operational decisions at some step well within the limits of a specific business process. In tactical decision-making, contextual analytics is essential. However, the smarter is to delegate tactical decisions that are structured, procedural, and "routine" to the underlying framework.

Executive management makes strategic decisions. Business owners need long term data that will help them set future company direction, help them analyze the effect of changes in strategy and predict the implications of these changes. In strategic decision-making, scenario modeling, predictive analysis, and collaboration is essential.

In short, a smart city needs a smart infrastructure. A smart infrastructure can identify patterns of routine processes that can be automated. This practice results in optimization of the subsystems of the city. Once patterns are identified, they can be used for multiple purposes:

a. patterns in existing processes can be optimized to save time and money.
b. patterns can become a basis to set quality standards and be used to define new social processes.
c. patterns can be used to define a baseline to compare against in case of intrusion and deviations.

Pattern detection and subsequent process automation is fairly well understood. Rules can be extracted using efficient algorithms on logs and traces of activity. Specific methods that can be used for this purpose are outlined in the section "Research and Scientific Issues." After patterns are discovered, and rules are designed according to these patterns, the system can perform these routine and structured tasks automatically. However, the challenge is to automate holistic behavior involving dynamics that are hard to detect. In fact, human beings can hypothesize intrinsic and holistic dynamics. After the hypothesis formulation, the system provides a platform to test the hypothesis.

Hypothesis Testing and A/B Framework

Testing of a hypothesis is done through controlled experiments, also called randomized experiments, A/B tests (and their generalizations), split tests, Control/Treatment tests, and parallel flights. Controlled experiments embody the best scientific design for establishing a causal relationship between changes and their influence on user-observable behavior. Controlled experiments typically generate large amounts of data, which can be analyzed using data mining techniques to gain deeper understanding of the factors influencing the outcome of interest, leading to new hypotheses and creating a virtuous cycle of improvements. Organizations, companies, and cities that embrace controlled experiments with clear evaluation criteria can evolve their systems with automated optimizations and real-time analyses. Consider the following experience for demonstrating the effectiveness of the split test methodology: Greg Linden at Amazon.com created a prototype to show personalized recommendations based on items in the shopping cart (Kohavi et al., 2007). An item is added to the cart, recommendations show up; another item is added, different recommendations show up. Although the prototype looked promising, a marketing senior vice-president claimed it would distract people from checking out. Greg ran a controlled experiment, and the recommendation feature won by such a wide margin that not having it live was costing Amazon a noticeable amount of money. As top priority, shopping cart recommendations launched live on Amazon.

City officials have to embrace the culture of experimentation similar to how Amazon as a company embraced it for revolutionizing Online Shopping experience. The adoption of a culture of experimentation starts with a simple assumption: intuition is misleading, and data is everything. Therefore, a system that makes it easy to run controlled experiments allows organizations and countries to innovate quickly and effectively.

More important than being able to run experiments, knowing what to measure and the ability to measure are keys to success. Once these hooks are placed in correctly, the methodology is straightforward: (i) setup experiment, (ii) define metrics, (iii) run experiment, (iv) monitor measurements, and (v) wait until enough data accumulates or until there is statistically significant evidence accrued. Using this methodology, hypotheses can be tested and validated or invalidated on a systematical basis.

In short, before taking a concrete action, city officials and policy makers have to think of how to measure the effectiveness of the solutions they are proposing. All other variables kept constant, only one variable is changed to have value A or value B. Figure 4 shows an example of the methodology being applied: two different versions of a website landing page are tested against the customer base. The difference between two versions is where the signup form is laid out on the page. The success metric is the total user signups. The metric is quantifiable and measurable.

Test results show that layout version A leads to more user signups, and therefore is the winner.

Consider the Bosphorus Bridge in Istanbul, Turkey. The Bosphorus Bridge, also called *Boğaziçi Köprüsü* in Turkish, spans the Bosphorus strait and connects Europe and Asia. The bridge is located between Ortaköy (on the European side) and Beylerbeyi (on the Asian side). It is a gravity anchored suspension bridge with steel pylons and inclined hangers. Its aerodynamic deck is hanging on zigzag steel cables. Assume that the objective of a city official at the city of Istanbul is to solve the traffic congestion problem during rush hours in the city. She chose to test the effect of the toll-pricing model on the bridge. Therefore, the variable to test is the toll collection model on the Bosphorus Bridge. Value A denotes that the model is the variable pricing model, and Value B denotes that the model is the fixed pricing model. For your information, currently on the Bosphorus Bridge, the toll is collected in two ways: there is the automatic toll collection through wireless card readers mounted on vehicle windshields, and the other method is the on-demand toll collection through prepaid cards.

One of the many KPIs in the above toll collection model selection scenario is the time it takes for vehicles to cross the bridge. A suit of break beam sensors, motion detectors, and video cameras are installed on the bridge and toll collection booths. These sensors are used to measure the average trip time and its variance, and transmit these readings through a base station to a backend server. The data on the backend server is statistically analyzed and an outcome is published. The outcome could be any of:

a. the variable pricing model reduces the average trip time.
b. the fixed pricing model performs better than the variable pricing model.
c. the difference between average trip times on using the above two models is negligible.

Figure 4. A/B Testing for a website layout selection. (Adapted from http://media.smashingmagazine.com/cdn_smash/wp-content/uploads/2010/06/abtesting.gif).

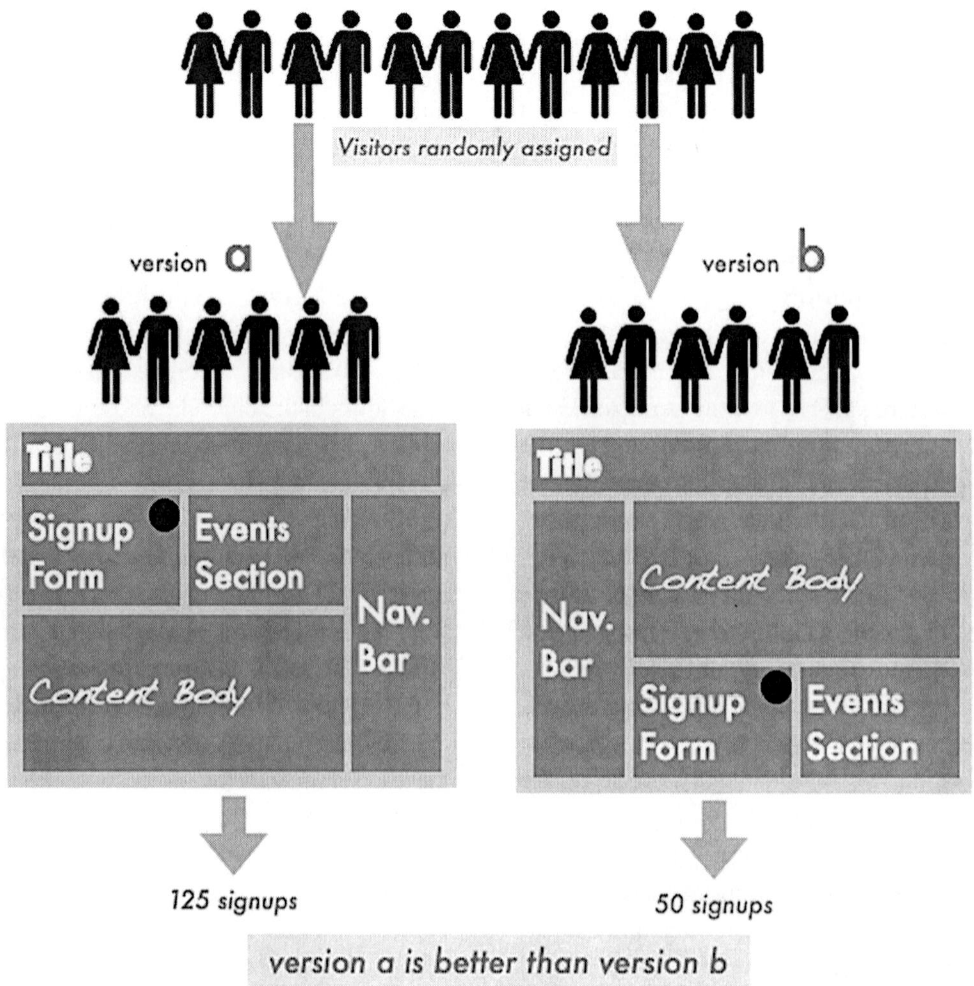

In case of (a) or (b), the action to take is to adopt the winning model. In case of (c), an experiment is devised to test an alternative model, or an experiment that tests a totally different variable, e.g., "the effect of alternating use of all lanes in one traffic direction", is setup. As can be seen, using the i3 methodology progressive actions can be taken. These actions ultimately can lead to a winner idea that does alleviate the traffic congestion problem. An interesting consideration here is to run multiple of these experiments so long as the infrastructure and the city itself allows it. The reason is that the methodology is inherently sequential, and it does take time to test each possible variable & value combination. If there are multiple bridges in the city, then there are multiple choices to test. First, two parallel tests, one for each bridge, can be devised. Each experiment tests the same variable & value combination. If there are no relative differences between the dynamics of the environment where these bridges are located, then:

a. a truly parallel experiment can be devised where each experiment tests a totally different variable.
b. the original test is permitted to run until its statistical significance threshold is reached.

Both of these alternatives address the same issue. They are both designed to expedite the experiment duration to arrive at a conclusion. Since time has intrinsic monetary value, the faster the decisions are made, the quicker the corresponding actions are taken; and therefore more OODA loops are completed, and the faster a regular city transforms to a smart city, and the faster a smart city transforms to an even smarter city.

The idea is to encourage and enforce the culture of experimentation at every decision level, i.e., both tactical and strategic. From an infrastructure perspective, the easier the experiments are setup and run, the faster is the adoption of experimentation among the city officials. The i3 framework advocates the use of instrumentation for data collection, interconnection for discovering associations and intelligence acquisition as the guideline to build a standardized frame of operations. Instrumentation allows measurability; interconnection allows holistic assessment and comparability. The intelligence designates the treatment (Menzies, 2003). The overarching goal of the system in design is to automate the discovery and application of best practices as much as possible so that the framework can take over the responsibility of the bulk of management operations in a city, i.e., operations that are routine, well-structured, and process centric operations. This is possible as long as an automated process does not throw a race condition. If a race condition occurs, then human intervention is necessary.

Interconnection through Service Oriented Architecture (SOA)

In order to interconnect the seemingly disparate subsystems of a city for holistic assessment, ideas

Figure 5. Service Oriented Architecture (SOA) Dynamics. (Adapted from http://www.w3.org/2003/Talks/0521-hh-wsa/soa.png).

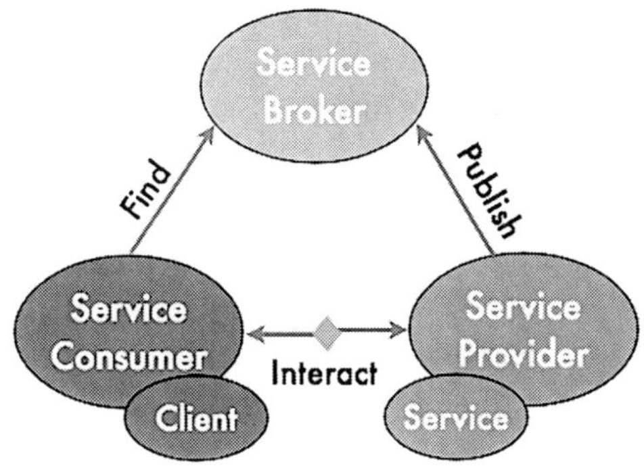

from SOA can be used. These principles have successfully been applied in abridging diverse set of business services in the past to close the loop in the service industry. The same methodology can be followed in the smart city initiative. A system based on SOA will package functionality as a suite of interoperable services that can be used within multiple separate systems from several business domains. Figure 5 depicts the architecture and the typical interaction between components. A service broker maintains a directory of services. Each individual service either consumes another service, or provides service to another service.

The city and its subsystems operate similar to how an SOA based supply chain operates. In order to interconnect various subsystems freely, each subsystem's service component has to be designed with interoperability in mind. The interservice communication has to be through Simple Object Access Protocol (SOAP), and the data exchange has to be in a structured format as XML to enable SOA based orchestration.

Reconsider the traffic congestion control example given in "Hypothesis Testing and A/B Framework" above. The toll collection model is proposed to regulate the rush hour traffic over the bridge. At major highway entrances, the one-car-per-green (ocpg) stop light regulates the traffic merging onto the highway that connects to the bridge. The timing of the ocpg stoplight directly affects the traffic that the bridge sensors will be exposed to. Therefore, a combinatorial solution that controls both the timing of the ocpg stop light at the highway entrance and the pricing model used by the toll collection system on the bridge requires a standard framework to synchronize. SOA based command control fits well in this regard to coordinate activity between these two individual traffic regulation services.

Publish-Subscribe System

After the ubiquitous instrumentation of the city with sensors and actuators, each device starts generating events that describe the current status quo in its surroundings. Following the standard log levels commonly found in Information Technology (IT) systems of today, events can fall into the one of the following five descriptive levels: (i) debug, (ii) warn, (iii) fatal, (iv) info, and (v) trace. Some events are regular info and debug events, which are informative messages that report regular system operation or report measurements. For example, “the average trip time measured” is transmitted to the base station as an event at info level. Warn events usually precede events that are fatal. For example, “the average trip time starts to move beyond its bi-monthly moving average” is a warn-level event. Average trip time climbs to an intolerable value during accidents and total road blockage. These are examples of events at fatal level.

Stakeholders can subscribe to individual events. Tactical decisions are usually given in the light of individual and local events. For example, a fatal local event causes an automatic dispatch of the traffic department officials to regulate traffic manually. Strategic decisions are given in the light of global events that are defined by the co-occurrence of a set of local events within a reasonable time frame or in the light of local events reoccurring systemically. If the average trip time fluctuates unpredictably for months, a new toll collection-pricing model is introduced to alleviate the problem.

For the stakeholder that designs the ocpg stop light solution to ease the merge-on traffic to the highway, the average trip time indicator on the bridge is a useful performance metric to monitor and measure effectiveness of this solution. If the ocpg solution does work indeed, then the decision to widely deploy the solution at other local areas is strategic. The stakeholders at the strategic decision level monitor and subscribe to the ocpg performance metrics for their informed decision making.

The above scenarios highlight the continuous monitoring in the system, where all stakeholders that are subscribed to events get notified when events of interest get published.

Furthermore, the events that are generated during the operation of subsystems in the city need to be stored for offline analysis. The types of analysis that need to be performed are statistical data mining and association rule mining. These operations require access to traces of activity at all levels over an extended period of time (Xu et al., 2004). In the context of i3 framework, association rule mining is important for discovering rules that underlie intrusions. The details of this analysis logic are the topic of the next section.

Research and Scientific Issues

Association rule mining is used to discover interesting relations between variables in very large systems. Interestingness of rules is evaluated according to various quality metrics. One of these metrics is regularity (Piatetsky-Shapiro, 1991). Regularity is defined using support and confidence. Support measures the applicability of a rule to a set of data items. Confidence measures the satisfiability of the rule among the applicable data items. For example, an association rule can describe co-purchase regularity between products in large scale transaction data recorded by point-of-sale (POS) systems in a supermarket: “if onions and potatoes are purchased, then a burger is also likely to be purchased" is a rule discovered in the sales data of a supermarket. These rules are used to drive marketing activities such as promotional pricing and product placements.

The association rule discovery works in two phases: first, all frequent itemsets with minimum support are found in the database. Then, frequent itemsets and the minimum confidence constraint are used to form rules. The search can be expedited by using anti-monotonicity property: all subsets of a frequent itemset are frequent, and all supersets of an infrequent itemset are infrequent. A-priori algorithm (Agrawal et al., 1994) uses this property.

In searching massive numbers of possible associations to look for collections of items that appear to be associated, there is a risk of finding spurious associations. This means that there are collections of items that co-occur with unexpected frequency in the data; however, this may happen purely by chance. In order to clarify this concept, consider a collection of 1,000 items. The objective is to find all rules that contain two items in the left-hand-side and 1 item in the right-hand-side. There are approximately 1,000 x 1,000 x 1,000 = 1,000,000,000 such rules. A statistical test for independence with a significance level of 0.01 implies 1% chance of accepting a rule while there is no association. If there are indeed no associations, as many as 10,000,000 false associations are made. Specifying custom significance levels can control this risk.

Once rules are discovered, and a rule engine with a rule base is constructed, the Rete algorithm can be used to implement the matching functionality (Forgy, 1982). The algorithm has a match-resolve-act cycle that supports forward chaining and inferencing.

Prototyping and Simulation

In the smart city initiative, the effectiveness and the feasibility of i3 framework has to be tested first in a controlled environment. This is necessary to fine-tune the framework before moving onto a city scale. Most importantly, the ideal controlled environment has to reflect the actual city environment, because that is where the framework will ultimately be used within.

International Airports stand as a promising candidate as it allows a rich set of physical and social system interactions. Another advantage is that it reflects the complex dynamics that occur in a metropolitan city. In an airport, there are many stakeholders in an international airport, and each stakeholder has a different perspective:

a. Passengers wish to travel safely.
b. Airlines need to ensure the security and safety of its passengers, employees, and assets.
c. Airport operators need to control all access to facilities and services.
d. Airport vendors and partners need access to secure facilities.
e. Employees need to be assured that their workplace is safe, and their colleagues do not misuse their access privileges to commit a crime.
f. Governments need more effective ways to secure their borders.

Air travel industry, law enforcement, and border management all require an integrated system that necessitates cooperation among all stakeholders, where data and processes of each organization is shared and tightly integrated (IBM, 2008).

RELATED WORK ON SMART CITIES

Siemens AG has been sponsoring research on strategic planning for city governance conducted by third party research organizations (Siemens, GlobeScan, & MRC McLean Hazel, 2007). The research findings indicate that greener mass transit solutions, active governance rather than passive administration, holistic view to problem solving rather than departmentalization and "silo mentality", and privatization of city services operations while retaining public leadership and control are prime avenues to consider and pursue further in building a sustainable and working solution for city management.

IBM is working with Kitson & Partners to build a smart city called Babcock Ranch outside of Fort Myers, Fla (Vizard 2010). The concepts being applied at Babcock Ranch can scale up to millions of people. In fact, Kitson plans overseas projects with a budget in the upper million-dollar range. IBM's responsibility in the project is to manage all the systems involved in building the city along and to create a dashboard system for ongoing management. Most of those systems will be powered using solar energy. The main design goal in the project is to avoid all the management complexity associated with cities that is a result of little or no planning. In many cases, the roads in those cities were built over cow paths, rather than planned to reduce costs or for long-term sustainability.

In order to make cities smarter, presence awareness technology can be used. It allows users to define how they would like to be contacted at different times of day, or in specific situations. A pilot smart city program in Masdar City in the United Arab Emirates presents a successful application of this technology: commuters will soon use mobile devices to order driverless electric cabs. Built-in presence awareness capabilities based on GPS and context-based awareness technologies will ensure that residents of this planned smart city will get to their destinations on time (IBM, 2009).

In another smart city initiative led by the EnergyAustralia consortium, IBM is playing an active role as the systems integration partner for the USD100m Smart Grid. The technology will enable residents to see real-time analysis of their household electricity usage and usage per individual appliance, which in turn can help them make better decisions about energy efficiency in their homes and minimize their environmental impact (IBM, 2010a).

On a muddy man-made island in the Yellow Sea, New Songdo City is rising to become a template for 20 new cities across China and India. From the trunk lines running beneath the streets to the filaments branching through every wall and fixture, Cisco is going to supply digital plumbing to the city of Songdo. Every square inch of the city is going to be wired with synapses so that the whole city runs on information. This initiative is potentially a multi-billion dollar opportunity. This revenue number includes revenues gained from installation of the basic infrastructure as well as revenues gained from selling the consumer-facing hardware and revenues gained from selling the services layered on top of that hardware. It is quite possible to imagine a Cisco-built digital infrastructure wired to Cisco's TelePresence videoconferencing screens mounted in every home and office, with engineers listening, learning, and releasing new Cisco-branded bandwidth-hungry services in exchange for modest monthly fees. Just like service oriented businesses, such as software as a service, hardware as a service, and infrastructure as a service, Cisco's intentions are to offer cities as a service, bundling water, power, traffic, and telephony into a single, Internet-enabled utility (Lindsay 2010).

In Amsterdam, Cisco and IBM will help develop an energy management system using 500 smart meters installed in participating households, giving each a glimpse of their individual energy consumption (Greenbiz Staff, 2009). IBM will establish the smart energy network, while Cisco will oversee the IP-based communications infrastructure that will enable the energy system to securely communicate with household appliances in real-time. Nuon, an Amsterdam-based energy company, will also take part by creating applications for the system. Other projects underway include connecting docked river cruisers and commercial sea vessels to the electric grid and the creation of a "climate street" that connects tram stops and street and store lighting with smart meters to help store owners and municipal workers manage energy use along a busy shopping route.

Japan is at the forefront of using of wireless technology to better organize societal processes. Cell phones have chips that turn them into "smart cards" that allow subscribers to pay for tickets, food and other items at shopping stores. Japan's DoCoMo plans to expand this service so commuters can use phones with special chips as train passes (Belson, 2005). Furthermore, DoCoMo's new generation of handsets can scan two-dimensional bar codes pasted at bus stops, which allows customers to get a bus schedule instantly, and an estimated time of arrival. Once on the bus, they can receive coupons sent to their cellphones from stores along the bus route.

CONCLUSION

Progressive policymakers around the world are emphasizing the use of smart systems based on i3 that can capture data from a myriad of instrumented and interconnected processes, devices, and objects, and then apply intelligence to this welter of information to detect patterns and take action in real time. The communication and rapid sharing of information help garner a common operational picture on the view of multiple stakeholders and let each stakeholder act in concordance with the rest of the law enforcement units. The goal is to leverage technologies such as modeling and associative data analysis to help form an understanding of how seemingly uncorrelated data in time and space in the form of witness statements, crime reports, newspaper articles, and web pages are interrelated. This ability empowers public safety agencies to take preventive action rather than after-the-event emergency response (IBM, 2010b).

MEMS and RFID technology are driving the transformation of dump cities of today into the smart cities of tomorrow. Smart and ubiquitous use of sensors and real time connectivity makes "how social processes and physical systems work and whether or not they perform up to the expectations" more transparent. By being able to measure the performance of city's subsystems systems and processes, it is easy and to leave the guesswork behind and to make informed executive decisions for improving management policies.

The i3 framework is proposed to accelerate the evolution of cities of today into the "smart" cities of tomorrow by providing a standardized frame of operations for city officials to make policy improvements and to make informed decisions on issues concerning the welfare of the society. The framework depends heavily on using data instead of intuition as the guide in the decision making process. The three key components of the framework are (i) instrumentation for data collection, (ii) interconnection for discovery of associations and relationships, and (iii) quick hypothesis testing for making informed decisions, and for cultivating intelligence.

In i3, it is straightforward to set up and run experiments for testing hypothetical policies, the decisions to transformation a regular city to a smart city can be given quickly because a large number of policy improvement iterations can be completed in a short amount of time.

In order to support the widespread use of i3 on majority of city management operations, the framework is proposed to be designed in such a way that it can discover routine and well-structured patterns; only to turn them into best practices that can be automatically executed without repetitive manual reasoning.

City officials can increase the competitiveness of their city in many possible directions. The beauty and advantages of i3 come from its applicability in organizing policy-making process. Using controlled experimentation to make informed decisions is a rational approach. It is purely scientific. It emphasizes using data to guide specific policy actions. Furthermore, it is complementary to other disciplined approaches that can be used to make a city more digitized and therefore smarter.

The power of i3 is the vision of universal knowledge. Subdividing knowledge into various fields fragments the totality of knowledge. However, a discovery in a subfield may have far reaching consequences in every aspect of our society, and it would be a big mistake to have a local perspective on outcomes and observations and evaluations, and therefore miss the global picture. Universality of knowledge is achieved by using heterogeneity at the center stage of data collection and data analysis processes. Each piece of data is analyzed with respect to all other available information.

The i3 framework emphasizes the urgency of action-orientation. A city official that has a global perspective on issues regarding the city has to act on her hunches and ideas. The regular methodology is to first observe, then orient the observed data through synthesis & analysis, then decide on a course of action after forming a hypothesis, and finally act. These actions will change the environment; new set of observations will appear. These will either support the original hypothesis or reject it. In anyway, it will let the officials build knowledge over time. This natural process is engraved in the framework. Therefore, city officials can make this methodology a mandate in their own and their subordinates policy operations. This will prove extremely useful in increasing the competitiveness of the city.

The i3 framework stresses the power of communication by sharing discovered knowledge. The discoveries have to be applied in society. The more they are applied and used, the more they get integrated into the belief system of the society. These usually take the form of best practices. These best practices are ethical, methodical, and efficient new ways to do everyday things while living in mega cities of today.

REFERENCES

Agrawal, R., & Ramakrishnan, S. (1994). Fast algorithms for mining association rules. In J. B. Bocca, M. Jarke, & C. Zaniolo (Eds.), *Proceedings of the 20th International Conference on Very Large Data Bases (VLDB), Santiago de Chile, Chile* (pp. 487-499). Waltham, MA: Morgan Kaufmann Publishers.

Belson, K. (2005, June 6). Technology: Fast phones in Japan- But no pot of gold. Retrieved from http://query.nytimes.com/ gst/fullpage.html?res= 9D04E7DA1538F935A 35755C0A9639C8B63

Berry, J. K., Delgado, J. A., & Khosla, R. (2008). *Precision farming and conservation advances agricultural sustainability.* Fort Collins, CO: USDA Soil Plant Nutrient Research Unit (SPNR). Retrieved from http://www.ars.usda.gov/ research/publications/ publications.htm?seq_no_115=223229

Boyd, J. (1995). *The essence of winning and losing* [Presentation]. United States Air Force. Retrieved from http://www.danford.net/ boyd/essence.htm

Dirks, S., Keeling, M., & Dencik, J. (2009). *How smart is your city?* [Report]. Somers, NY: IBM Institute for Business Value.

Djurdjanovic, D., Liu, J., Marko, K. M., & Ni, J. (2010). Immune systems inspired approach to anomaly detection, fault localization, and diagnosis in automotive engines. In Schumann, J. M., & Liu, Y. (Eds.), *Applications of Neural Networks in High Assurance Systems (Studies in Computational Intelligence 268)* (pp. 141–163). Berlin: Springer. doi:10.1007/978-3-642-10690-3_7

Elgamal, A., Conte, J. P., Masri, S., Fraser, M., Fountain, T., & Gupta, A. …El Zarki, M. (2003). Health monitoring framework for bridges and civil infrastructure. In F. K. Chang (Ed.), *Proceedings of 4th International Workshop on Structural Health Monitoring*, Stanford University, Stanford, CA (pp. 123-130). Lancaster, PA: DEStech Publications.

Forgy, C. (1982). Rete: A fast algorithm for the many pattern/many object pattern match problem. *Artificial Intelligence*, *19*(1), 17–37. doi:10.1016/0004-3702(82)90020-0

Greenbiz Staff. (2009, July 14). *IBM and Cisco to help Amsterdam become a 'smart city'*. Retrieved November 30, 2010, from http://www.greenbiz.com/ news/2009/07/14/ ibm-and-cisco-help-amsterdam- become-smart-city

Han, T., & Yang, B. S. (2006). Development of e-maintenance system integrating advanced techniques. *Computers in Industry*, *57*(6), 569–580. doi:10.1016/j.compind.2006.02.009

IBM. (2006). *Santa Barbara sheriff's department wields innovation to fight fraud and inefficiency*. Retrieved November 30, 2010, from ftp://ftp.software.ibm.com/ software/solutions/pdfs/ ODB-0141-00A_ITM.pdf

IBM. (2008). *Unlock opportunities for innovation while addressing security risks in air travel*. Retrieved November 30, 2010, from IBM http://www.ibm.com

IBM. (2009). *New strides in mobility signal opportunity calling*. Retrieved November 30, 2010, from http://www-304.ibm.com/ businesscenter/cpe/download0/ 175785/FV0905_Mobility.pdf

IBM. (2010a, October 10). *IBM chosen by EnergyAustralia for smart grid, smart city initiative*. Retrieved November 30, 2010, from http://www.tmcnet.com/ usubmit/2010/10/08/ 5057234.htm

IBM. (2010b). *IBM smarter public safety solutions*. Retrieved November 30, 2010, from http://www.ibm.com

IBM. (2010c). *Optimize your software with IBM Cognos 8 business intelligence*. Retrieved November 30, 2010, from http://www.ibm.com/ cognos

In-Stat. (2006). *RFID tags and chips: Opportunities in the second generation*. Retrieved November 30, 2010, from http://www.instat.com/ abstract.asp?id=3& SKU=IN0502115WT

Jardak, C., Rerkrai, K., Kovacevic, A., Riihijarvi, J., & Mahonen, P. (2009). *Email from the vineyard*. Paper presented at 5th International Conference on Testbeds and Research Infrastructures for the Development of Networks & Communities and Workshops (TridentCom), Washington, DC.

Kang, J. M., Shekhar, S., Wennen, C., & Novak, P. J. (2008). Discovering flow anomalies: A SWEET approach. In *Proceedings of the 8th International Conference on Data Mining (ICDM)*, Pisa, Italy (pp. 851-856).

Karaki, J. N., & Kamal, A. E. (2006). A taxonomy of routing techniques in wireless sensor networks. In Ilyas, M., & Mahgoub, I. (Eds.), *Handbook of Sensor Network Protocols: Compact wireless and wired sensing systems* (pp. 1–24). Boca Raton, FL: CRC Press.

Kohavi, R., Henne, R. M., & Sommerfield, D. (2007). Practical guide to controlled experiments on the Web: Listen to your customers not to the HIPPO. In *Proceedings of the 13th ACM SIGKDD international conference on knowledge discovery and data mining* (pp. 959-967). New York: ACM.

Lindsay, G. (2010, February 1). *Cisco's big bet on New Songdo: Creating cities from scratch*. Retrieved January 17, 2011, from http://www.fastcompany.com/ magazine/142/ the-new-new-urbanism.html

Menzies, T., & Hu, Y. (2003). Data mining for very busy people. *Computer*, *36*(11), 22–29. doi:10.1109/MC.2003.1244531

Niu, P. (2010). *MEMSIC adds structural health monitoring capabilities to wireless sensor platform*. Retrieved from http://investor.memsic.com/ releasedetail.cfm? ReleaseID=462923

Piatetsky-Shapiro, G. (1991). Discovery, analysis, and presentation of strong rules. In Piatetsky-Shapiro, G., & Frawley, J. (Eds.), *Knowledge Discovery in Databases* (pp. 229–248). Menlo Park, CA: AAAI Press.

Reuters. (2007). *U.S. blames airline, FAA for deadly Florida crash*. Retrieved November 30, 2010, from http://www.reuters.com/ article/ idUSN30280 19120070530

Rice, J. A., & Spencer, B. F., Jr. (2009). *Flexible smart sensor framework for autonomous full-scale structural health monitoring* (NSEL Report No. 18). Urbana-Champain, IL: University of Illinois at Urbana-Champaign. Retrieved from http://hdl.handle.net/ 2142/13635

Siemens. (2010). *ICT for city management: Using information and communications technology to enable, engage, and empower city stakeholders*. South Africa: SIS Global Center of Competence for City Management, sponsored by Siemens AG.

Siemens, GlobeScan, & MRC McLean Hazel. (2007). *Megacity challenges: A stakeholder perspective*. Munich, Germany: Siemens.

Sweedberg, C. (2010, November 18). Driscoll's monitors its berry shipments in real time. *RFID Journal*. Retrieved from http://www.rfidjournal.com/ article/view/8011/

Tapia, E. M., Intille, S. S., & Larson, K. (2004). Activity recognition in the home using simple and ubiquitous sensors. In A. Ferscha & F. Mattern (Eds.), *Pervasive computing: Proceedings of the Second International Conference on Pervasive Computing* (pp. 158-175). Berlin: Springer.

Vizard, M. (2010, June 8). *IBM helps build smarter city in Florida*. Retrieved November 30, 2010, from http://www.itbusinessedge.com/ cm/blogs/ vizard/ibm-helps- build-smarter-city-in-florida/

Wilson, D., & Atkeson, C. (2005). Simultaneous tracking & activity recognition (STAR) using many anonymous binary sensors. In *Pervasive computing: Proceedings of the Third International Conference on Pervasive Computing* (pp. 62-79). Berlin: Springer.

Wireman, T. (1990). *World class maintenance management*. New York, NY: Industrial Press.

Wyatt, D., Philipose, M., & Choudhury, T. (2005). Unsupervised activity recognition using automatically mined common sense. *American Association for Artificial Intelligence*, *1*, 21–27.

Xu, W., Bodik, P., & Patterson, D. (2004). *A flexible architecture for statistical learning and data mining from system log streams*. Paper presented at the Workshop on Temporal Data Mining: Algorithms, Theory and Applications at the 4th IEEE International Conference on Data Mining (ICDM '04), Brighton, UK.

KEY TERMS AND DEFINITIONS

A/B Testing: Controlled experimentation.

Association Rule Mining: Discovery of behavioral and structural relationships in data.

Counterterrorism: The act of protection against unlawful and hostile activities.

Homeland Security: Security rules and procedures set for protecting a homeland.

Monitoring Framework: Information science infrastructure that is used to monitor host system dynamics.

Publish-Subscribe System: A system where published information is automatically sent to a subscriber of that particular information.

RFID: Radio frequency identification.

Sensor Network: A network of low cost, batter powered, wireless devices with antennas.

Smart City: A city that is managed preemptively rather than correctively.

Chapter 2
City Competitiveness and Spatial Location

Diana Sarita Hamburger
Federal ABC University, Brazil

ABSTRACT

This chapter discusses the importance of Information Technology to spatial location and city competitiveness. Understanding spatial location in urban networks can be informed by economic geography concepts, especially those with insights into how urban areas form and develop. Relative distance to markets and the flow of goods and services influences the spatial position of cities, shapes how urban settlements evolve, and helps explain their distribution. Concepts like accessibility and centrality—and strategies for measuring them—can be used to determine a good place to locate a business or transportation hub. This chapter makes a case for the importance of considering information utilities, especially telecommunication networks, as important part of economic geography, and ultimately the growth and competitiveness of cities.

INTRODUCTION

This chapter addresses one component of the geographical models used to evaluate and measure cities' location advantages. The approach derives from the spatial analysis discipline, which emphasizes models and the importance of spatial location. The following discussion analyses how to describe and measure the spatial location in a manner that will contribute to understand cities competitiveness.

Cities are inter-related subsystems influencing each other. The economic analysis assumes inter-regional trade between them as a main component in competition analysis. The exchange of merchandise involves elements like goods and people′ transportation, traffic infrastructure - considering

DOI: 10.4018/978-1-61350-174-0.ch002

traffic routes, networks, stations, bus terminals, airport terminals, etc.

New processes and information technology provide a continuing source of innovation requiring organizational changes. The importance of information leads to a condition where relationships between urban centers are important in determining competitiveness. A wide variety of interactions must be considered to define the capacity of an economy (national or local) to transform the opportunities for innovation in technological change in fact (Bradford, 1994).

The three concepts strongly linked to the traditional definition of the region (distance, contiguity in space and friction) have undergone changes with impact on the modern region (and the urban system). The reduction in transportation costs reduces the importance of these factors and results in the increasing of communication as a location factor. In the past, the size of the regions was an important criterion in its definition. The size also entered into obsolescence as a criterion. What matters today is the structural complexity of an organized territory (Boisier, 1994).

Distance is a relation between two points, considering a geometric space. A distance is defined in a metric system. The metric can be the Euclidian distance between two points or take a more complex metric system, like the energy needed to surpass this distance (Brunet et al, 1993). The metric defined in a space can indicate if two points are so close that the distance between them can not be noticed. In this case, those points are contiguous. A description of space as point's connecting lines (that is usual for transportation and electric networks) correspond to the *graphs* mathematical model (as referred in Getis, 1989 and Jiang et al, 1999). In this conception, distance can be described by the effort needed to surpass it (as in Fotheringham, 1981 and 1983), sometimes expressed as friction of this space, resulting in a description of distance that can be called functional distance and can be measured in cost or time (Noronha, 1992).

To represent the urban system distance, contiguity in space and friction, the urban system has been diagrammed as a network. In this *graph* model, the transportation routes as lines and urban centers as nodes constitute the *graph*. In those models the spatial location of each urban center is considered one of the main factors. In the quantitative analysis, the distance can be measured in kilometers, time or cost to transpose the distance. The market size dimension is the population or income in the urban center.

The prerequisites for a modern region will be: flexibility and elasticity while facilitating factors in and out of networks with speed and timeliness. These important criteria suggest the structure of pivotal regions and organized complex territories identified in the scale of political-administrative division history (Boisier, 1994).

To have a better understanding of the urban system, there is a need for criteria and tools to evaluate those exchanges. In the competitive environment it is important to consider and measure information flows.

Economic conditions of global cities are based on the ability to store and process information and generate knowledge as stated by Castells (1998) and rescued by Abrahamson (2004). The argument presented is that the network to be described should be the connection between electric, telecommunication and transportation networks.

This approach can be applied in logistics, when deciding the best place to stablhish a deposit or to transportation procedures when describing the best place for a hub. It deals with a new spatial understanding where telecommunication flows are easier, more intense and should be considered in a more explicit way.

The information science advances and the evolution of spatial technology changed the spatial location and the spatial information. They became an asset and merchandise. Spatial location can be easily achieved through GPS devices, geoprocessing software's and geospatial databases (e.g. GoogleEarth).

Those technical and social changes that bring the knowledge and representation to the spatial location are also responsible for change in the importance of each location. The spatial characteristics change due to information technology and its social consequences. When considering the geographical space as an expression of the social relations, the meaning of the spatial location has to change altogether with social transformation.

The form in which social changes derives from innovation in information technology and can be expressed in the analysis of spatial locations is the object of this chapter, this requires a conceptual update of the spatial location.

The urbanization process is still transforming the world. The urban growth and expansion are part of a complex set of social and economic changes. The city is the place of concentration of activities and the center of regional integration. They are the nodes that link regional components.

Last decades technological changes created a new condition for interaction between distant agents. This change in the quality of spatial interaction requires an integrated analysis approach. There is a need for a conception that embraces the new forms of spatial interaction to comprehend urban systems. These changes are present in the infrastructure networks (especially electric energy, telecommunications and transportation). Quantitative and qualitative tools are required to apprehend those networks and their importance. In this sense the spatial separation concept seems like a nice manner to represent these variations.

The concept of spatial separation always refers to the various elements that can be exchanged, like people, goods, ideas, orders, etc. but the bibliography about it has failed in including variables that represent those factors.

GEOGRAPHIC APPROACH TO CITY COMPETITIVENESS

When dealing with urban systems it is clear that the connection between urban centers occur through the utilities and transportation networks. Those technological intensely dynamic devices are the crucial basis for production and consumption that subsidize the urban development.

The description of the urban locations considering their position in the utilities networks and their relation is essential to understand the urban system.

The cities size, measured by their population, GDP (Gross Domestic Product) or public equipments, articulated by the transportation networks infrastructure are criterion used to understand functionality, hierarchy and polarization between cities. Advances in telecommunications systems have transformed the regional connections. There is a thickening of the technical networks, enabling new forms of interaction between urban centers. The importance of the geographical position of the various urban centers changed to reflect these new connections.

In the current process, global circulation prevails over the production and the flows become increasingly important in comparing positions.

The existence of fluidity to the movement of ideas, messages, products or people, is interesting to the hegemonic cities. Objects are created and designed to encourage fluidity: pipelines, canals, highways, airports, etc. This fluidity is increasingly responsible for changes in the value of objects and places (Santos, 1996).

City competitiveness has been described in geography as related and dependent on spatial location. In this sense, spatial location should be considered as a relative position that is defined by characteristics and distribution between cities.

The travel distance has been proposed as the measurement to evaluate distance to consumption markets, distance from raw material origins or concurrent distribution. The model was improved with distinctions in components like market segmentation, a more or less specialized market, analysis of the spatial distribution considering distances through the transportation network instead of the geometric distance. In the evaluation of an industrial or commercial location the transportation characteristics were considered.

Acquire and maintain spatial information include the need to measure, conceptualize, represent spatial features and spatial interactions that explain geographical relations.

To represent the new spatial relations that result in the inter-dependence of the telecommunication and transportation accessibility a new model should be developed.

The Meaning of Distance

Space can be understood in, at least 3 different ways (Sui, 1999): geographical space where social relations are a constituent, concept that expresses where we live; mathematical space that usually correspond to the representation of the geometrical space; and the subjective space, an individual psychological experience. The spatial analysis that subsidizes the city competitiveness considers geographical models that use the mathematical space to represent the geographical space.

So, there is a need for a new form to measure distances, based in the separation through the utilities networks like telecommunication, transportation and electric networks.

The Euclidean distance as studied in school geometry is adequate to most of the spatial location references, but when dealing only with spatial location. But when considering the spatial interactions, as in the geographical models to represent city competitiveness and business competitiveness among enterprises, the main issue is the effort needed to surpass the distance.

In the case of obstacles, a first solution is the distance through the transportation network, (also called *Manhattan* distance) (Goodchild, Maguire and Rhind, 1991). This diagram considers the transportation network, with prohibited directions and where symmetry may not occur. So when travelling by land area, not always the smallest distance is the Euclidean distance.

When considering urban environment, the straight line between two points may not be possible to drive through because of obstacles (Gatrell, 1991). Another expression used to describe this effort to surpass distance is "functional distance" meaning the amount of effort, money, or energy necessaries to overcome it.

Spatial interactions tend to occur over shorter distances. The conventional concept of "friction space" is based on the notion that distance usually requires some amount of effort, money, and / or energy to be overcome. For this reason and because of this "friction", the spatial interactions tend to be more common among smaller distances, the intensity of interaction declines with distance. The distance will have a greater dampening effect on some types of interaction than in others. A related concept is that of a "zone of lower friction" in which the distance and cost (any measure) to overcome does not seem important (Slobodian, 2000). The distance in this perspective can be described as spatial separation. In this sense, the telecommunications transformation result in a complete change in city competitiveness.

The abstract description of the surface that subsidizes this analysis is understood as follows. The distribution of elements in the surface usually is irregular, being concentrated in some areas. Those areas are surrounded by other surfaces. In the case of urbanized areas; the interaction usually is presented as a network with nodes connected by lines. This description considers the concentration of economic activities, population or markets connected through transportation and telecommunication links. Any degree of urbanization can be described this way. A region might be defined by its points and lines or nodes and connections (George, 1965).

The first property of the networks is connectivity. Transportation and telecommunication networks express the potential mobility and are a condition to an increase in circulation. Through the connections or absence of them, infrastructure networks can connect or isolate (Dias, 1995). Geographical connections are present everywhere and every time because they are an attribute of space. Those attributes are essential in the new reality where time/space relations have been accelerated (Souza, 1996).

The insertion of new centers to an economic system through the interchange of goods, people and information is referred as realizing the spatial integration process where urban centers depend on the transportation network (as in Corrêa, 1997; Machado, 1997). This concept relates to both the idea of development poles and with the theory of central locations. To characterize those networks there should be measures that make it possible to evaluate access, connectivity and to establish indicators to understand those measures.

The linkages among cities became the node to an international system of cities, surpassing the nation's boundaries, which now depend upon the linkages among cities (Campbell, 1991).

The increase in the interdependence in economic exchanges among cities and nations can be noticed from the bird's eye but the massive flows of information and the numerous connections between them can not be viewed as clearly (Abrahamson, 2010).

Telecommunications and the New Meaning of Distance

The Internet was presented as the most equitable and most transformative new technology. It is often placed as a technology that enables the "shortening distances" and establishing a limit to the concept of place ("placeless"), or at least to its importance. In fact, there is a lack of empirical evidence to evaluate the real effect of these new technologies in the socio-economic development.

Understanding the Internet as the most equitable of new technologies, contradicts the assertion that only largest cities in a country have capacity to access (BNDES, 2000). Indicators are needed to evaluate: if the internet is an urban phenomenon or underground, it represents an effective "shortening distances" and how this influence the telecommunications means urban development, and which socio-technical factors that influence variation in telecommunications development (Moss et al., 1997).

There is a new question regarding the concept of distance. References are increasing in bibliographies considering the effect of telecommunications in reducing the meaning of distance. This shows the transformation in the possibilities of interaction between distant places, especially in the velocity of those relations (Gould, 1996; Joly, 1999; Miller and Wul, 2005).

This change in the spatial interactions has two aspects. One is the technological change that generates many new opportunities for information exchange and the other is the social transformations reducing national and regional cultural restrains, generally called globalization.

Likewise the "*death of distance*", as put by Kolko (1999) does not occur; it is a true transformation in the role of space in social relations. These changes are not, of course, only technological but also institutional and social. Still, the role of technology in these changes is indisputable, as shown in Castells (1998). Literature is increasingly demonstrating the inadequacies in considering the internet as having an equal distribution (Castells, 2003; Abrahamson, 2004).

The concept of space based in the social relations that occur in space, cannot be in the form of representation of space, free to seek ways to address these changes. One of the difficulties presented in this sense is the lack of common theoretical framework for different approaches to space.

Even though we understand that those changes are not independent, here we will be referring only to the technological component.

There is a need to redefine the traditional measures that emphasize location of potential destinations, from one origin zone that don´t consider the informational dimensions that change accessibility (Hanson, 1999). Even the consideration of cost or time measurement to evaluate distance does not represent this new condition. We can call this new spatial characteristic that has different possibilities to evaluate, spatial separation. Spatial separation is how we can measure as the relative spatial position between urban centers. It can be evaluated considering different utilities networks as electric, telecommunications and transportation networks. The integration on those networks is described in the studies that discuss the new characteristics of cities (Mitchell, 2005, 2010; Reader, 2009).

Each technical network requires specific spatial separation measurement to evaluate the biggest or smaller proximity between spatial locations. At least two different networks should be considered: telecommunications and transportation. To each of them, the parameters should be studied. It can be the velocity to the telecommunication data and the amount of energy available (in kilowatt/h, for example) or other. The tendency is to use the measure of cost as the unit that makes it possible to integrate the analysis into a single measure.

Telecommunications Data

The need for consistent and reliable data is a requirement for the development of these indicators. In telecommunications there are few sources of reliable data, because of the competitiveness of the sector, especially sources of spatially referenced data.

A good discussion on Internet data can be rescued from Moss et al. (1997). Providers have reasons to hide data for economic reasons. The Internet operates, primarily, with an infrastructure network in existence, which increases the difficulty of obtaining data with comparable levels of geographic aggregation and the few elements that can be measured have limited relevance. It is difficult to establish a theoretical framework that applies to urban research. Number of computers connected full time, name of the location of the domains with domain registers, and the capacity of backbone networks where tested as empirical measures.

The array of computers connected to the Internet full time, located by address, can find the number of hosts, by county. This measure indicates the location of major equipment corporate network nodes. It is a useful measure. The results confirm the expectation of high concentration in a group of metropolitan areas and the presence of academic and technological centers. This measure seeks to identify the adoption of these technologies in a wide range of industries and population groups; however, it fails to differentiate the various types of equipment. It is a measure based purely on the technical organization of the Internet to infer their adoption in different regions. The location of the institution, given from the zip code, represents the social phenomenon. But different sizes of organizations cannot be noticed. The geographic data are not reliable, since they do not always relate to the main location of the user's domain, quantifying the institution, but not the scales. Nodal cities showed high concentrations of Internet domains. The expansion of internet in the last years still resulted in isolated areas.

As concerns the Backbone networks intended to quantify the flow of information between cities. Through lines such as highway networks and channels to these networks are the physical manifestation of the track information. The aggregate analysis of the topology and capacity of the backbone network indicates a high degree of centralization in employment and use of Internet technologies. While the count of hosts described a phenomenon and complete physical count of domains and presented a logical and virtual phenomenon, the measure of capacity of the backbone is a hybrid (Moss et al., 1997).

Theoretically, any city in a given network has access to all locations served by networks.

However, there are traffic congestion, packet data connections and indications that providers have established direct links in routes of greatest traffic. These patterns indicate the superiority of direct connections and their importance to the ability to import and export information. To these authors, the number of computers connected full time is not very useful as an indicator, since it expresses the purpose or function of the system, domain name registration and location are only effective for institutions, and especially to their headquarters, and propose the ability of network backbones as the most conclusive indicator. In the United States, among metropolitan areas, most of the capacity of backbone network is used to connect these regions with each other, rather than connecting the cities of less hierarchy (Moss et al., 1997).

These areas have developed an urban system highly interconnected and complex. In fact, much of the literature of the Internet phenomenon highlights its dissociation with other cities. An important factor is the allocation of network capacity. As a result, larger cities have access to greater volumes of data, faster and at less cost to the infrastructure of the Internet that cities of comparable size, but less well-leased. Some prerequisites influence the allocation of these technologies, such as producing areas of information for the Internet and consumers of Internet information. Most of the telecommunications technologies have a physical infrastructure (antenna, satellite, etc..) that can be evaluated and rated and that determine the interaction potential of a given location with distant sites. The same happens with other network utilities. It is important to mention that the soft infrastructure (backbone and virtual domains), or electronic equipment easily reconfigured (host computers and fiber optic networks), enables very rapid changes. By definition, the Internet is very volatile and in permanent changes (Moss et al., 1997).

Urban utility networks tend to be largely ignored or taken for granted in urban economic policy. Studies of urban utility networks tend to focus on isolation, dealing with simple networks, rather than adopt an integrated perspective and cross-over periods of time, or that indicate the relationship of the development of networks with broader political and economic trends (Graham & Marvin, 1999).

Even though the telecommunication connections could allow a more complex urban system it still did not generate equitable access to information for large geographic areas.

Evidently, the network and flow characteristics are related to the nature of the element that goes through it. Likewise, the spatial form and type of connection form a complex set of interrelated relationships.

The network structure is the shape of their interconnections in its qualitative and quantitative aspect. The connection types, the form of exchange, the existence of reciprocity, density and intensity of flows, etc..

All the connections between distant points on the surface should be described. They can be described considering the components that should be exchanged and the networks that make it possible. Each network should be described as much as the relations between them. Some networks that should be evaluated in urban systems: transportation network (considering the intermodal connections), electrical networks; pipelines; the conventional telephone network, fiber-optic cables; and other links through the transmission of electromagnetic radiation. As it was used to transportation data all networks must be studied in their main components: form and structure, element transmitted, connection type and flow characteristics.

These networks can provide a more detailed classification according to the capacity of the system, the material that composes or its management system, for example. Likewise, other networks can be described. An example is the network of sewers that can be understood as a particularity of the water network, and the water distribution system (Mahoney, 1991).

Analysis and Synthesis Needs to Synchronize Urban Subsystems

The emergence, growth and deepening of functional specialization between urban centers depend on a connection between them. The development of information technologies and communication involves a wide variation in the forms of connection between the urban centers (telex, telephone, internet, mail, etc.). The potential transformer (or not) of these forms of connection between the cities is also on the ability of municipalities to absorb those elements to improve competitiveness.

Economic sectors develop their activities using different means. The methods used vary depending on the production of goods stage or completion of a service. The position in the networks enables, or not, the development of certain activities in certain cities, constituting each city potential, allowing it to turn into a central decision-making city, a financial center, or a production center, etc. For example, a newspaper can use a plane for distribution, but information should be collected in central areas and that is where those who control the networks usually are located. There's an argument that the decline of the relationship between rate and distance for telecommunications, would result in a scattering of human settlements, which should be imminent. There is decentralizing cities spreading and becoming complex systems. Even though, telecommunications allow the evolution of an increasingly complex system, with multiple connections and hierarchies, the place remains important.

The position of cities in the global economy depends on their ability to generate process and exchange information and, therefore, to study the emerging structure of systems for exchanging information electronically, allows knowing the characteristics of cities.

The deployment of networks of utilities requires investment and is the basic infrastructure for the functioning of modern economic and social system. The deployment of electricity networks or a pipeline requires the reproduction of a pattern of private space of consumers and a supply system that includes a significant investment in site (Graham & Marvin, 1999). The capital intensity and reliance on networks of local utilities are of fundamental importance in understanding their relationship with cities. The geographical scale and extent of the network operation is also essential for this understanding; they indicate how the network can overcome the vulnerabilities arising from dependence on the conditions of the local economy. Although dependent to individual cities and towns, utilities are involved in programs or development regional projects.

The relationship between networks of utilities and urban system can be summarized considering: how extensive technical and institutional systems are related to the society of which they are components. The infrastructure of utilities is an important factor for economic development and social space of cities, regions and economic spaces. The cost, quality, availability and reliability of network services, and how they vary in space, the technologies they employ, and how their development is regulated are factors that influence the development and restructuring of cities and regions (Graham & Marvin, 1994). The recent renaissance of the economic policy analysis of cities has failed in identifying the importance of infrastructure for utilities and their regulation and failed to recognize the crucial relationships between cities and utilities.

Urban utility networks tend to be largely ignored or taken for granted in urban economic policy (Graham, 2009). Studies of networks tend to focus on isolation, dealing with simple networks, rather than adopt an integrated perspective and cross-over periods of time, or that indicate the relationship of the development of networks with broader political and economic trends. As a result, it is difficult to understand the links between utilities and city development. There is a need to include utilities in the assessment of the changing political economy of cities, because

Figure 1. Network connection representations

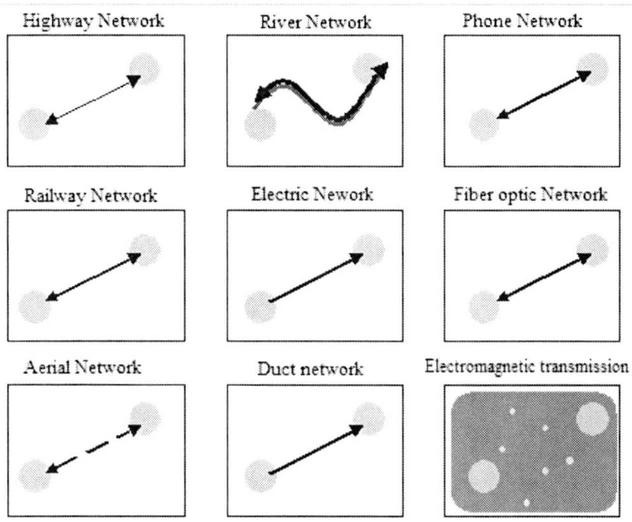

of its symbiotic role in the economic, social and geographical range of cities and their role as actors in urban policy arena.

SPATIAL LOCATION FOR CITY COMPETIVENESS

The spatial location does not appear isolated in contributing or avoiding cities competitiveness. It is a component of the economic relations that is so integrated in the daily lives that we only notice the need for logistic solutions that are used to send systematic messages, transport goods or people and to acquire needs.

Some elements are required for an interaction between non contiguous places occur. There should be accessibility allowing people or goods to flow from one place to the other. There should be mobility conditions for the interaction to exist. Those are conditions for any business opportunity. The form we have to analyze if those conditions are present is mapping and representing spatial infrastructure to each urban center.

Any location and any city should be evaluated considering its accessibility. The accessibility being the spatial position measured in the articulated networks. The proposal on how to represent network connections is presented in Figures 1, 2 and 3. These figures represent that the lines connecting cities networks are bi-directional and others have only one direction (represented by the arrow points). Moreover, some lines do not present a physical way, like the aerial transportation (Figure 1).

On the nodes characteristics presented in Figure 2, it can be observed that not all urban centers (small dots) represent nodes (large dots) in all networks. Transmission of electromagnetic radiation with no defined channel does not constitute a network, as defined above. The networks present purposes and forms of spatial interaction. Networks of transporting goods and passengers predominantly use the highway, airway, railway and waterway.

The drainage systems and electrical, as well as pipelines, are unique to certain elements. Although various forms of connection are being leveraged by investments in infrastructure, the major recent changes is given in telephone networks, fiber optics, and the routes that go through the transmission of electromagnetic radiation.

Figure 2. City representation

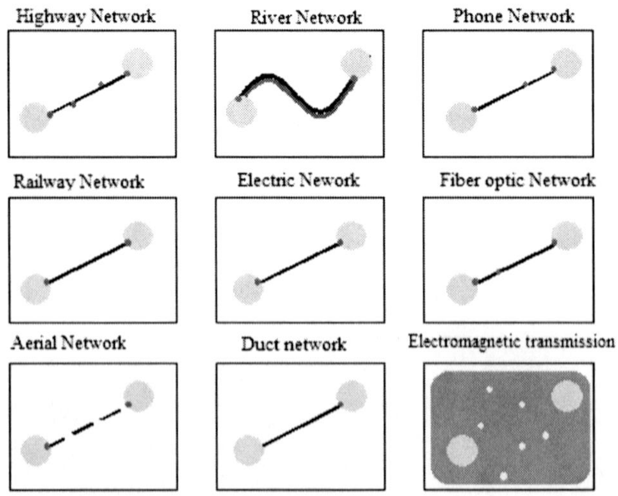

Figure 3. Integrated urban system network

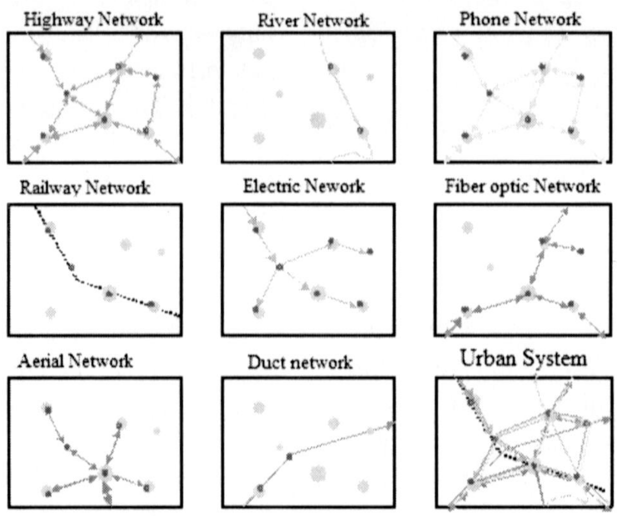

The links that go through the transmission of electromagnetic radiation do not connect necessarily fixed points in the territory. In this sense they do not have a format, as the established network. However, by allowing a link between distant points in the territory, they can be considered network in another classification system. Technological development makes possible a growing exchange between the forms of connection. Combination products are generated in networks that were previously autonomous.

From the viewpoint of the process of transmission through space, there are organizational forms that use the links separately and others that use them in an articulated and operational way.

The networks are usually driven in an integrated manner. The use of fiber optic cables often requires electricity, for example. The interconnected networks comprise technical composites. There are technical networks that match the networks made, such as internet, for example.

Trade in goods and trade in services often rely on a physical channel to be displaced. It seems possible and necessary to dismember goods and services according to their potential for transport and movement across communication systems more or less agile. The separation of transport systems and communications has promoted relations between distant points, but do not have a role in regional integration as it allows an action of a center "independent" of its surroundings. This condition is evident in offering distance learning courses, emergency doctor by phone, etc.

The same city might have a different meaning in each network. Cities potential must be represented as geographical networks. Urban centers are the main focuses of the various spatial interactions necessary for social reproduction, so preponderant interactions, are intended. The urban network is a complex synthesis of geographical networks (Correa, 1997). The presentation on how those networks can be represented is shown (Figure 3). Each network is presented with its specific characteristics: directional flows or not (indicated by arrows), including all urban centers (dark dots), or not and defining spatial connections.

Characteristics of space can be expressed in terms of attributes associated with specific spatial units. The features that characterize a space-saving high technical level, with strong industrial impact, depend largely on the density of land use. Urban areas have a dense spatial distribution. The spatial location of urban cities considering those networks is the form to evaluate spatial location for city competitiveness.

CONCLUSION

The integration of different areas of the globe in an interconnected economy and political system is associated with a denser network. The development of different activities in one place or location tends to attract other activities and deepen those relations. Cities centralize many networks.

Many flows occur in an integrated form. There is a discontinuity between the city and its surroundings, explicit in its physical aspect, which can be observed by the change between discrete activities and forms of land use. It can be verified that there are substantial exchanges between urban cores discontinuity. These exchanges can take place between near or far cities. The exchanges that characterize the financial system take place between centers without limits of distance or physical restrictions, other than those of the communications network.

The analysis of utility networks, study and planning of cities, is important in assessing the ideal density for urban areas. This analysis starts from the contrast of two components: the quality of life and environmental quality, as opposed to a greater or lesser extent infrastructure, which represents rising costs for the municipality. On the connection between urban centers, the greater or lesser extent and density of activities that enable a fruitful interaction, depending on the extent of necessary infrastructure, is also a complex issue.

Technological development has enabled a significant change in the speed of communication between different regions. This high speed has changed the meaning of distance when it comes to financial flows and information worldwide. The space seems limited when it comes to financial flows or communication ideas.

There is a need to improve the ways to describe and measure the utilities networks, considering their form, connections, the element they transport and the flows characteristics.

But it is still questionable to what extent this spatiality was overcome when one considers, for example:

- the decision of large firms and corporations still located in major centers, a sense of origin-destination for orders;

- major producers and transmitters of information are located in major centers, resulting in a sense of origin-destination for information;
- there are major financial centers, resulting in a sense of origin-destination of choice for financial flows.

With respect to services, an evaluation of the bond or not with the spatiality becomes more complex face of the diversity of this activity. But health services and education have been considered as indicators. So services that depend on the movement of people and goods still have a higher spatial compared to services that depend on information exchange.

The urban networks study traditionally emphasizes the movement of goods and people. Especially with the technological advances, meaning a change in this direction, the transportation of persons and property remains the aspect of the urban network where the overcoming of distance remains more complex. In this sense, the spatial model is appropriate, first, with emphasis on municipal transportation. Later we will evaluate the adequacy of this model to other aspects that make up the urban network. With the advancement of new technologies, the issue of networks emerges with a new form, involves an analysis that considers its relations with the urbanization process, with the territorial division of labor and the increasing differentiation between cities.

ACKNOWLEDGMENT

This work was supported by FAPESP, whom we gratefully acknowledge (Grant 02/13834-2).

REFERENCES

Abrahamson, M. (2004). *Global cities*. New York, NY: Oxford University Press.

Boisier, S. (1994). *Los escenarios del desarrollo descentralizados en America Latina (Serie Investigación, documento 94/09)*. Santiago, Chile: ILPES.

Bradford, C. I. Jr. (1994). The new paradigm of systemic competitiveness: Why it matters, what it means and implications for policy. In Bradford, C. (Ed.), *The new paradigm of systemic competitiveness: Toward more integrated policies in Latin America*. Paris, France: OECD Development Centre Documents.

Campbell, J. (1991). *Map use and analysis*. Dubuque, IA: Wm. C. Brown Publishers.

Castells, M. (1998). *The rise of the network society* (*The information age: Economy, society, and culture,* vol. 1). West Sussex, UK: Wiley-Blackwell Publisher.

Castells, M. (2003). *The Internet galaxy. Reflections on the Internet, bussiness and society*. Oxford, UK: Oxford University Press.

Chrisman, N. (1992). *Exploring geographic Information Systems*. New York, NY: John Wiley & Sons.

Dias, L. C. (2003). *Redes: Emergência e organização*. Retrieved from http://www.cfh.ufsc.br/~imprimat /artigos/leila-dias.htm

Fotheringham, A. S. (1981). Spatial structure and distance-decay parameters. *Annals of the Association of American Geographers. Association of American Geographers*, *71*(3), 425–436.

Fotheringham, A. S. (1983). Some theoretical aspects of destination choice and their relevance to production-constrained gravity models. *Environment & Planning A*, (15): 1121–1132. doi:10.1068/a151121

Gatrell, A. C. (1991). Concepts of space and geographical data. In Maguire, D. J., Goodchild, M. F., & Rhind, D. W. (Eds.), *Geographical Information Systems* (pp. 119–146). New York, NY: Longman Scientific & Technical.

George, P. (1965). *Cidades, redes urbanas, região. Conferências no Brasil*. Rio de Janeiro, Brasil: Instituto Brasileiro de Geografia FIBGE.

Getis, A. (1989). A spatial causal model of economic interdependency among neighboring communities. *Environment & Planning A*, (21): 115–120. doi:10.1068/a210115

Gould, M. (1996). *Modeling third wave (Virtual) accessibility*. Retrieved from http://www.ncgia.ucsb.edu/ conf/BALTIMORE/authors /gould/paper.html

Graham, S. (2009). *Disrupted cities*. London, UK: Routledge.

Graham, S., & Marvin, S. (1996). *Telecommunications and the city: Electronic spaces, urban places*. London, UK: Routledge. doi:10.4324/9780203430453

Hanson, S. (1999). Reconceptualizing accessibility. In Janelle, D. G., & Hodge, D. C. (Eds.), *Information, Place, and Cyberspace* (pp. 267–278). Berlin, Germany: Springer-Verlag.

Jiang, B., Claramunt, C., & Batty, M. (1999). Geometric accessibility and geographic information: Extending desktop GIS to space syntax. *Computers, Environment and Urban Systems*, *23*, 127–146. doi:10.1016/S0198-9715(99)00017-4

Joly, O. (1999). Geographical position: *State of French art of spatial accessibility indicators* [Working paper]. France: SPESD (Study Program on European Spatial Development) working group 1.1. Retrieved from www.nordregio.se.spespn/files/1.1annex5

Kolko, J. (1999, November). *The death of cities? The Death of distance? Evidence from the geography of commercial Internet usage*. Paper presented at 27th Annual Telecommunications Policy Research Conference 1999, Alexandria, VA.

Machado, L. O. (1992). Sociedade urbana, inovação tecnológica e nova geopolítica. Boletim de Geografia Teorética, AGETEO. *Rio Claro*, *22*(43), 398–403.

Mahoney, R. P. (1991). GIS and utilities. In Maguire, D. J., Goodchild, M. F., & Rhind, D. W. (Eds.), *Geographical Information Systems* (pp. 101–114). New York, NY: Longman Scientific & Technical.

Miller, H. J., & Wu, Y. (2005). Gis software for measuring space-time accessibility in transportation planning and analysis. *GeoInformatica*, *4*(2), 141–159. doi:10.1023/A:1009820006075

Mitchell, W. J. (2005). *Placing words symbols, space, and the city*. Cambridge, MA: The MIT Press.

Mitchell, W. J., Borroni-Bird, C. E., & Burns, L. D. (2010). *Reinventing the automobile: Personal urban mobility for the 21st century*. Cambridge, MA: The Mit Press.

Moss, M. L., & Townsend, A. M. (1997, November). *The role of the real city in cyberspace: Understanding regional variations in Internet accessibility and utilisation*. Unpublished manuscript, Taub Urban Research Centre, New York University, NY.

Noronha, V. T., & Goodchild, M. F. (2005). Modeling inter-regional interaction: Implications for functional regions. *Annals of the Association of American Geographers. Association of American Geographers*, *82*(1), 86–102. doi:10.1111/j.1467-8306.1992.tb01899.x

Poltras, C. (2009). Downtowns, past and present. *Urban History Review. Revue d'Histoire Urbaine*, *37*(2), 3.

Reader, J. (2004). *Cities: A swift examination of the pleasures and problems of city life*. New York, NY: Grove Press.

Santos, M. (1985). *Espaço e Método*. São Paulo, Brasil: Nobel.

Santos, M. (1996). *A natureza do espaço, técnica e tempo, razão e emoção*. São Paulo, Brasil: Editora Hucitec.

Slobodian, B. (2000). *Spatial analysis*. Retrieved 9/4, 2000 from http://www.mala.bc.ca/~slobodian /concepts/GC_20_8.htm

Sui, D. Z. (1998). GIS based urban modeling: Practices, problems and prospects. *International Journal of Geographical Information Science*, *12*(7), 651–671. doi:10.1080/136588198241581

Chapter 3
The Role of GIS in City Competitiveness

Khan R. Rahaman
University of Minho, Portugal

Júlia M. Lourenço
University of Minho, Portugal

ABSTRACT

Virtually every city and region is engaged in activities to improve their relative global competitiveness. The Geographic Information System (GIS) is one of the powerful tools of information storage and information access, providing spatial data to different stakeholders and cities across the world. This chapter will highlight the role of GIS technology in empirical assessment of the competition among cities or regions, using a variety of data assembled by many different individuals, businesses, and institutions. This valuable information can be used in decision-making by stakeholders who are taking part in the competition and can be disseminated, accessed, and updated in a dynamic way. This chapter discusses the origins of urban competitiveness, dynamics and functions of competition, and current and future research possibilities made possible by GIS.

INTRODUCTION

While the economic literature has a presumed "objective" – for example some measure of competitiveness or economic product, the "governance" literature, characterized by institutional arrangements, fails to assess the outcome of those changes (Prosperi & Lourenco, 2010). Information technology has brought opportunities both for private, public and individual actors within the city life. Actors now can easily access the information and become more competent in the global market (Turok, 2004). In the early sixties, cities were competing mostly with industrial and labor inputs within the city territory. The situation has changed because of the globalization and now cities are mostly competing with information, skills of human resources and macro economic

DOI: 10.4018/978-1-61350-174-0.ch003

aspects. A question remains (Sim et al., 2003), if telecommunications have not made cities obsolete, have they at least altered the economic function of cities in a global economy? For this, the major dynamics have been shifted over last twenty years based on some notions as: i) command points in the organization of the world economy, ii) key locations and marketplaces for the leading industries of the current period finance and specialized services for firms and iii) major sites of production for these industries, including the production of innovations in these industries (Sim petal., 2003). Private sectors were predominantly competing among them in the early ages. However, public sectors have come into action from the late 1980s and early 1990s to brand their cities. In the age of internet, telecommunication and digital media world, nature of competitions among cities is changing rapidly. Both private and public sectors are working together to brand their cities in this globalization era so that both the parties can take economic, environmental and social advantages of investments and innovations. Cities have to ensure that their institutional structures cater to the rising demand from business and industrial corporations. There is an increasing need for institutional thickness and free market systems. In the modern world, cities and regions are competing with each other to prove the potentialities they have to offer. For example, they do compete for assembling a skilled and educated labor force, efficient modern infrastructures, a responsive system of local governance, a flexible land and property market, high environmental standards and a high quality of life. Competitiveness in urban regions has emerged from different directions. Some researchers have argued the influential factors as: infrastructure, memberships in networks, institutional and effective policy-networking, quality of living environment, human resources as well as existing number of firms and industries are the major determinants of competition (Lengeyl, 2003; Begg, 1999). Besides, interactions among actors, institutions and municipalities determine the level of competition among cities or urban centers. These competitions have both positive and negative impacts in terms of society, economy and environment. Common descriptions of the competitive process as 'displacement' and 'zero-sum' imply on one hand, it is unproductive and to be discouraged, since one area's success may only come at the expense of others. On the other hand, pressure on local bodies may prevent complacency and encourage timely delivery of suitable economic infrastructure, services and skills. The pyramid model of competitiveness shows sustainability at the top, essential conditions that can bring a positive outcome of the competition among cities, in the middle layer and at the bottom, there are different policy inputs (Lengyel, 2003). Business environment, physical and knowledge infrastructures are some major policy inputs that should be considered to achieve the sustainability of cities. To implement these policy instruments, essential conditions are considered to be business performance, productivity, prices and products, and available labor supply. Geographic Information Systems combine geo-data with other types of information and generate maps for better recognition of local problems and more effective solutions. GIS directly supports a variety of local activities (land-use and urban planning, economic development, infrastructure, transportation and emergency management, educational planning, tax assessment, environmental monitoring, public information services) to know the extent of competition among cities and regions (Pawlowska, 2001). GIS is a set of tools that enables combining digital map with non-spatial characteristics and relationships associated with geographic objects like:

- **Automated mapping technology:** manipulating map information;
- **Database management:** managing attribute data;
- **Land records information:** providing the cartographic and attribute data;

- **Topological data structures:** Providing definitions of spatial relationships among points, lines and polygons
- **Spatial analysis capabilities:** Retrieving, manipulating and displaying map and location-related attribute data, etc.

The role of such user-friendly and interactive technology can bring more dynamic and accurate information based on spatial and attribute data to measure the competition among cities, regions and urban areas in the world. More important, GIS can help assemble sets of factors or criteria and can be worked together upon different scenarios. Besides, GIS can offer the opportunities to compete among cities according to the needs and values of each specific society. It considers each and every individual criterion and can measure the performance. The service GIS can offer is very much useful for individuals, society, industries and businesses to rationalize the competition among cities or urban areas based on real time data. This chapter will highlight the role of GIS technology to assess empirically the competition among cities or regions based on different criteria and values assembled by different individuals, businesses, institutions, etc. Apart from individual activity, location of firms and investment opportunities a GIS system can be prepared to provide interactive information on, for example, topics such as existing infrastructures, the quality of life and living environment. In sum, GIS can provide accessibility to valuable information for decision-making by stakeholders who are taking part in the competition and can disseminate information about the actors and the attained level of competition in a dynamic way.

ORIGINS OF COMPETITIVENESS

Since the late 1980s, Paul Krugman has rapidly established himself as amongst the most prominent of economists, particularly for his work on 'new trade theory' and competitiveness (Boddy, 1999). Two of the main reasons for the notion of competitiveness gaining currency are the increasing international mobility of capital and more open national markets – globalization for short (Turok, 2004). Due to the process of globalization, cities and regions have been confronted with increasing competitive pressures, and this made policy-makers at the regional and urban level think about how to respond (Begg, 2002). Concepts like industrial districts (Becattini, 1990), innovative milieus (Camagni, 1991), learning regions (Asheim, 1996) and regional innovation system (Cooke, 1998, 2001) have described the crucial importance of externalities in maintaining competitiveness that extend or cross the boundaries of individual firms but which operate within the boundary of a territory. The emergence of new economic powers, particularly in Asia, and generally more competitive markets for products and services have intensified the pressure of costs and increased economic insecurity and the risk of instability. A variety of micro-economic, supply side measures have been put forward to improve the efficiency of firms' internal processes and to enhance the quality and value of their products, and thereby secure their share of world markets and jobs. Competitiveness is not an end itself rather more an indication of the drivers and dynamics of economic success over time. Competitiveness thus encompasses three important variables such as:

a. Ability of local firms to sell their products in contested external markets (trade)
b. Value of these products and the efficiency with which they are produced (productivity) and
c. Utilization of local human, capital and natural resources (employment).

Nowadays, competitiveness is being measured through different criteria. World Economic Forum (Global Competitive Report, 2009) has complied twelve pillars to assess the global competitiveness among countries in the world. Those pillars are as follows:

1. **Institutions:** The institutional environment is determined by the legal and administrative framework within which individuals, firms, and governments interact to generate income and wealth in the economy
2. **Infrastructure:** Extensive and efficient infrastructure is an essential driver of competitiveness. It is critical for ensuring the effective functioning of the economy, as it is an important factor determining the location of economic activity and the kinds of activities or sectors that can develop in a particular economy.
3. **Macro economic stability:** this is concerned about the government policy and business environment within a country or territory.
4. **Health and primary education:** It ensures the healthy workforce to add values in labour supply to receive more investments in the city territory.
5. **Higher education and training:** In today's globalization system, economy requires economies to nurture pools of well – educated workers who are able to adapt rapidly to their changing environment.
6. **Goods market efficiency:** Healthy market competition, both domestic and foreign, is important in driving market efficiency and thus business productivity.
7. **Labour market efficiency:** Labour markets should have the flexibility to shift workers from one economic activity to another rapidly and at low cost, and to allow for wage fluctuations without much social disruption.
8. **Financial market sophistication:** An efficient financial sector allocates the resources saved by nation's citizens as well as those entering the economy from abroad to their most productive uses.
9. **Technological readiness:** Information and Communication Technologies (ICT) have evolved into the general purpose technology of time, given the critical spillovers to the other economic sectors and their role as efficient infrastructure for commercial transactions.
10. **Market Size:** The size of the market affects productivity because large markets allow firms to exploit economies of scale. Traditionally, the markets available to firms have been constrained by national borders.
11. **Business sophistication:** It is conducive to higher efficiency in the production of goods and services.
12. **Innovation:** It is particularly important for economies as they approach the frontiers of knowledge and the possibility of integrating and adapting exogenous technologies tends to disappear.

Michael Porter (2007) has also emphasized four different aspects of competitiveness as:

a. The quantity and quality of available business inputs—human resources, for example, or transportation infrastructure;
b. The rules and incentives that govern competition—such as policies that protect intellectual property, ensure transparency, safeguard against corruption, and encourage investment;
c. The size and sophistication of local demand, influenced by such things as standards for product quality and safety, consumer rights, and fairness in government purchasing; and
d. The local availability of supporting industries, such as service providers and machinery producers

Competitions can only be visible when the rapidly changing dynamics are being considered. The above criteria are being considered to assess the global competition among countries of the world and based on this information Switzerland was the number one in a list of 136 countries.

India was placed on the 50th position and, at the bottom, there was Burundi.

Geographic maps can plot this competition visibly better based on socio-economic information, environmental aspects and business possibilities around the world. Political chaos often denies the investors to put their money into danger. These criteria can be easily incorporated in a GIS platform to take the advantage of digital media.

CITIES IN THE INFORMATION ERA

By the end of 1980s, it had become clear that the rise of what was then called the 'information sector' would have a big impact on the functioning of cities and regions (Berg & Braun, 1999). According to Beniger (1986), the transition to an information society had already started at the turn of the century. Castells (1989, 1996) concludes that we have entered the information age or era. There are different analyses of the implications of the information era, both in general and for cities. The information era symbolises a whole set of changes, that we consider as the start of a transition to a new stage of urban development. Hall (1995) points to a set of major economic, political and social factors (globalization, transport infrastructure, integration, information technology, the opening up of new areas) that affect the urban system. One of the driving forces behind the information age is the rapid and continuous development of information and telecommunication technology. Moreover, the combination of information and telecommunication changes society fundamentally (Berg, 1999). What are the implications of these technological developments on the use of space: the spatial behaviour of people, firms and other organizations? This question remains as usual to know the physical location and its importance. Clearly, these technological developments are all the more important to economic activities that rely heavily on information and knowledge. ICT has boosted the informational economy and the exchange of information, but other trends have also reinforced the importance of information. For a large group of entrepreneurs, information has become their end product or a key element in their services. But information has also become a very important asset in other economic value chains such as: transport, banking, marketing and sales, tourism etc. (Berg, 1999). These activities need to be accessible for, and have access to, information at both ends of their production process. An example can be found in the banking sector where parts of bank operations have been moved (back-office activities) to cheap labour countries. ICT enables these types of firms to follow a more cost-efficient location strategy, moving activities for which physical proximity to other business units is not essential, to cheaper locations. New development very often emerges in an urban environment. Inner cities in particular offer a climate where trend – setters like to be. Businesses that need to respond to and spot trends in music, fashion, design and graphics all benefit from an urban location. Furthermore, cities historically have been the place where information/knowledge and educational institutions like universities and other research institutes have located, giving these larger cities another advantage over the periphery. In such a competitive environment, the old 'certainties' no longer exist: although the geographical situation is still relevant – cities in the centre have an easier time than cities in the periphery – it is no longer as dominant as it used to be. The quality of the living environment, the level of the cultural services and access to knowledge are examples of important location factors in such an environment in cities of today. On one hand, small cities in deprived regions face problems in adapting to the demands of the new knowledge economy. An urgent planning challenge is 'finding a place' for such cities and their regions in this economy. As they no longer attract investment which is increasingly drawn to mega-cities and core scientific regions (Bennewoth, 2006). Universities can play an animator role in these new broad based governance networks that is mobilising through initiation and vision building. GIS allowing manipulation of data may act as a branding tool of remote locations facilitating communications and information availability.

On the other hand, administrations of mega cities face multiple challenges including informal settlements and air and water pollution control. Adequate mega city management requires a well functioning spatial data infrastructure (SDI). The investigation of current applications of state-of-the-art SDI technology in the world's existing mega cities including Network SDI development in their home countries shows a large diversity in terms of stage of development. Whilst for some countries and cities almost no useful information was retrieved from the web others are in the conceptual phase of SDI development. Often NSDI progress and urban SDI development is correlated. In some regions primary and secondary data production is in progress. The most advanced SDI implementations are to be found in some countries and cities where web based services for broad access to comprehensive distributed spatial data pools are already in operation (Boos, 2009). The internet has transformed information access in the first world – and also in the cities of the developing world. On the web, there are a large number of websites that allow users to query geographic information related to urban characteristics (Díaz, 2009). Once government officials start accessing GIS data from their desktop and integrating multiple data sources, there is no going back.

URBAN DYNAMICS AND REGIONAL COMPETITION

Cities and towns are waking up to the fact that an entrepreneurial and anticipatory policy is called for to cope with urban and regional competition. In the past two decades, a growing number of European cities and towns have acknowledged that marketing could be a powerful instrument in such a new style of urban management (Berg & Braun, 1999). Research (Hall & Hay, 1980; Berg, 1987) has shown that European towns and cities tend to develop according to a fixed pattern. In that pattern, three stages can be distinguished. The first is the stage of spatial concentration, also called the urbanization stage. It is the stage of development of countries and regions where the agricultural society is passing into an industrial one. At the stage of urbanization, to virtually all intents and purposes the fast-growing towns from a spatial function entity. Such socio-economic activities as living, working and shopping proceed for the most part within the municipal border. Spatial interactions are intra-municipal rather than inter-municipal.

It changes considerably at the next stage (second stage) of urban development, that of suburbanization or urban sprawl. It is basically a period of spatial deconcentration, when the suburban municipalities on the outskirts of the large cities or towns find their population growing fast, while the large central cities or towns of agglomerations are losing quite a lot of inhabitants. By far the greater part of suburban communities has their job in the central town. Voluminous commuter flows between the suburbs and the central town develop as a result.

The ultimate consequence is that the large agglomerations move from the suburbanization stage into that of counter-urbanization – the third stage distinguished in the pattern of urban development. Counter urbanization means that the entire agglomeration is losing inhabitants and employment, mostly to the smaller agglomerations some distance away. By this stage, the migrant population's functional ties with the agglomeration of departure have mostly been cut. The new settlers are pursuing most of their socioeconomic activities (working, shopping, sports, leisure, etc.) within the borders of their agglomeration of settlement.

Based on these notions, definition of urban competition is changing over time. It has been experienced that in the beginning, the dynamics were rather physical and they become virtual in the end. Dynamics are depending heavily on information nowadays for city's competition. People can use internet, cellular phone and the media information for knowing latest news and information regarding their expected activities. Goods and services can be distributed within desired time line if information becomes available to the stakeholders. Similarly, labour market can be more competitive and can get more skilled persons based on available information. It has been experienced that spatial patterns of cities, urban sprawling and counter urbanism are being considered as major dynamics nowadays for regional competition. Evidently, then, the degree of which cities are competitive depends mainly on their stage of development. At the urbanization stage, towns function as independent units, and their functional connections with other municipalities are relatively few. Suburban sprawl means the formation of a functional urban region or agglomeration. Competition comes from farther afield in counter urbanization. Large towns are not only in peril from their own suburbs, but find themselves competing also with the alluring living environment of medium-sized towns farther away. Urban competitiveness mainly refers to a city's abilities to more quickly create more and better wealth with fewer resources than other cities and to provide welfare to its inhabitants in the course of competition and development (Pengfei, 2006).

Based on client interactions, global city competitiveness gets major key determinants of location choice. According to the market report of NeoIT, there are two major categories those are working as prime determinant of location choices of firms in cities. They are: Generic or basic factors and Enterprise specific factors. Both of them are highly depending on revenue potentials and time factors.

Generic factors are those which are independent of the country from which the work is offshored, and the industry of the offshoring enterprise. Factors that are very much dynamic and fall into this category are: human capital, costs, infrastructure, environment and risks. Any enterprise looking at globalizing its services would typically look in detail at each of these factors in this information era to create a shortlist of options. Specific factors are those which are dependent on the source country, industry of the enterprise, or the specific processes being offshored. These include language compatibility, physical proximity, socio-economic affinity and sector specific expertise availability. All these factors are depending on four major pillars in terms of revenue potentials over time.

FUNCTIONS OF COMPETITION BETWEEN CITIES AND REGIONS

Competitiveness has been a highly studied topic in recent years, calculating indices by countries, cities, through this described a set of variables to approximate the degree to which a society is capable, in addition to generating economic development, guarantee the social rights of individuals (Bustillos, 2008). Cities and regions are competing while each other to attract investors (economic) and customers for creating job opportunities and updating the services (environment) for the citizens. Competition between firms believed to have two main benefits for economic development (Carlin et al., 2001; Beath, 2002).

First, it provides a selection mechanism; firms with outdated products or inefficient processes do not survive, while new entrants introduce better products and techniques. Selection by exit and entry reallocates resources from inefficient producers and declining sectors to more efficient and growing ones (H.M. Treasury, 2001). Second, it provides strong incentives to existing firms to improve their technology and organization. The threat posed by rivals encourages them to become more innovative and efficient, which increases their market share, lowers the average cost of production in the industry and reduces the price to consumers.

Competition between places can not operate in a similar way. The agents and their powers are different and competition is moderated by other resource allocation mechanisms. Considering selection first, cities can not bankrupt if they are uncompetitive, unlike businesses. New entrants also emerge infrequently and insignificant compared with most markets in which firms operate, since building new urban economies is obviously costly and slow. Most countries have public finance system that cushion the impact of economic decline. Yet, places still stagnate and decline, even if they do not 'close down'. There may be local reasons, such as demand for their products. New urban areas also emerge over time, showing a dynamic process of change at work that must affect the functioning of the economy in the long term. Edge cities and new towns may grow on the back of the amenities, services and markets of their neighbouring cities, without their physical and financial costs. Growth pressures and property shortages in successful regions mean that local authorities are often encouraged to respond to market demand and allow new development, not try to steer private investment through planning restrictions or strategic use of public infrastructure (Turok & Bailey, 2004).

Nevertheless, a competitive political system, electoral pressures to create and safeguard jobs, and lobbying by selected business interests to help them grow mean that all except the most prosperous dormitory suburbs and towns normally make some effort to maintain or enhance their economic position. It is also likely to vary in different countries depending on the link in the financial system between a city's economic performance and its tax receipts. In general, one might expect cities with inherent economic advantages at the outset to be more inclined to participate in competitive activities than those in weaker positions, because they stand more chances of success (Turok, 2004). One of the difficulties financing local decision – makers is uncertainty about what policies to pursue. Their choice may be influenced by central government controls, the rules of other external funders, the pressure for visible actions, fashionable ideas or advice from consultants. They may also be swayed by special interest for whom the payoffs from development are more clear cut, such as major property owners or other business groups that stand to benefit from increasing demand and rising land prices and rents (Logan & Molotch, 1987). Other economic interests with larger numbers, smaller individual stakes and more diverse concerns tend to find it harder to organize for collective action. This raises the obvious danger that the selected policies favour narrow interests. Overall, there is little doubt that there has been a steady growth in sub-national policies in the UK as in other European countries that are explicitly or implicitly competitive, even if the rationale is sometimes questionable and the incentives are not clear cut, as they are in North America (Turok, 2004).

URBAN PLACE MARKETING PERSPECTIVE

More than in the past, urban management needs to be strategic market oriented, and responsive to the trend towards increasing competition and interdependence among cities that ensures from internationalisation and the associated spatial-scale enlargement of the cities' economic relations. The policy implication of the simultaneous co-existance of urban and regional economy is: the economic globalization and scientific development are imposing opportunities to grow as well as threats of decline for all the cities in the world, no matter if it is big or small, developed or underdeveloped, emerging or declining (Pengfei, 2006). However, it is not enough for a town to be more attractive than its competitors as a location for enterprises. The mission of urban management is guiding the harmonious development of the city or region. The key principle is that urban managers need to be more entrepreneurial. That does not imply that their role is identical to those of managers of firms and agencies supplying ordinary products and services. To manage a town as 'enterprise', an entirely different approach is needed. City administrations have little experience of entrepreneurship, as they are prone to depend and rely on higher authorities for their actions and are characterized by a rather different organization culture. Marketing could be an important instrument in urban management; by adopting marketing principles, the municipal organization may become more customer-oriented, ready to give service to, and mind the interests of, the town's citizen. Ashworth and Voogd (1994) claim that it is not so much the idea that activities, including those provided by government agencies for social rather than economic reasons, operate within a market, but the conscious application of marketing approaches that is new. To some politicians and city administrators in Europe, the use of marketing as an instrument for urban policy is at least controversial (Berg and Braun, 1999). Philo and Kearns (1993) talk of a process of manipulation whereby urban bourgeoisies are seeking to mobilise segments of culture, history and locality in the competitive selling urban places both to outsiders (to attract capital) and to insider (to legitimate redevelopment). The culture, history and locality are partly the result of the evolution of markets. They associate the 'selling of place' with intensive 'commodification of places'. Lastly, cities have their origin in the concentration of human and hence trading activities and it cannot be denied that the concentration of economic and social and cultural potentials has made cities into what they are today.

To entrepreneurs, marketing is a useful instrument; looking at their business and their proposition through the eyes of the customer helps them to reach the company's objectives (profit, market share, etc.). Cities – or, rather, city governments, city managers and other public (semi) institutional – have different objectives (Berg & Braun, 1999). It is important to realize that one cannot with impunity promote prosperity in terms of income and employment without giving heed to the quality of the urban living environment and living conditions of underprivileged groups in the community. But that does not exclude the use of marketing principles to attain these objectives.

Urban place marketing network works very efficiently when information is available in hand for the firms, labours and for the investors. What characteristics make the city's image and what is their effect on separate goods and services within a network of urban place marketing? First, the marketing of cities and towns is a question of creativity and good ideas: what are the cities target groups, the position of the city, the communication strategies that are needed, etc. There are many dimensions to the marketing of cities (many different customers, many products and combinations of products, and the city as a product) that bring about the involvement of a great number of producers and managers (not necessarily the same). In the case of cities, there are numerous possibilities. The product can be the activities of the elected city government, a public utility company, a public development corporation and/or other government agencies, a project developer, an investment corporation, citizens, etc. But it is also possible for the city government to establish a private company that is concerned with urban marketing issues, or that a private organization like the chamber of commerce or the local entrepreneur's association may contribute to the marketing of a place. Moreover, in some cases, a combination of private actors in one sector, like a combination of promoters of tourist attractions, could market the city or its products. Kotler asserts that, increasingly, authority and responsibility for marketing and public overseeing are delegated to private organizers (Kotler, 1993).

Information is one of the foremost instruments in this issue for urban management and selling cities to the investors for bringing more employment opportunities in the society and for upgrading the living environment of the community. Geographical Information System (GIS) can help considering all the dynamics that a city can offer to the stakeholders who are investing and doing businesses.

MODELS OF COMPETITIVENESS

At present, more and more experts dealing with regional science are interested in the analysis of regions as functional regions or nodal regions, also examining factors influencing competitiveness (Maskell, 1998). There are three main approaches and models which are important to assess the regional competitiveness. Those models are being considered dynamic in this information era. GIS can incorporate those models very easily putting the attribute and spatial data together to plot them in a customized manner for peoples of different interest groups.

Pyramid Model of Regional Competitiveness:

Measuring regional competitiveness has been traced back to four related economic categories: income generated in the region, labour productivity, employment rate and the openness. The notion of competitiveness obtained in this way cannot be used, however, to identify factors responsible for regional competitiveness or areas which are to be strengthened or developed by regional development policies and programmes for improved competitiveness (Lengyel & Lucovics, 2006). The pyramidal model of regional competitiveness seeks to provide a systematic account of these means and to describe the basic aspects of improved competitiveness (Gardiner, Martin & Tyler, 2004; Lengyel, 2003, 2004).

Factors influencing regional competitiveness can be divided into two groups of direct and indirect components. Of particular importance are programming factors with a direct and short-term influence on economic output, profitability, labour productivity and employment rates (Lengyel, 2003). But social, economic, environmental and cultural processes and parameters, the so-called 'success determinants', with an indirect, long-term impact on competitiveness are also to be taken into account.

Three levels can be distinguished with regard to the objectives of regional development programming and the various characteristics and factors influencing competitiveness:

- **Basic categories of regional competitiveness (ex post indicators; revealed competitiveness):** these categories measure competitiveness and include income, labour productivity, employment and openness.
- **Development (programming) factors of regional competitiveness (ex ante factors; improving competitiveness):** factors with an immediate impact on basic categories. These can be used to improve regional competitiveness by means of institutions in short-term programming periods.
- **Success determinants of regional competitiveness (social and environmental conditions; sources of competitiveness):** determinants with an indirect impact on basic categories and development (programming) factors.

These determinants take shape over a longer period of time and their significance reaches beyond economic policymaking When characteristics determining competitiveness are placed on a chart one obtains the 'pyramidal model' of regional competitiveness (Figure 1): the components of long-term success are to be found in the base, the middle layer is constituted by the development (programming) factors, the basic categories included in the standard definition of revealed competitiveness are located one level higher, while the standard of living and welfare of the region's population, the ultimate objective, forms the peak of the pyramid.

Considering that GIS aims, essentially at manipulating, and displaying while storing data in a

Figure 1. Pyramid Model of Regional Competitiveness. (Source: Lengyel, 2003)

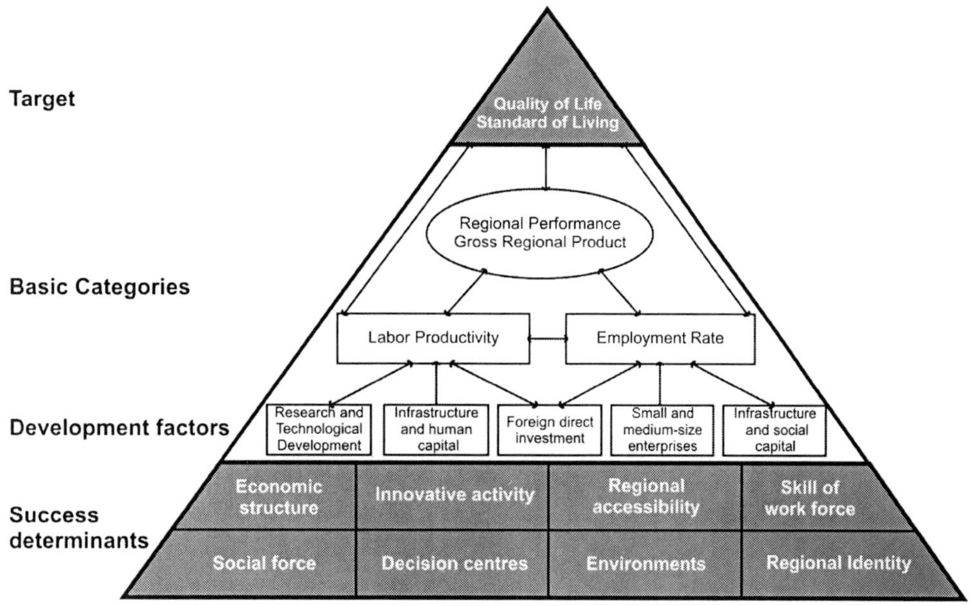

sort of cartographic sandwich, one can use a GIS to relate various types of demographic or socio-economic or other data-say, for example, incidence of poverty, number of registered voters, and measured levels of certain environmental pollutants-to a standard geospatial data set of points, lines, and bounded areas - such as a city map (Jackson, 2008). This model certainly gives idea about the data need and about the possibilities GIS can offer to analyse the quality of life index at the end of the pyramid. The final outcome GIS based maps offers the way competitions can be measured based on primary and secondary sources of information in specific locations of interests.

TREE MODEL OF COMPETITIVENESS

The scientists group ECORYS proposes a new model of regional competitiveness – Competitiveness Tree Model (ECORYS, 2010). This model was developed in order to illustrate how the complex factors that influence competitiveness are related to positive outcomes such as social inclusion, welfare and sustainability. The organic nature of the tree emphasizes the cyclical character of the concept of competitiveness. The quality of the soil and effective functioning of the system of roots, trunk and branches determine the strength of the tree and the capacity to bear fruit. This is a dynamic process as the fertility of the soil is determined partly by the extent to which it is revitalized by fruit from the tree. Using this metaphor, the Competitiveness Tree Model is envisioned in Figure 2. Regions compete in attracting and retaining firms, qualified labour, private capital and investments, public funds and also visitors and customers.

In the ECORYS model research regions are divided into 4 types:

- Production site regions;
- Increasing return regions;
- Knowledge hubs;
- Less competitive regions.

Figure 2. Competitiveness Tree Model. (Source: ECORYS, http://www.ecorys.com/competitiveness/resources/competitiveness_remi.ppt#1)

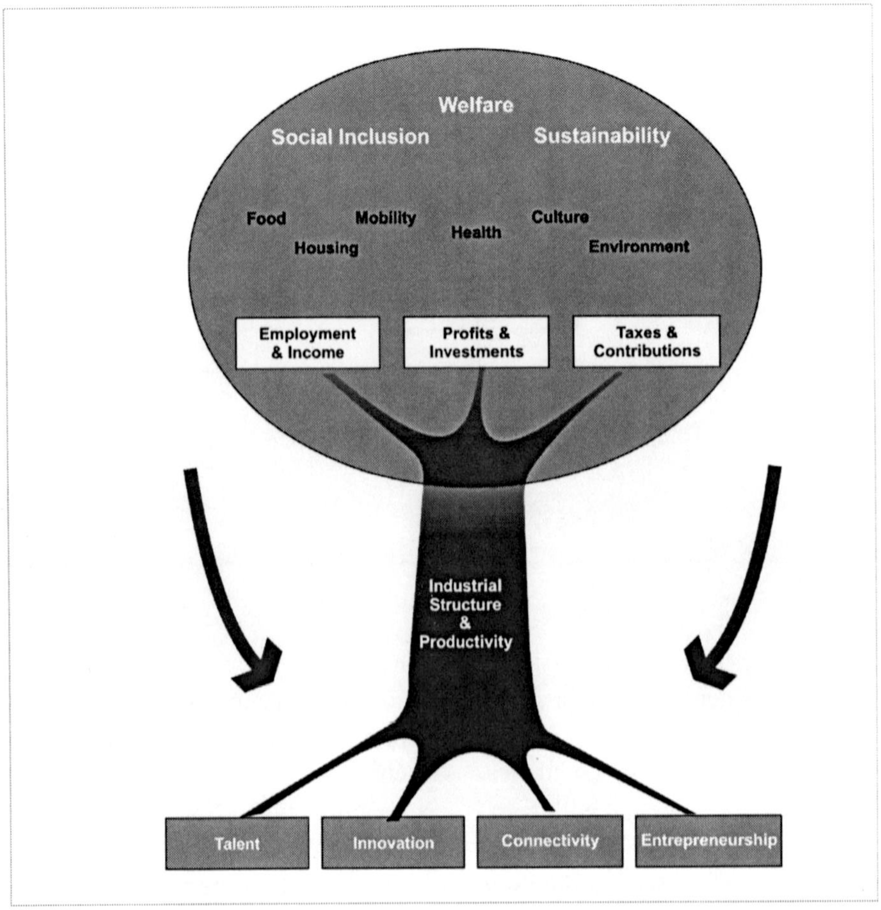

Each of these regions has its own drivers of competitiveness and therefore it is necessary to work out different development scenarios.

Webster and Muller (2000) are of the opinion, that urban regions control only some of the factors which determine their competitiveness. National policy frameworks and socio-economic conditions are also very important; these are related to national taxation, human resource development, tariff, macro economic, industrial incentives, policies, etc. However, in many countries, national factors are becoming relatively less important because of global forces, for example trade liberalization which makes tariff policies less important, or because of internal changes - decentralization which may result in devolution of responsibility for critical competitiveness factors. It is very difficult to forecast how competitiveness of a given region will change; even knowing what competitive advantage will exist in the near term.

As discussed in this model, four different pillars have different characteristics which can be modelled in GIS platform to produce maps that show the characteristics of competitions. This approach allows both aggregation of geometric components and aggregation of measures associated to those components, defined in GIS fact tables. Spatial data in the different thematic layers of a GIS system can be mapped univocally to each other using a common frame of reference (Gomez et al., 2009), in this case, the given four pillars can be considered as different layers and they can be brought under a common framework to show the level of competition among cities and locations.

Four Assessment Models of Competitiveness

Economic structure has traditionally been the focus of competitiveness assessment, key elements usually considered include economic composition, productivity, output and value added, and investment – foreign and domestic. Territorial endowment refers to the non-tradeables associated with a given place, such as location, infrastructure, natural resources, amenity, cost of living and doing business, and an urban region's image. Human resources refer to the skill levels, availability, and costs of labor in an urban region. However, it is increasingly becoming clear that the value of human resources is highly related to the environment in which they are deployed, just as the same plant will grow very differently in different soil regimes. Thus, there are enormous differences in returns to human resources depending on the institutional milieu and place where they work. Human resources, along with the Institutional milieu are probably the most important factors in explaining competitiveness. Human resources determine the extent to which activities in cities can move up value chains. Human resource development must "fit" emerging economies in cities if human resources are to be a competitive asset. The quality of education and training is at least as important as the number of students graduated. The institutional and cultural milieu refers to business culture, governance and policy frameworks (including incentive structures), and network behavior. It is increasingly clear that development and competitiveness is largely a product of networking, which is based on, and creates social capital (Granovetter, 1985; Putnam, 1993).

Figure 3. Four Assessment Categories Model of Regional Competitiveness. (Source: Webster & Muller, 2000)

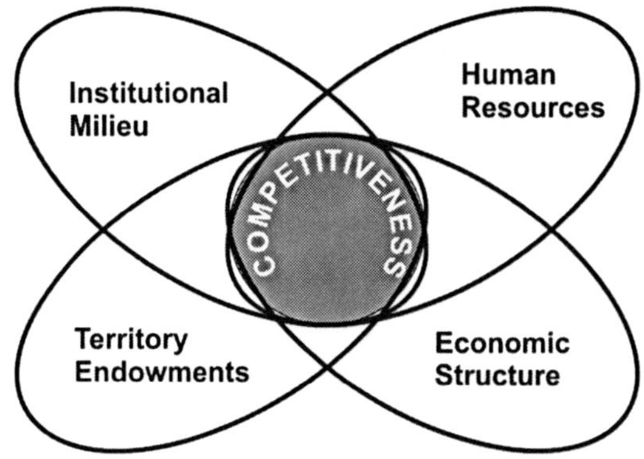

Figure 3 shows that competition remains in the center of four different pillars. Available information about those four pillars can offer more competitive regional competition with more opportunities for stakeholders in the society too.

This model is very important also to imply into GIS based analysis. In this case, competitiveness is the pivotal criteria to be assessed based on other four criteria (humans resources, economic structures, territorial endowments and institutional milieu). The richer, more complex data holdings generated by GIS data sharing can benefit organizations and society (in this case the farms and industries) grappling with specialized or unique problems (Bishop, 2010). Based on this notion, available data can be structured according to the four assessment categories model which definitely shows the competition in a clearer view with the help of GIS techniques.

GIS AND SPATIAL ECONOMICS

Geographical information systems (GIS) are widely used in business, government, and a growing number of academic disciplines (Overman, 2010) especially in city and country planning. Traditionally, spatial data has come from two main sources. Ground survey (what goes on at a particular location) and census (what does the distribution look like across location). These are familiar to researchers interested in spatial economic and spatial planning (Overman, 2010). Examples include aerial photography and, increasingly, data from satellites (either locational or radiometric). The latter are said to provide remote sensing data, although sometimes that term is applied more generally to any data collected indirectly. Geospatial Information Systems (GIS) are growing and the effective management of our resources requires a greater level of insight. Increasingly that greater clarity is being provided by GIS applications (Kwok, 2008). Remote sensing typically provides large amounts of data on the earth's surface. These data are of inherent interest but, may be of particular interest to researchers seeking sources of exogenous variation to use as instruments. The means of inputting data into the GIS depend on the source. One can move from analog maps to digital data by either digitizing (manually tracing over the map) or scanning on the map. Both of these processes require a considerable amount of skill and most researchers want to outsource these tasks (on the basis of either absolute or comparative advantage). One proviso, however, is that it can be very important to understand the sources of measurement errors that can occur during the inputting process. Some of these are specific to the way that GIS handles data (e.g., a polygon representation a lake has a gap in the edges that represent it so the polygon is open, when it should be closed; lines on a transport network should intersect at junctions that allow interchange between routes but incorrect positioning of routes may mean they do not) and some are so general as to be not worth considering in detail (e.g., forgetting to input a line of data). Locating and characterizing observations is the most basic form of analysis in GIS. Competitiveness can be measured based on spatial information to prove which cities or areas provide attractive packages for investment and for future opportunities. GIS can analyse these information based on locational or spatial analysis which can be easily visible through web support system in the internet for everybody. When looking for a good location for a new shop, one of the main considerations would be proximity to a sufficiently large group of potential customers. Obviously, one approach to this problem would be to compute for each location the number of people within a reasonable distance, say 10 minutes by bicycle (Ritsema & Jong, 1999). How far are people willing to travel to visit a facility? In such cases, it is a good idea to do some sensitivity analyses for the range if it is considered for specific type of competition of labour markets in some cities. As researchers become more familiar with using GIS to integrate data from different sources, they will be increasingly exposed to new sources and types of data that can help increase understanding of spatial economic phenomena. Competitiveness of urban locations in this information age might be a good example in spatial economic analysis in this case. Municipalities need access to data, such as parcels, owners, buildings, road and infrastructure information, to provide for public administration (Alkan, 2010). The development and approval of a comprehensive municipal land use plan is normally a long and arduous process that requires great effort from local administration, particularly in small municipalities with a shortage of resources and expert staff. Besides, because it is necessary to engage citizens in the plan development process, they should be provided with access to all the documentation and means required to participate (Díaz, 2009). Applications based on Web and Geographic information System technologies have been designed to promote the dissemination of urban information and to enable citizens to register objections to the land use plan during the different phases of the planning process, as well as to answer those objections. That applies to citizens as well who now have information in their screens which otherwise would only have been available at City Hall, threw an appointment.

This represents efficiency for local organizations and citizens are now able to access GIS information online. So we are starting to add value simply by drawing information off a web site, and in so doing we are starting to break down barriers (Kwok, 2010).

IMPLICATIONS OF CHANGING GOVERNANCE ARRANGEMENTS FOR GIS CREATION AND USE

Urban management may be defined as incorporating a corporate policy and resource allocation strategy dealing with health, education, transport, local economic development, poverty reduction, infrastructure provision, labour market's availability, land use, translated into programmes and projects for implementation. Operations involve the management of assets and services, involving planning and allocating responsibility for service delivery, maintenance, regulation and revenue generation. In an urban management system, a number of potential roles for GIS may be envisaged, both as technical tool and as a communication medium (Rakodi, 2003). Some of them are as follows:

- The storage, manipulation, retrieval and display of spatially referenced data, including land use, linear services and infrastructure, population, house types, labour markets, shopping areas, etc. For example, remote sensing is being used to map land use change, land suitability and environmental sensitivity in different cities of the world to a scale suitable for city or regional strategic planning. If linked to other databases, which are typically zonal (administrative or census areas), in this case of competitiveness assessment, thematic maps can be produced (Kasturirangan & Rao, 1997). The potential has been demonstrated in an analysis Saharanpur City (population 378,000 in 1991) where satellite images, aerial photographs and field checks have been used to measure the loss of agricultural land to urban extension between 1988 and 1998 (Rakodi, 2003).
- Automated data handling (e.g., payroll administration; valuation, billing and collection; storage; tracking and retrieval of development applications) is often the first attempt at computerisation made by local governments. Although not necessarily spatially referenced, there is potential for linking the data available from records to a GIS.
- Improving the availability and flow of information in and amongst organizations responsible for urban management, with a view to improved monitoring of change, anticipation of issues and monitoring of implementation.
- Land suitability analysis and hazard mapping, by linking population and environmental health data to identify areas at risk from environmental pollution; and by linking the degree of risk to population distribution and infrastructure in order to identify priority areas for preventive activities (Paulsson, 1992).
- Predictive and prescriptive (or optimising) models of urban spatial interactions can be used to forecast the impact of changing land use and transportation infrastructure on the distribution of population and other activities and vice versa (Batty & Densham, 1996). The former include, for example, shopping models and transporation models and the latter models for locating facilities.
- Increased access to information via increased access to computers and the development of web means that potentially a wider range of actors in urban areas can be informed about planning and management issues. As the potential for using web-based information resources including GIS

(and models) interactively develops, there is also potential for using them in planning and decision making to increase opportunities for consultation during decision making and planning, through the development of on-line discussion forums. As the systems are further developed, there might also be potential for people to redesign what is viewed on screen remotely, either individually or in discussion with others.

GIS are only tools – their value depends on high level managerial, political support and institutional capacity to define information needs, set priorities and then make appropriate use of the information generated. In this article, the information age and GIS can help the decision makers to compete their urban or regional territory before people who are feeling interests to see the exact spatial and attribute information in time around the world. However, corporate policy formulation and decision making capacity may be weakly developed, especially in local authorities organized on traditional sectoral lines, with an administrative rather than developmental view of their role and operating to short political time horizons.

FROM GOVERNMENT TO GOVERNANCE

To govern is to steer, guide, direct, control, regulate, influence or determine. In particular, it is to rule with or exercise authority and administers the affairs of a state. Government, therefore, refers to the formal institutional structure and location of authoritative decision making, embracing the legislative and executive branches of the state apparatus and those who control them (Hewison, Rodan & Robison, 1993). Governance in contrast, refers to the action, manner or system of governing in which the boundary between organizations and public and private sectors has become permeable… the essence of governance is the interactive relationship between and within government and non-governmental forces (Stoker, 1998). It is about relationships between the state and civil society, rulers and ruled, government and the governed (Radoki, 2003). Until the 1980s, the predominant assumption was that governments had the authority and capacity to govern: to formulate and implement policy, and to realise development goals. Translated into an urban context, the stated-led approach to development implied that public sector organizations could plan and manage urban development and the debate focused on an appropriate allocation of roles and responsibilities between central and local government and between the administrative departments of government and semi-autonomous public sector agencies. The tasks urban government is expected to perform are similar almost everywhere (Davey, 1996) such as:

- Provide infrastructure for the efficient operation of cities;
- Provide services which develop human resources, improve productivity and raise the standard of living of residents;
- Regulate private activities that affect community welfare and the health and safety of the urban population; and
- Provide services and facilities that support productive activities and allow private enterprise to operate efficiently.

The challenges to traditional public sector approaches to urban government have had an impact on various aspects of governance and management in the 1990s, including political arrangements, decision making processes, ways of ensuring accountability, the allocation of roles and responsibilities between public sector agencies and other potential service providers and between different levels of government, and the mobilisation and allocation of financial resources for urban development.

The participatory creation of maps, above and beyond their interpretation, started in the late 1980s. Participatory mapping has spread like a pandemic with many variants and applications not only in natural resource management but also in other domains (Chambers, 2006). Participatory GIS is an emergent practise in its own right; developing out of participatory approaches to planning and spatial information and communication management (Rambaldi & Weiner, 2004 in Rambaldi, 2006).

Considering the information society, available documentations and updated approaches of good governance, the World Bank (WB) has fixed criteria for good governance in 2000. Those criteria are also important to be available before citizens and stakeholders who are competing with each other such as: cities and regions. Good governance has been comprised in this competitive information age as:

- Accountability, transparency and integration of local government;
- Local government institutions sensitive to the needs of the poor;
- Cost effective fulfilment of local government service objectives, through sound financial management and access to market-based credit; and
- A social compact with users for fees-for-services.

The key regional issue appears to be exploiting these new initiation activities to produce a genuine shift in regional governance arrangements which embody a shared and cross regional vision of a confident region well-adapted more generally to the demands of the knowledge economy (Benneworth, 2006).

HARD AND SOFT NETWORK FOR COMPETITIVENESS

The concept of competitiveness has grown significantly in the past decade and it is now common for cities, regions and nations to assess, improve and publicise their competitive standing vis-à-vis other places. With the blessings of information and internet, people can easily know about their present status in the world market and immediately they can improve the drawbacks within limited time to overcome the situation. Yet it remains rather unclear just what competitiveness is when applied to places, which have objectives distinct from those of private sector organizations, for whom share price and profitability serve as common metrics. To be effective, networks must operate at the global, national, regional and local scales, gathering knowledge via social interaction, which are being considered as soft infrastructure in networking. In another way, soft networks mean people related networks and hard network means infrastructure dependent networks. Both of them offering information so that cities and regions can supplant each other by offering updated information to its stakeholders.

Both the purely business aspects of private enterprise – and the interaction of firms with the public sector – are sustained by networks (Maleki, 2002). It is increasingly the case that these networks must operate at the global, national, regional and local scales. It is critical that links to outside, even global, networks be established and maintained. In the information age, this is heavily depending on the internet facilities as well as on better visualization. GIS plays a good role in it to visualize the spatial information into real world for easily accessibility to people who are interested in the specific information.

The need for complex internal and external networks is now well-established in transnational corporations. The internal network focuses on routine intra-firm communication and on learning by sharing internalizing what is obtained from

Figure 4. Convergence of Information Network. (Source: King & Kraemer, 1995)

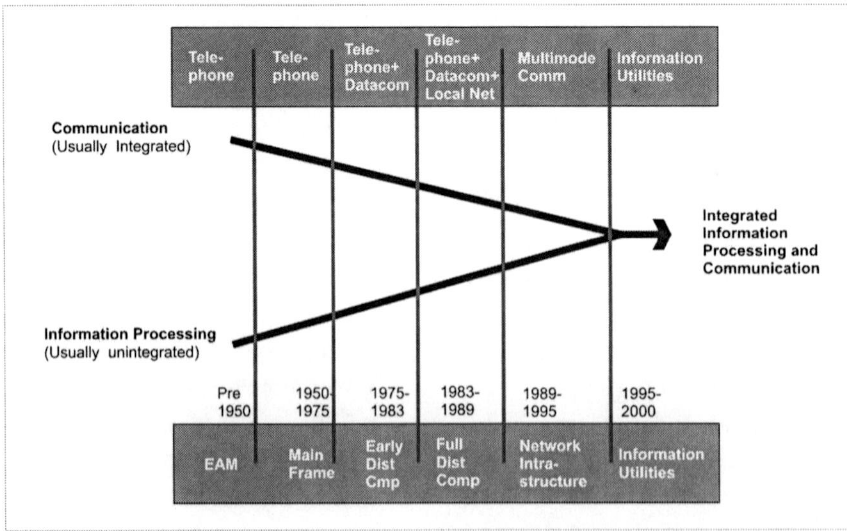

the external network (Nonaka &Takeuchi, 1995; Zanfei, 2000). This involves both co-operation with local firms and a global network. Translated to the context of places, cities must develop and maintain links to knowledge sources outside the community – what Flora and Flora (1993) call 'vertical links', developed and maintained by active, entrepreneurial gatekeepers. Successful city–regions like–successful companies–maintain connections with other places in the global economy (Kanter, 1995).

The internet is a new hard network that is, in principle, ubiquitous. Any telephone can connect to the internet as can, increasingly televisions and other devices. However, the convergence of computing, telecommunications and information media into digital form masks great inequalities in the availability of bandwidth for data transmission. The recognition of digital indicators of competitiveness reflects the fact that all of us and our organizations are compelled to live in at least two dimensions of reality: the old and the new. The relationship between the space of flow in global telecommunications networks and cities is less than clear. So GIS can play important role to put them together and to bond soft and hard networks together.

Figure 4 shows how the communication is being transmitted to the network society. This opportunity gives more people better access to information with up-to-date evidences of competition in terms of business, marketing, investment as well as city structures and governance.

The Internet also has imposed new requirements on companies and the changes it has brough are 'more pervasive and varied than anything that has gone before' (Cairncross, 2000). Competitive firms are not only those whose managers are under excessive time-pressure, living 'in a permanent state of emergency' (Thrift, 2000). Essentially, a new threshold of performance is demanded, incorporating speed and short product and service life-cycles. In addition, management by Web or e-management means both the internet and a web links with partners, employees, external contractors, suppliers and customers in interdependent collaborations (Byrne, 2000). The e-government of Singapore suggests that this is not merely a North Atlantic phenomenon. However, the virtual city is a not a substitute for a real city and trends toward exclusion are all too evident.

Figure 5. Quality of Life: Ranking in Major Portuguese Cities. (Source: Lourenco, Alves, & Mendes, 1998)

		Climate	Shopping service	Crime	Unemp-loyment	Housing	Mobility	Built heritage	Purchase	Pollution	FINAL
Rank	Cities	Score	Score	Score	Score	Score	Score	Score	Score	Score	SCORE
1	Lisbon	0.93	1.54	0.24	0.39	-2.69	-0.91	3.26	3.31	-1.86	0.38
2	Guarda	-0.18	-0.08	0.88	0.75	0.91	0.05	-0.58	-0.51	0.76	0.26
3	Coimbra	-0.07	0.58	0.80	0.66	-0.62	0.55	0.41	0.06	-0.36	0.23
4	Bragança	-0.64	-0.10	1.15	0.22	1.23	-0.31	-0.37	-0.54	0.41	0.16
5	Castelo Branco	-0.09	-0.02	1.20	0.03	1.00	0.27	-0.65	-0.49	-0.15	0.15
6	Santarém	-0.07	-0.41	0.54	0.50	0.44	-0.09	0.04	-0.50	0.48	0.12
7	Aveiro	0.39	0.18	-1.04	1.05	0.20	0.13	-0.61	0.02	0.16	0.05
8	Viana do Castelo	-0.05	-0.60	0.13	0.72	0.40	0.12	-0.20	-0.67	0.38	0.04
9	Évora	-0.09	0.01	0.10	0.00	-0.66	-0.27	1.28	-0.23	0.32	0.04
10	Leiria	0.39	-0.51	-0.69	1.29	0.56	0.23	-0.65	-0.33	-0.01	0.03
11	Faro	0.93	-0.01	-1.36	0.31	-0.12	0.42	-0.58	0.32	-0.17	-0.06
12	Porto	-0.05	0.97	-0.03	-0.58	-1.66	-0.19	0.52	1.76	-1.08	-0.07
13	Braga	-0.51	-0.16	-1.06	0.41	0.37	-0.04	0.15	-0.20	-0.18	-0.13
14	Vila Real	-0.64	-0.24	0.29	-0.48	0.37	0.04	-0.47	-0.65	0.25	-0.15
15	Viseu	-0.51	-0.51	-0.60	-0.03	0.45	0.24	-0.40	-0.48	0.29	-0.15
16	Beja	-0.09	-0.13	1.03	-1.21	-0.64	-0.22	-0.41	-0.41	0.44	-0.18
17	Setúbal	0.93	-0.37	-1.90	-1.08	0.16	0.18	-0.31	0.08	-0.13	-0.32
18	Portalegre	-0.53	-0.13	0.32	-2.96	0.30	-0.19	-0.41	-0.54	0.46	-0.41

But the need for bricks and real people is just as important for e-governments as it is for e-business. All these hard and soft networks can be connected well in GIS platform to make the information available for everybody if successful competitive measures and instruments be properly applied.

EVALUATING URBAN QUALITY OF LIFE USING GIS: AN EXAMPLE IN PORTUGAL

Proper information can depict the actual competition figures if it is available to all stakeholders. In Portugal, a research work has been done by the researchers of the University of Minho to find out the living standard scenario in major Portuguese cities. GIS plays an important role to gather information and to produce scenarios using attribute and spatial data from the municipalities.

Figure 5 shows that Lisbon has been ranked in the top considering climate, shopping services, crime, unemployment rate, housing, mobility, built heritage, purchasing power and pollution as major indictors to assess the quality of life in major Portuguese cities.

Competition of cities can be found in different dimensions as, quality of life, attracting firms and investments, employment opportunities, business environment, supply and production of goods and services etc. The most powerful tool to assess the competition is the information. If information is available, the decision to choose places and cities for investment, living and opportunities can be easily accessible. With the help of GIS, information technology and relevant tools in hand, competitiveness can be measured following the tradition and contemporary models described in this chapter.

HOW DOES PARTICIPATORY SPATIAL PLANNING (PSP) RELATE TO GIS, GOVERNANCE AND COMPETITIVENESS?

Participation in spatial planning is clearly related to legitimacy as a governance criterion, but a strong participatory approach also supports other governance imperatives of equity and respect for people (McCall, M., 2003). Four intensities of PSP, from the least to the most participatory level of the ladder can be found as recognizable:

PSP as 'information sharing' implies one or two way communication between outsides and local people, and is primarily technical information, such as need assessment.

In PSP as consultation, external agents refer certain issues to local stakeholders for refinement or prioritizing, but it is the outsider who pre-define the salient problems, and analysis is controlled by outside.

If all local and external actors are involved in decision making, they jointly identify priorities, analyze current status, assess alternatives, and implement. Participation is seen as a right, not just as the means to achieve project goals.

PSP as 'initiating actions' means that independent initiatives are made and owned by empowered local people, e. g. people self-mobilize to perform community activities; a different situation from simply implementation with their own labor inputs.

There are critical differences in the underlying purposes or intentions of the parties (external or internal) which are pushing PSP as a strategy and/or promoting P-GIS (McCall, 1988):

Facilitation: PSP is promoted in order to ease outside interventions and interests, to improve external project efficiency, or to pass a share of the cost burden onto the 'beneficiaries'.

Mediation: PSP is promoted to link (mediate) outside demands and local people's priorities in order to increase programme effectiveness, to build up local community capacity, or to modify outside interventions towards local aspirations and needs.

Empowerment: PSP is promoted to reinforce local decision-making and responsibilities towards community empowerment, to support equitable social redistribution, and to empower weak groups in resource access and control.

Considering the above notions, it can be said that PSP is an important tool that can help promoting urban competitiveness in the information era. GIS can plot the real information available for all (if good governance exists with equitable access to the information by all) and cities can learn from each other to promote its own activities with the blessings of information technology.

It is shown in Figure 6 how several data input opportunities are available in the mechanism to float the information to its user's worldwide through internet.

The GIS platform offers this immense opportunity to analyze regional and urban management data which are useable for all stakeholders.

CREATING ONLINE MONITORING COMMUNITIES

The development of online data visible for all interest groups may generate the creation and development of virtual community of volunteers, scientists, environmental technicians and elected officials (for ensuring better governance) that are involved within its limit. The creation of online monitoring communities should enable all the stakeholders to share their views on the state of the GIS data and available information of major cities of regions in the world. It implies the existence of tools to support communication, information exchange as well as teamwork among stakeholders (Gouveia, et al., 2004). Communication of data findings is one of the major issues involved within volunteer monitoring. Not only citizens need to publicize their findings, but they also need to receive feedback and information from other stakeholders, namely from decision-makers or scientists. The existence of feedback

Figure 6. Finding results of urban competitiveness based on criteria and World Wide Web input

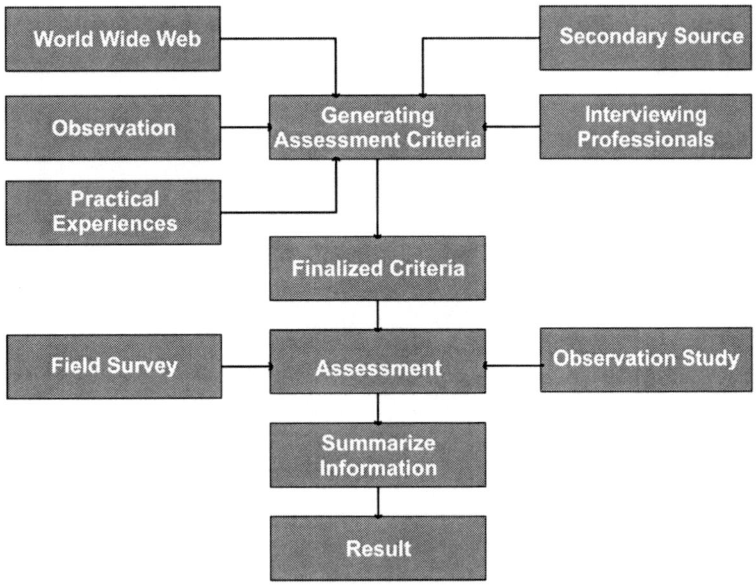

mechanisms is a key issue in any public participation support system since it promotes citizen's commitment to participate. Moreover, volunteers also need to communicate among themselves to support the organization of monitoring activities. Another important component within the creation of online communities is the need to have tools for information exchange among the stakeholders. The client – server architecture available on the WWW supports different mechanisms for data exchange being the ftp protocol one of the most known. The collaborative monitoring can use the mechanisms available but should prefer mechanisms that facilitate multimedia data exchange access. The creation of online monitoring communities should be backed up by a database management system that allows users to register in the system and create users profiles that help the system to target user needs and preferences.

FURTHER RESEARCH GUIDANCE REGARDING CITY COMPETITIVENESS

There are still various opportunities to work further considering information, GIS and urban competitiveness. Every city is competing with others to attract tourists, investors so on and forth. Interactive maps with information are available online. However, while most of the people have access to the cell phone either from the most developed countries to the poorest countries of the world, most people do not yet have open access to personal computers. People can take advantage using the cell phones to know more about places of attractions within a minute if the information is available for them. More important, people can input information according to their needs. Scientist should come forward to make the information available for cell phone users so that more people can take part in the system of monitoring, contributing and taking advantages of the information system. By seeking ways to build multiple perceptions of reality into a competitiveness analysis through GIS (e.g., using various stakeholder perceptions of what is liked and disliked about a specific location), there is an increased ability to interconnect "fact" and value, the separation of which is a common critique of conventional GIS. One way to do this would be to combine the evaluations of individual stakeholders into a composite, value-based layer in GIS that is then compared to more "objective" GIS representation.

CONCLUSION

The production and use of information technology (IT) is increasingly recognized in national leaders as a vital component of global competition. This technology, both in processing and communications, is now woven into the fabric of the economic and social life of developed countries and is a major focus of national investment in newly industrialized countries. Developing countries are playing catch-up by making significant increases in spending for IT in the hopes of not being left behind in the global struggle for competitive advantage. In the old days GIS was managed by engineering and IT didn't get very heavily involved, but now GIS is becoming an IT pillar (Ball, 2010). GIS does not do a lot of simulation or pre-planning but it is basically becoming part of the information infrastructure, at the same time, the smart grid system is producing a great deal of new information, but it's not an analysis or visualization tool. GIS becomes the visualization tool.

Competitiveness of urban regions is heavily depending on important factors such as: Infrastructures, firms, human resources, quality of living environment, institutions and effective policy-networks and membership networks (Linnamaa, 1999). A quick look at the cutting edge of GIS analysis and modelling suggests that it is grappling with a very different set of issues. A quick rundown of the list of highlighting topics are included as: Cellular automata and agent based modelling, visual analytical tools for environmental and urban systems, analysis of human movements data, spatiotemporal data mining, hierarchies, scaling and fractal structure of geographic patterns, modelling vehicle dynamics and crowd behaviour, patterns of human spatial behaviour and migration, urban high resolution morphology, small world modelling and spatial interactions, etc. GIS can incorporate most of the attribute data to present them according to the need of spatial analysis. This can improve the information to all to promote competition among cities in the world with relevant and dynamic information.

Where multiple actors operate, a clear framework is needed for the allocation of powers, function and resources and for the distribution of legislative and executive responsibilities to see how competition among countries, cities or regions is functioning. In practice, these are often unclear, contested and unpredictable. The multi-sectoral nature of urban management implies that a number of action programmes are needed, each likely to involve more than one agency forgetting the boundary of its own. However, the ability of agencies, even local government, to devise a coherent and binding strategic development framework is limited. There is, therefore, needs of simultaneous develop capacity within agencies and to improve collaboration between them with up to date information they can offer. Neither is easy: modes of operation are entrenched in institutional culture and resistance to change widespread; rivalry, conflicting aims and priorities, and political competition may prevent conflicts being resolved and consensus reached. GIS are a potential tool for improving the information available to all governance actors and thus the information base for decision making. As they develop, they might provide additional scope for both more informed decision makings (through predictive or prescriptive modelling and visualisation) and more inclusive and interactive participation in policy formulation, planning and projects design, locational importance, measuring competition, producing real life maps of resource allocation, but plans for their use must be financially, institutionally and politically realistic.

REFERENCES

Alkan, M., & Bulut, G. (2010). GIS and remote-sensing-based urban-information system design and development: A case study for Kozlu, Zonguldak. *Scientific Research and Essays*, *5*(19), 2889–2899.

Asheim, B. (1996). Industrial districts as 'learning regions': A condition for prosperity. *European Planning Studies*, *4*(1), 379–400. doi:10.1080/09654319608720354

Ashworth, J., & Voogd, H. (1994). Place promotion: The use of publicity and marketing to sell towns and regions. *Place Branding*, *1*(2), 39–52.

Ball, M. (2010). *Exploring the Role of GIS in the Enterprise*. Retrieved June 24, 2010, from http://vector1media.com/dialog/ interviews/7320-interview-exploring-the-role-of-gis-in-the-enterprise.pdf

Batty, M., & Densham, P. J. (1996). *Decision support, GIS, and urban planning*. London, UK: Center for Advanced Spatial Analysis, University College London.

Beath, J. (2002). UK industrial policy: Old tunes on new instruments? *Oxford Review of Economic Policy*, *18*(2), 221–239. doi:10.1093/oxrep/18.2.221

Becattini, G. (1990). In F. Pyke, G. Becattini, & W. Sengenberger (Eds.), *The Marshallian industrial districts as a socio-economic notion*. Geneva, Switzerland: International Institute of Labor Studies, IIL Publisher.

Begg, I. (2000). *Cities and competitiveness*. London, UK: South Bank University.

Begg, I. (2002). *Urban competitiveness*. Bristol, UK: Policy Press.

Beninger, J. (1986). *The control revolution: Technological and economic origins of the information society*. Cambridge, MA: Harvard University Press.

Benneworth, P. (2006). *The role of university spin-off firms in strengthening regional innovation systems in weaker places*. Paper presented at the 6th European Urban & Regional Studies Conference, Roskilde, Denmark.

Berg, L. (1987). Urban system in a dynamic society. Avebury, UK: Aldershot.

Berg, L., & Braun, E. (1999). Urban competitiveness, marketing and the need for organizing capacity. *Urban Studies (Edinburgh, Scotland)*, *36*(5), 987–999. doi:10.1080/0042098993312

Bishop, S. (2010). Building programmatic capacity at the grassroots level: The reactions of local nonprofit organization to public participation geographic information system. *Nonprofit and Voluntary Sector Quarterly*, *39*(6), 991–1013. doi:10.1177/0899764009334984

Boos, S., & Mueller, H. (2009). Evaluation of spatial Information Technology applications for mega city management. SDI Convergence. Research, Emerging Trends, and Critical Assessment. In van Loenen, B., Besemer, J. W. J., & Zevenbergen, J. A. (Eds.), *Nederlandse Commissie voor Geodesie (Netherlands Geodetic Commission), 48*.

Bustillos, B., Urista, V., Rentería, G., Vega, G., Vazquez, E., & Delgadillos, C. (2008). A model to measure the degree of competitiveness of medium-sized cities in the Chihuahua region of Mexico: *Annual Statistics Report 2005*.

Byrne, A. (2000, August 28). Management by Web. *Business Week*, 84-96.

Cairncross, F. (2000). Inside the machine: A survey of e-management. *The Economist, 11*, 8-10.

Camagni, R. (Ed.). (1991). *Innovation networks: Spatial perspectives*. London, UK: Bellhaven.

Carlin, W., Haskel, J., & Seabright, P. (2001). Understanding the essential fact about capitalism: Markets, competition and creative destruction. *National Institute Economic Review*, *175*(1), 67–84. doi:10.1177/002795010117500107

Castells, M. (1989). *The informational city: Information Technology, economic restructuring and the urban-regional process*. London, UK: Basil Blackwell.

Castells, M. (1996). *The information age: Economy, society, and culture. The rise of the network society* (*Vol. 1*). London, UK: Blackwell.

Chambers, R. (2006). Participatory mapping and geographic information systems: Whose map? Who is empowered and who is disempowered? Who gains and who loses? *The Electronic Journal on Information Systems in Developing Countries*, *25*.

Cooke, P. (2001). Regional innovation systems, clusters, and the knowledge economy. *Industrial and Corporate Change*, *10*(4), 945–974. doi:10.1093/icc/10.4.945

Cooke, P., Uranga, M., & Extebarria, G. (1998). Regional innovation systems: An evolutionary perspective. *Environment & Planning A*, *30*(9), 1563–1584. doi:10.1068/a301563

Davey, K. (1996). The structure and functions of urban government. *Urban Management: The challenge of growth*. Avebury, UK: Aldershot.

Díaz, P., Parapar, J., & Touriño, J. (2009). *Web-GIS based system for the management of objections to a comprehensive municipal land use plan*. Research Study of the Department of Agroforestry of the University of Santiago de Compostela, 2009.

ECORYS. (2010). *Competitiveness of regions: What works and what doesn't work?* Retrieved May 16, 2010, from http://www.ecorys.com/ competitiveness/resources/ competitiveness_remi.ppt#1

Feldman, M., & Audretsch, D. (1999). Innovation in cities: Science-based diversity, specialization and localized competition. *European Economic Review*, *43*(2), 409–429. doi:10.1016/S0014-2921(98)00047-6

Flora, C. B., & Flora, J. L. (1993). Entrepreneurial social infrastructure: A necessary ingredient. *The Annals of the American Academy of Political and Social Science*, *529*(1), 48–58. doi:10.1177/0002716293529001005

Gardiner, B., Martin, R., & Tyler, P. (2004). Competitiveness, productivity and economic growth across the European regions. *Regional Studies*, *38*(9), 1045–1067. doi:10.1080/0034340042000292638

Gomez, L., Haeseroets, S., Kuispers, B., & Vaisman, A. (2009). Spatial aggregation: Data model and implementation. *Information Systems*, *34*(6), 551–576. doi:10.1016/j.is.2009.03.002

Gouveira, C., Fonseca, A., Camara, A., & Ferreira, F. (2004). Promoting the use of environmental data collected by concerned citizens through information and communication technologies. *Journal of Environmental Management*, *71*(2), 135–154. doi:10.1016/j.jenvman.2004.01.009

H. M. Treasury/DTI. (2001). *Productivity in the UK: 3 – The regional dimension*. London, UK: H. M. Treasury.

Haal, P., & Hay, D. (1980). *Growth centers in the European urban system*. London, UK: Heinemann Educational Publishers.

Hall, P. (1995). *Towards a general urban theory: Cities in competition*. Melbourne, Australia: Longman.

(1993). Introduction: Changing forms of state power in South East Asia. InHewison, K., Rodan, G., & Robison, R. (Eds.), *Southeast Asia in the 1990s: Authoritarianism, democracy and capitalism* (pp. 2–8). Sydney, Australia: Allen and Unwin.

Jackson, S. (2008). The city from thirty thousand feet: Embodiment, creativity, and the use of geographic information systems as urban planning tools. *Technology and Culture*, *49*(2), 325–346. doi:10.1353/tech.0.0039

Kanter, M. (1995). *World class: Thriving locally in the global economy*. New York, NY: Simon and Schuster.

Kasturirangan, K., & Rao, P. (1997). *Urban planning and management: Space technology options-Computers in urban planning and urban management*. Bombay, India: Narosa.

King, J., & Kraemer, K. (1995). *Information infrastructure national policy*. Irvine, CA: Center for research on Information Technology and organizations, University of California (UCI).

Kotler, P., Haider, D., & Rein, I. (1993). *Marketing places*. New York, NY: The Free Press.

Kwok, A. (2008). *City Hall and GIS: Government GIS, local government, technology*. Retrieved October 2010, from http://www.futuregov.asia/articles/2008/oct/10/ city-hall-and-gis/

Lengyel, I. (2003). The pyramid model: Enhancing regional competitiveness in Hungary. *Acta Oeconomica*, *54*(3), 323–242. doi:10.1556/AOecon.54.2004.3.3

Lengyel, I., & Lucovics, M. (2006). An attempt for the measuring of regional competitiveness in Hungary. In *Proceedings of the 46th Congress of Regional Science*, Volos, Greece.

Linnamaa, R. (1999). Competitive advantage and leadership of urban regions in information society, Suomen Kuntaliitto, Helsinki, Finland, Acta-sarja 106.

Logan, J., & Molotch, H. (1987). *Urban fortunes: The political economy of place*. Berkeley, CA: University of California Press.

Lourenco, J., Alves, E., & Mendes, J. F. G. (1998). *Evaluating urban quality of life in Portugal*. Braga, Portugal: University of Minho.

Malecki, E. (2002). Hard and soft networks for urban competitiveness. *Urban Studies (Edinburgh, Scotland)*, *39*(5–6), 929–945. doi:10.1080/00420980220128381

Maskell, P., Eskelinen, H., Hannibalsson, I., Malmberg, A., & Vatne, E. (1998). *Competitiveness, locational learnings and regional development: Specialisation and prosperity in small open economies*. London, UK: Routledge. doi:10.4324/9780203276921

McCall, M. (1988). The implications of Eastern African rural social structure for local-level development: The case for participatory development based on indigenous knowledge systems. *Regional Development Dialogue*, *9*(2), 41–69.

McCall, M. (2003). Seeking good governance in participatory-GIS: A review of processes and governance dimensions in applying GIS to participatory spatial planning. *Habitat International*, *27*(4), 549–573. doi:10.1016/S0197-3975(03)00005-5

Neo, I. T. (2006). Offshore Insight. *Market Report Series*, *4*(8), 4–14.

Nonaka, I., & Takeuchi, H. (1995). *The knowledge-creating company*. Oxford, UK: Oxford University Press.

Overman, H. (2010). GIS a job: What use geographical Information Systems in spatial economics. *Journal of Regional Science*, *50*(1), 165–180. doi:10.1111/j.1467-9787.2009.00644.x

Paulsson, B. (1992). Urban applications of satellite remote sensing and GIS analysis, World Bank/ UNCHS/ UNDP. *Urban Management Program*, *DP*(9), 1-40.

Pengfei, N., & Qinghu, H. (2006). *Comparative research on the global urban competitiveness*. Beijing, China: Chinese Academy for Social Sciences.

Philo, C., & Kearns, G. (1993). *Culture, history, capital: A critical introduction to the selling of places*. Oxford, UK: Pergamon Press.

Porter, M., & Kramer, M. (2006, December). Strategy and society: The link between competitive advantage and corporate social responsibility. *Harvard Business Review*, 2–15.

Rakodi, C. (2003). Politics and performance: The implications of emerging governance arrangements for urban management approaches and Information Systems. *Habitat International*, *27*(4), 523–547. doi:10.1016/S0197-3975(03)00004-3

Rambaldi, G., Kyem, P., McCall, M., & Weiner, D. (2006). Participatory spatial information management and communication in developing countries. *The Electronic Journal on Information Systems in Developing Countries*, *25*, 2006.

Ritsema van Eck, J., & Jong, T. (1999). Accessibility analysis and spatial competition effect in context of GIS-supported service location planning. *Journal of Computers. Environment and Urban System*, *23*(2), 75–89. doi:10.1016/S0198-9715(99)00016-2

Schwab, K. (2010). *The Global Competitiveness Report 2009 – 10*. Geneva, Switzerland: World Economic Forum.

Sim, L., Ong, S., Adeesh, A., Parsa, A., & Keivani, R. (2003). Singapore's competitiveness as a global city: Development strategy, institutions and business environment. *Cities (London, England)*, *20*(2), 115–127. doi:10.1016/S0264-2751(02)00119-1

Stoker, G. (1998). *Public-private partnerships and urban governance*. Basingstoke, UK: MacMillan.

Thrift, N. (2000). Performing cultures in the new economy. *Annals of the Association of American Geographers. Association of American Geographers*, *90*(4), 674–692. doi:10.1111/0004-5608.00217

Turok, I. (2004). Cities, regions and competitiveness. *Regional Studies*, *38*(9), 1069–1083. doi:10.1080/0034340042000292647

Turok, I., & Bailey, N. (2004). Twin track cities: Competitiveness and cohesion in Glasgow and Edinburgh. *Progress in Planning*, *62*(3), 135–204. doi:10.1016/j.progress.2004.04.001

Webster, D., & Muller, L. (2000). *Urban competitiveness assessment in developing country urban regions: The road forward*. Washington, DC: The World Bank.

Zanfei, A. (2000). Transnational firms and the changing organization of innovative activities. *Cambridge Journal of Economics*, *24*(5), 515–542. doi:10.1093/cje/24.5.515

Chapter 4
City Competitiveness and Airport:
Information Science Perspective

Maria Nadia Postorino
"Mediterranea" University of Reggio Calabria, Italy

ABSTRACT

Transport systems allow people to take part in social and working activities, which are variously spread in space and time. Cities where transport systems are widely developed benefit from a high-quality accessibility, which makes them more attractive both to users and investors. As a consequence, they can gain in competitiveness and economic growth. Particularly, an airport can play an important role in assuring the development of the city best linked to it, as it enlarges its external accessibility and attractiveness. As information science applications can be used to obtain more reliable, efficient and safer systems, with reference to these latter aspects this chapter describes the importance of improvements concerning surface transport systems linking cities to their nearest airports as well as improvements assuring safer operations and capacity enhancement at the city airport airside.

INTRODUCTION

Competition is defined as "a situation in which people or organisations compete with each other for something that not everyone can have" (Oxford Dictionary, 7th Edition). Given the current globalisation tendency, not only firms traditionally compete for the market but also cities now compete for investors (e.g., companies, organizations, agencies), visitors and inhabitants and then demand a suitable planning. In this context, "city planning" is the control of the development of cities in terms of their buildings, roads and services, so that they can be agreeable and convenient places

DOI: 10.4018/978-1-61350-174-0.ch004

Figure 1. Relationship between airport and city competitiveness

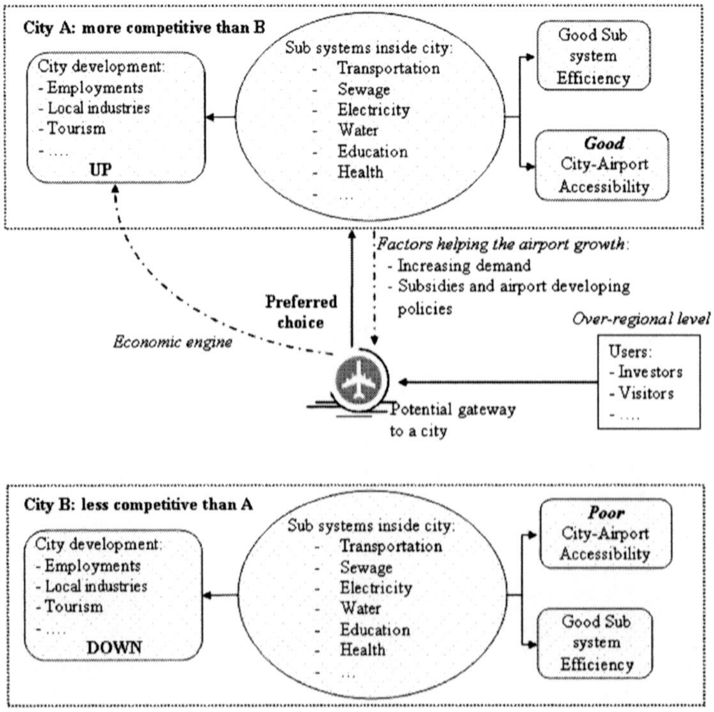

for people to live in, as well as attractive areas for investors and visitors.

Two important aspects of city competitiveness are city efficiency and accessibility. In terms of efficiency, a competitive city is a place where people, living there or visiting it, do not waste time or money. In terms of accessibility, defined as a measure of potential opportunities for interaction (Hansen, 1959), a competitive city is well connected and also offers services and activities (such as shops, firms, education centres, health centres, and so on) easily reachable.

It is worthwhile to note that the concept of (time) efficiency is linked to that of accessibility, as a good accessibility also means that people do not waste time to move inside the city itself or to reach the city from outside. Accessibility is assured by transport systems, linking not only city districts among them but also the overall city to its surroundings.

While several chapters of this book focus on city competitiveness by examining synchronization of subsystems inside it (then from the perspective of a single urban area), the aim of this chapter is to enlarge that point of view in order to examine the city competitiveness from the wider national and over-national perspective, beginning from transport considerations. Then, the problem is now reversed and the question is: given an efficient synchronization of city subsystems, how can city competitiveness, helped by Information Science[1] (IS) applied to transports, increase in a over-regional perspective? Or, in other words: at over regional level, why could people (as investors, visitors, and so on) prefer to go to that city?

Figure 1 tries to provide an answer from the transport perspective. All the other things being the same, external investors and visitors prefer destinations well linked to the over-regional transport networks; particularly, airports potentially represent the gateway to a city, as they can guarantee quick links towards the remaining parts of the over-regional territory.

There are some studies in the literature trying to examine how developing policies aimed at improving efficiency and accessibility, included the airport role in this context, relate to city competitiveness (see for example Crouch and Ritchie; 1999; Williams, 1999; Turok and Bailey, 2004; Chorianopoulos *et al.*, 2010). Then, the easiest answer to the question: why could people prefer to go to that city? is that the preferred city should be more attractive than another one at least from two points of view: efficiency and accessibility.

As Figure 1 summarizes, cities A and B have the same subsystem efficiency, but different city-airport accessibilities, therefore A is the preferred choice as it is better linked to the airport and then to the over-regional areas. Generally, surface transport links between an airport and a city are assured by express bus and/or metro services; to be efficient, these systems should integrate with the urban services (e.g., direct bus lines to/from the airport are connected to the urban bus main terminals; express metro lines are connected to the urban metro main stations).

The relationship between cities and airports is still more complex; given some efficient urban subsystems, not only a good city accessibility improves the overall city competitiveness, but also affects the airport growth. In fact, increasing air travellers, whose destination is that more competitive city, require more air services and facilities; particularly, the aircraft traffic growth calls for better safety measures to manage both ground movements at the airside and airport capacity. On the other hand, good airport services, together with efficient urban subsystems, attract investors and visitors, and then facilitate the city development; in other words, the airport potentially is the economic engine that facilitates the city development.

Finally, although the current tendency is towards privatization, many airports are still owned by local governments that can develop suitable policies bringing benefits to a city. As an example, municipality start-up subsidies[2] can attract airlines, so that the airport can: *i*) increase its air supply; *ii*) attract more people, both incoming and outgoing ; *iii*) help the economic development of the city. Generally, start-up subsidies are part of more general urban planning policies concerning land use and activity organization as well. In fact, the activity system, and its development, depends on general policies both at local and national level (as fiscal policies, tax system reforms, macro-economic scenarios, and so on), that in turn affect the transport system, mainly in terms of transport demand. Then, the airport growth generally is a consequence of more general planning actions, because air transport demand depends both on transport service and socio-economic characteristics. However, these issues are outside the scope of this chapter.

As Figure 1 shows, the starting point of this discussion is the airport rather than the city itself, because airport-related improvements (particularly, surface transport connections and management of ground movements at the airside), in turn, lead to city competitiveness improvements. As there is a mutual relationship between airport and city, urban services play the role of city attractive factors; as an example, good city accessibility but poor urban subsystem efficiency do not attract people, or, in other words, do not make a city "competitive". Urban subsystem efficiency is largely discussed in some other chapters, while this one focus on the airport role to help city competitiveness.

Keeping these considerations in mind, and according to the over-regional perspective, the focus of this chapter is on the use of IS to help city competitiveness by facilitating the development of "its" airport, particularly in terms of surface transport links requirements and management of the airport ground movements. That general problem will be considered under two perspectives: the use of IS applications outside airports (surface transport links requirements) and inside airports (management of the airport ground movements).

Preliminary, an overview of the role of the air transport system (and mainly of regional airports) to increase the competitiveness of an urban area is presented.

THE ROLE OF THE AIR TRANSPORT SYSTEM TO HELP CITY COMPETITIVENESS

It is well known that transport systems represent the vital structure of a country; in fact, economic and cultural exchanges arise only if there is a transport network linking distant areas among them.

In transport engineering terminology, a "transport network" is defined as the set of "nodes", "links" and "travel cost functions" allowing people to move within a region (Sheffi, 1984; Patriksson, 1994; Cascetta, 2001). Nodes represent specific points of the transport system (e.g. intersections, bus stops, train stations, airports, ports). Links are services and/or facilities linking nodes between them (e.g. roads linking intersections; transit lines linking bus stops; railway lines linking train stations; air/ship lines linking respectively airports/ports). Performance on each link is defined in terms of link transport cost, a function of travel times and monetary costs depending on the kind of link (Akcelik, 1988; Highway Capacity Manual, 2000; Cascetta, 2001; Ortuzar and Willumsen, 2001).

With specific reference to the air transport system, airports represent the nodes of the air transport network while the air services realized by airlines represent the links connecting airports among them.

As the technological development has facilitated the production of faster means of transport, the economic and cultural exchanges have become quicker too. The increase of both cultural and economic relationships is a stimulus to the development of cities; particularly, the life quality of people living in those areas can improve due to financial investments attracted there. All the other things being the same, investments are attracted where transport services and facilities are better; further, the more the transport services and facilities increase, the more the economic and cultural exchanges increase and generate activities (e.g., events, meetings, exhibitions and so on). To summarize, cities can start a positive trend that generates and attracts investments if urban planning policies take into account both activity and transport system aspects.

In this context, the air transport system can give an useful contribution to increase city competitiveness.

From an economic perspective, air transport potentially generates employment and wealth. Following some recent studies (ATAG, 2004), the air transport industry has generated nearly 14 million jobs (direct, indirect and induced) and it is one of the most important industries in the world. Another important economic benefit of the air transport is its spin-off effect on other industries, mainly tourism. There is a reciprocal relationship between air transport and tourism, as tourism depends on transport to carry visitors, whereas the transport industry depends on tourism to generate demand for its services. Further, the air transport is one of the greatest contributors to the progress of modern societies: as an example, many technical improvements have been obtained in the air transport industry (e.g., aircraft engines, avionics, but also airports facilities as handling vehicles, baggage sorting mechanisms and so on) or have been stimulated by it so as to achieve greater safety and security.

According to its primary role, air transport not only makes possible quick transfers of people in general, but facilitates connections to remote areas and fast delivery of emergency and humanitarian aids as well as medical supplies.

Airports are important elements of the air transport system, and most of the economic benefits already discussed are due to them. Airports need employees at several levels of expertise, facilitate the growth of industries producing goods that require fast transports, help the tourism and

Figure 2. City and airport accessibility

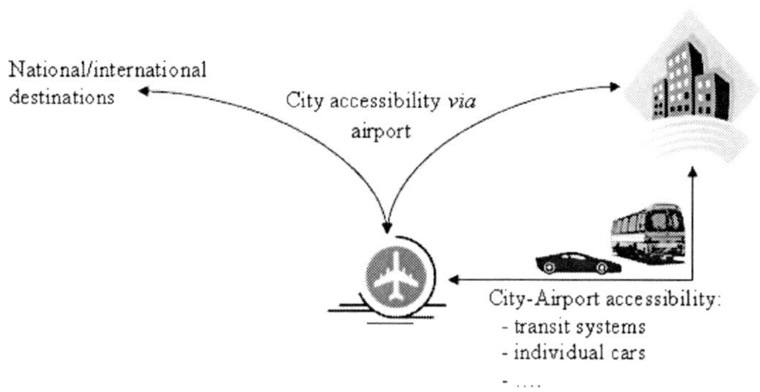

all its related activities (e.g., hotels, restaurants, leisure attractions). To summarize, it can generate employment and wellness in its surroundings.

On the other hand, the airport, as terminal of the air transport network, links cities among them and increase their accessibility in a over-regional perspective. Further, air travellers need to arrive at the airport, and to leave it, by using surface transport modes (e.g., individual cars, buses, trains and so on), then the airport must also be linked to the surface transport network in order to be accessible to air travellers (Figure 2)

Therefore, in the air transport system airports play a double role: 1) as part of the air industry, they are potential economic engines to help the development of a geographical region; 2) as nodes of the air transport network, they are gateways to a region and increase its accessibility.

According to the second point, air accessibility has the great advantage of enlarging the potential competitiveness range of a given city with respect to other cities, because the air transport main characteristic is the capability to realize fast links among distant areas.

In the European Union (EU) as well as in US the importance of airports is well recognized (Postorino, 2010a); many policies have been implemented to reach a satisfactory development of the air transport system and most of them take into account not only the traveller needs but also the environmental constraints to obtain a sustainable development.

Following the European classification (European Commission, 2005), large airports, grouped in national and community airports, have respectively from five to ten and more than ten million passengers per year, while regional airports, grouped in small regional and large regional airports, have respectively less than one and from one to five million passengers per year.

Surface transport networks linking the airport to the near cities are one of the most important factors helping the successful development of the airport (large or regional).

Apart from individual modes, small regional airports generally are linked to cities by bus lines; when passengers at the airports are between one and five million passengers per year (large regional airports), in addition they could be linked by rail services. Larger, national and hub airports are also linked by other surface public transport services (e.g., standard trains, high speed trains, and so on), often integrated each other.

Excluding socio-economic and demographic considerations, the best ranked European airports (ACI-Europe, 2007) have good accessibility, particularly in terms of railway/metro links availability and variety of modes (tab. 1). Similar considerations can be made for US, although in this case individual modes (as taxi, limousine

Table 1. Top ten large and regional European airports

Top ten large European airports				
Airport	Local/regional trains at the airport	Rapid/international trains at the airport	Suburban service to reference city	Local and regional buses
LONDON LHR	Yes	-*	Yes	Yes
PARIS CDG	Yes	Yes	Yes	Yes
FRANKFURT FRA	Yes	Yes	Yes	Yes
MADRID MAD	-	-	Yes	Yes (only local)
AMSTERDAM AMS	Yes	Yes	Yes	Yes
LONDON LGW	Yes	- *	Yes	Yes
MUNICH MUC	Yes	-	Yes	Yes
ROME FCO	Yes	-	Yes	Yes (only local)
BARCELONA BCN	-	-	Yes	Yes
PARIS ORY	-	-	Yes	Yes
* Railway network in Great Britain is a little different from continental railway networks, as it serves an island.				

Top ten regional European airports			
Airport	Bus to nearest city centre	Railway service to reference city	Local/regional public services
BUCHAREST OTP	Yes	Yes	-
BERGEN BGO	Yes	-	-
GIRONA GRO	Yes	-	Yes
IBIZA IBZ	Yes	- *	Yes
FUERTEVENTURA FUE	Yes	- *	Yes
PALERMO PMO	Yes	Yes	Yes
SEVILLA SVQ	Yes	-	-
GOTHENBURG GOT	Yes	-	Yes
BOLOGNA BLQ	Yes	Yes **	Yes
BASEL-MULHOUSE BSL	Yes	Yes **	Yes
* Railway network serves a small island ** Shuttle buses link the airport to the main station in the city centre			

services and so on) are mostly preferred by travellers also due to transport system policies quite different from EU.

While large airports are generally linked to main cities of recognized competitiveness (Eurocontrol, 2007), regional airports can represent an important element to increase the competitiveness of medium size cities (Button, 2010). In fact, as Figure 1 summarized, cities and airports develop together, because cities benefit from the airport transport and economic functions as well as airports expand if air travellers are attracted by cities around them.

In recent years, many policies have been implemented, particularly in EU but also in East Asia and South America, to help the development of regional airports (Postorino, 2010b); as stated before, at a local level the development of a regional airport potentially produces great benefits for a community, both in direct terms (employments and induced jobs, accessibility improvement and so on) and indirect terms (e.g., as engine to promote the development of tourism or other economic activities).

The growth of a regional airport, and then the benefits it produces on the community, depends on how attractive it is to both air travellers and airlines. It is worthwhile to note that air travellers really choose cities as destinations, the airport being only the gateway to cities. On the contrary, airlines choose airports because their function is to produce air services; anyway, the airline choice to operate flights at an airport depends on how many air travellers will use it.

Apart from compulsory destinations, air travellers choose a city as destination on the basis of some basic attractive factors already discussed: efficiency and accessibility. From the air traveller point of view, an airport is part of the city subsystems, then the concept of efficiency and accessibility refers to both the city and its airport.

As stated in the introduction, the starting point here is the airport as contributor to the city competitiveness, then the focus will be on it.

In terms of transport, air travellers choose the best travel options according to journey times and monetary costs; choices concern not only the air services but also the airport where the air services are offered. Particularly, user travel times depend on flight frequencies, structure of the air network (point-to-point *vs*. hub-and-spoke services)[3] and airport accessibility. This latter, in turn, depends on the trip performances of surface transport systems defined, again, in terms of journey times (e.g. bus/rail/car/taxi run time; bus/rail/taxi waiting time) and monetary costs (e.g. bus/train/car parking/taxi fares).

Generally, regional airports are more efficient than larger airports, because they reduce accessory times (as check-in, security control, baggage recovery and transfer times among the others); on the other hand, good surface transport systems linking the airport with the city increase the accessibility of both (Figure 2). Depending on the airport location, the transport services between city and airport can be considered either urban transport services (if the airport is very close to or inside the city) or extra-urban transport services (in the other situations). Anyway, also in this latter case, they must be integrated with urban transport services, because user elementary destinations are inside the city.

If the reciprocal, positive trend starts and traffic demand (both in terms of air travellers and aircraft) increases, some new problems could arise. In fact, regional airports generally have spare capacity as well as good safety levels because potential collisions among aircraft at the airside are near to zero due to low traffic. Anyway, increasing demand could generate a reduction in spare capacity and a worsening of the safety levels.

In the light of this overview, two important questions can be posed: 1) how cities and airports can be linked by efficient surface transport modes? 2) if the air traffic at the airport increases, due to the positive trend generated by the city-airport relationship, how safety and capacity at the airport should be guaranteed in order to maintain its efficiency? The following sections will try to give the answers by considering the use of IS applications outside airports (question 1) and inside airports (question 2).

INFORMATION SCIENCE APPLICATIONS OUTSIDE AIRPORTS: I.T.S. TO IMPROVE TRANSPORT LINKS BETWEEN CITIES AND AIRPORTS

Air travellers want reliable and quick surface transport systems linking cities and airports. Generally, they can choose between public (e.g., bus, metro, train) and individual (e.g., car, taxi) transport systems.

While individual transport facilities (e.g. roads, parking and so on) are generally available, public transport system improvements could represent a specific policy to reduce the share of individual modes, in order to obtain at least two positive effects: 1) the reduction of the pollution levels due to cars; 2) the reduction of car parking areas at the airport landside. Furthermore, public transport systems could be important for incoming passengers not having private cars.

People can decide to use public transport systems (mainly bus and metro/train) only if they are really competitive with respect to private modes, particularly in terms of coverage (availability in space) and reliability (availability in time).

Coverage refers to the public transport line paths and stop distribution within the examined area (Murray and Wu, 2003), then it depends on the transport network organization; on the contrary, reliability depends on the system management, because, to assure reliability, the difference between scheduled and observed travel time (as running time, but also expected arrival time at stops) should be nearly zero.

By using individual modes (e.g., cars, taxi), a unique trip needs to users to reach the final desired destination from their origins. On the contrary, public line paths are fixed and, for a given user, origin/destination pairs generally are joined by more than one line; then, user has to: *i*) change at interchange nodes (where two or more public transport services converge); *ii*) wait the desired service. Obviously, the less the waiting time at these nodes is, the more the user satisfaction is.

To increase the quality and efficiency of public transport services linking city and airport (both urban or extra-urban), and then give an answer to the question: how cities and airports can be linked by efficient surface transport modes?, IS can provide a useful contribution in terms of Intelligent Transport Systems (ITS).

ITS applications use telematics technologies and data processing methods to obtain an "intelligent" transport system control in terms of efficiency, safety and environmental aspects (see for example Ogden (Ed), 2003; European Commission, 2008; ITS Strategic Research Plan 2010-2014 for US). While the main function of an ITS is to guarantee an optimal system control, it may also be used to give user information about the system itself. According to the nature of the received information, users may modify their trip choices and then produce new transport conditions; these latter should be suitably taken into account by transport planners so as to maintain the system reliability.

As to public transit ITSs, their main aim is to process data collected in real time in order to maintain reliable conditions and also give on line information about the network situation; users may optimize their trip choices (as line, departure time, interchange and so on) on the basis of the received information.

It is known that even if travellers use individual modes, anyway the availability of public transit ITSs increases their overall trip satisfaction (Cascetta, 2001) and represents a city attractive factor. Then, a good public transport system contributes to both the airport and city attractiveness.

The following is a technical overview of the general architecture of an ITS for public transport systems.

For a given transport system, an ITS consists of three main parts:

- data detection and collection unit;
- data storing unit;
- data using unit.

The specific characteristics of these units depend on the goal of the ITS itself (e.g., user information; system management and user information; design and planning; and so on).

ITS specialized for transit systems are also identified as Advanced Public Transportation

Figure 3. General overview of a control and transmission system

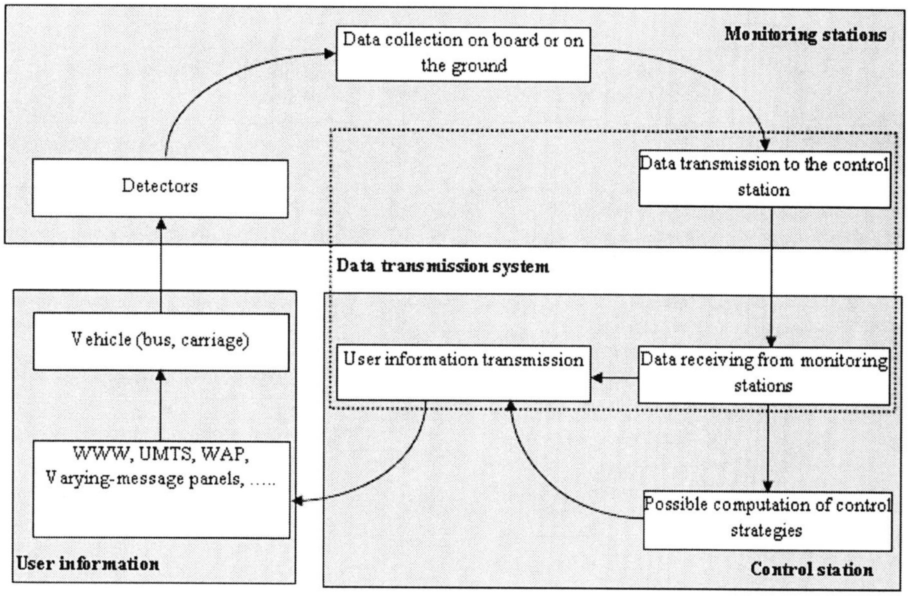

Systems (APTS). APTS focuses on the development, employment and evaluation of advanced technologies to improve the reliability, safety, efficiency and cost of public transport services. Similarly, public transport user information systems are called Advanced Traveller Information Systems (ATIS); they include a significant range of advanced computer and communication technologies designed to provide transit users pre-trip and real-time information; in this way, they can make more conscious travel decisions. Generally, information are provided at stops and via telecommunication systems (e.g., mobile phones for subscribers, Internet), but also on-board (e.g., monitors) and at main terminals. The most used user information systems are Variable Message Signs (VMS) located at the vehicle stops; information is updated in real time and may concern waiting time to the next passage, crowding degree on the next vehicle, and so on.

Inside an APTS, the Fleet Management Systems (FMS) involve the integration of fleet based communication, vehicle monitoring/location, automatic passenger counting and vehicle control technologies to improve the overall planning and operations of transit systems. Generally, they include a specific location technology and a method to transmit the location data from the vehicle to a transmission centre (TRB, 2008).

Control and user information (APTS/ATIS) systems consist of some main elements (Figure 3):

- automatic vehicle location (AVL) and monitoring stations[4];
- control stations for information processing;
- data transmission systems;
- user information communication systems.

Table 2 describes some of the most used AVL technologies together with their main advantages and disadvantages.

Data coming from the detection systems are sent to one or more monitoring stations, equipped to receive data and transmit them to the control station(s) where they are stored and processed. The importance of the data storing procedure is well known and larger countries as EU and US have adopted unified data storing procedure across their member States. For example, the EU supported the implementation of an abstract model

Table 2. Automatic Vehicle Location technologies (source: US Federal Transit Administration, 2000)

AVL technology	Main advantages	Main disadvantages
Signpost/odometer systems	Low in-vehicle cost No blind spots or interference Repeatable accuracy	Requires well equipped infrastructure (network of signposts) No coverage outside signpost network Frequency of vehicle location updates dependent on density of signposts
Dead-reckoning systems (based on odometer and compass readings)	Relatively inexpensive Any infrastructure costs Only odometer needed (assumes vehicle is on route)	Accuracy degrades with distance travelled Requires direction indicator and map matching system to track vehicles off-route
Ground-based radio positioning and paging systems (based on radio triangulation)	Low capital cost Moderate accuracy Low maintenance costs	Monthly service fees that can be relatively high with high frequency of use Signal attenuation by foliage, tunnels and high buildings
Global Positioning Systems	Reasonable accuracy Global coverage Reasonable cost per vehicle	Signal attenuation by foliage, tunnels and high buildings Subject to multipath errors

of common public transport concepts and data structures that may be used to develop several public transport information systems (Transmodel, 2001). Such architecture was formally adopted by the European Committee for Standardization (CEN) as a European standard, ENV 12896 (CEN/TC 278/WG 3). The implementation of a specific data base architecture may facilitate the interoperability among transit system operators, thus making possible an efficient use of the overall transit network.

It is worthwhile to note that since January 2009 the US Department of Transportation Research and Innovative Technology Administration (USDOT-RITA) and the European Commission Directorate General for Information Society and Media are cooperating to develop coordinate research programs specifically focusing on cooperative vehicle systems.

The control process requires the following recurring steps:

1. system information acquisition (by the monitoring stations);
2. information processing and definition of the optimal control strategies;
3. application of selected strategies and user information communications.

A control station generally consists of:

- monitors to see recordings coming from the remote stations;
- equipment to receive data coming from the monitoring stations;
- computers to process data and set control strategies as well as to receive and transmit information;
- specialized software to manage and control traffic;
- equipment to transmit information to users.

As to control aspects, the use of APTS/ATIS is intended to guarantee:

- service regularity;
- quick interventions in case of events modifying the scheduled service;
- high safety and security levels, both for users and staff;
- user information about the characteristics of the service in the short time.

The transmission system links together detection points, monitoring stations, control stations, user information systems. Information is transmitted by radio systems suitably arranged or by modems. Modems are linked to: *i*) fixed phone network if each station is equipped with fixed phones; *ii*) mobile phone network otherwise.

Information concerning the application of the control strategies set at the control stations will be communicated to users. According to the acquired information, users may modify their travel behaviour and then produce variations in the transport systems. Such variations are monitored thus closing the control process sequence.

To summarize, service regularity (and also safety and security levels) can be guaranteed by adopting fast specific actions if there are events modifying the scheduled service/performances (e.g., unexpected demand levels at stops and/or on-board, requiring more time to get on/off; road accidents or traffic congestion causing a slowdown in scheduled transit times; on-board security problems; and so on). Information about lack of scheduled performances can be obtained by comparing real time data transmitted by monitoring stations with time series data (recorded in the past). If the difference between real time and time series data concerning specific variables (as transit time at given points) is greater than a prefixed threshold, then operators should adopt specific actions in order to realign the offered service. In some cases, despite the adopted actions, service regularity cannot be guaranteed and the operators predict new service variable values (e.g., time to reach the next stop) by using time series data and current system conditions known by monitoring stations. Information is then given to users mainly in the form of expected waiting time at stops.

As an example, GPS equipped buses transmit information to a remote control centre where operators verify the alignment between the expected position at a given time (time series data) and the actual one. In case of significant differences with respect to a prefixed threshold, bus drivers are asked to speed up or slow down so as to match again the expected performances. If this is not possible, VMS at stops (or mobile phone messages) inform users about the system performance variations.

It is important to note that expected values depend not only on network conditions but also on demand; as an example, a bus arriving late at a stop meets a number of users greater than the expected one if users arrive continuously at stop. Larger demand produces greater boarding times and a further bus slowdown. Then, expected waiting times at next stops should take into account not only time series data related to running times, but also time series data concerning demand rates at stops.

In this situation, users reaction to information may produce new network conditions as they can choose another path/service, another transport mode or other. Particularly, user perception of trip performances (as waiting time) is very different in case of user information systems (Hickman, 1995; Golledge, 1997; Chorus *et al.*, 2006; Erke *et al.*, 2007) and then travel choices will be different. As a consequence, the effects of user information should be investigated so as to obtain more effective short time predictions about the network performances (Lee et al., 2009).

To conclude this section, it is worthwhile to note that travellers give great importance to public service quality, so ITS applications to public transport systems linking city and airport (but also to public systems inside the city) represent a good opportunity to increase the city attractiveness and competitiveness.

INFORMATION SCIENCE APPLICATIONS INSIDE AIRPORTS: MANAGEMENT OF GROUND MOVEMENTS AT AN AIRPORT

As Figure 1 described in the Introduction, there could be a positive relationship between city and airport, which helps the airport growth in terms of increasing demand. Anyway, in the medium-long run, a positive demand trend at the airport could generate problems in terms of safety and capacity and then pose the question: if the air traffic at the airport increases, due to the positive trend generated by the city-airport relationship, how safety and capacity at the airport should be guaranteed in order to maintain its efficiency? The question is crucial, because an ineffective airport also damages "its" city, as both airlines and air travellers give great importance to safety issues at the airport.

Information Science, mainly in terms of Information Systems, can help to give an answer.

First, an overview of the airport capacity concept and the safety problems arising in the airside during aircraft and land vehicles transfers are discussed, so as to better understand the implications of operational systems used to manage airport ground movements; then, the main characteristics of satellite-based Information Systems to solve both safety and capacity issues will be briefly described.

Readers already familiar with the concept of airport capacity may skip the first sub-section.

Airport Capacity: A General Overview

As well known, airports are divided into two main areas: landside and airside. The landside generally refers to operations and facilities concerning passengers and visitors. On the contrary, the airside refers to operations and facilities concerning aircraft; it includes runways, taxiway and aprons where aircraft and handling vehicles move.

For each airport, the airside layout is generally different, also depending on the expected movements at the airport. In smaller airports, runway and taxiway may also overlap and the aprons are generally close to the passengers terminal so that they move on foot between the aircraft and the gate. On the contrary, larger airports have complex layouts as generally there could be many runways and a taxiway network connecting aprons and gates to runways.

Between these two very different cases there are a lot of intermediate situations where layouts may have several degrees of complexity, although they do not achieve that one of largest airports. Particularly, this is a typical condition of many large regional airports and some small regional airports where the air traffic increase has required an airside expansion. In these airports, the expansion often is made without sufficient, available land, as airports are generally close to urban centres; then, solutions to obtain the capacity increase generally consist of two crossing runways, or crossing runway and taxiway, as the available land should be used as efficiently as possible.

Airport capacity due to infrastructure characteristics depends on elements assuring landing, take-off and taxiing operations (i.e., runways and taxiways) as well as on how such operations are realized. Further, number and location of aprons also contribute to define the airside parking capacity. Then, airport capacity generally depends on runway(s), taxiway(s), and aprons capacities: the smallest value among these three ones represents the airport capacity limit.

Usually, greater attention is given to runway capacity, as typically runways are the limiting element, but taxiways and aprons capacities could give an important contribution too.

Computation of the airport capacity is not an easy task, as it depends on many factors, some of them having a stochastic nature. For this reason, there are many capacity definitions, according to the point of view.

As stated by US Federal Aviation Administration (FAA, Airport Capacity and Delay, Advisory Circular 150/ 5060-5, 1983), the airport capacity is a measure of the number of landing/take-off operations that can be realized by an airport (or by one of its components) in a reference time equal to one hour. The capacity of each component is independent of the capacity of the other ones, so single capacities should be computed for each relevant component.

As general definition, the capacity of a generic airport element (as runway, taxiway, apron) is the maximum number of aircraft that can be handled safely in the reference time unit, without giving both air traffic controllers excessive workload and aircraft excessive delays.

As regards runways, Saturation Capacity is the average number of landing/take-off operations realized at a runway in one hour without violating the Air Traffic Control (ATC) separation rules, without considering the aircraft delay and by assuming continuous landing/take-off aircraft demand. More usual the concept of Practical Capacity, defined as the number of operations that can be accommodated in a given time period, by considering all constraints incumbent to the airport, and with no more than a given amount of delay (Wells, 2000; de Neufville and Odoni 2003). Generally, the runway capacity depends on the time interval between two consecutive landing/take-off operations at the runway itself: the less the interval the larger the capacity. Anyway, for safety reasons the time interval cannot decrease under a prefixed threshold depending on the aircraft characteristics, particularly aircraft weight and speed category. In fact, each aircraft produces an own "vortex effect", that can generate significant lift variations for a following aircraft being in that vortex. To avoid such situations, time intervals between two consecutive operations also take into account the vortex effects, depending on the characteristics of the leading and following aircraft. Generally, small aircraft following larger aircraft need greater time intervals so as to reduce the vortex effect generated by the larger ones. Finally, runway capacity depends on the runway occupancy time, i.e. the time required by an aircraft to free the runway; it includes the approaching and climbing times as well, respectively for landing and take-off operations. Optimal aircraft departure sequences at the runway (i.e., groups of similar aircraft so as to optimise separation times due to the vortex effects) and the planning of initial climb phases can help to increase the runway capacity.

As regards taxiways, capacity is defined as the number of aircraft the system is able to transfer from aprons to runway(s) and vice versa in a given reference time unit (generally one hour). From a practical point of view, well developed and efficient taxiway networks, without runway intersections and good designed rapid exit taxiways, do not represent a constraining element to the airside capacity. Anyway, as the taxiway configuration becomes complex, the number of conflict points increases (e.g., intersections among different taxiway links and among aprons and taxiway links).

As regards aprons, capacity is defined as the number of aircraft stopping there in a given reference time unit (generally one hour). It depends on the time interval between two consecutive seat occupancies by two different aircraft that, in turn, is linked to the Scheduled Occupancy Time (i.e., the expected aircraft occupancy time) and the Positioning Time (i.e., the time required for aircraft in/out positioning at the available seat).

The concept of capacity is linked to that of time delay; in fact, as the number of aircraft requiring an airport service increases up to the capacity value, the average delay for each landing/take-off aircraft increases, where the aircraft average delay is defined as the difference between the time it needs to complete the operation with capacity constraints and the time it would need without capacity constraints.

Ultimately, capacity can increase if time intervals referred to several operations are reduced. Then, starting from a given airside configuration, airport capacity can increase if transfers of aircraft, but also land vehicles, are managed in order to: *i*) avoid paths converging simultaneously towards intersections, as they require waiting times at the intersection itself; *ii*) reduce runway occupancy times, by assigning efficiently paths consisting of exit taxiway and taxiway links; *iii*) reduce separation distances among moving objects (aircraft and land vehicles) in the airside, particularly along taxiways.

Figure 4. Examples of layouts generating conflict points

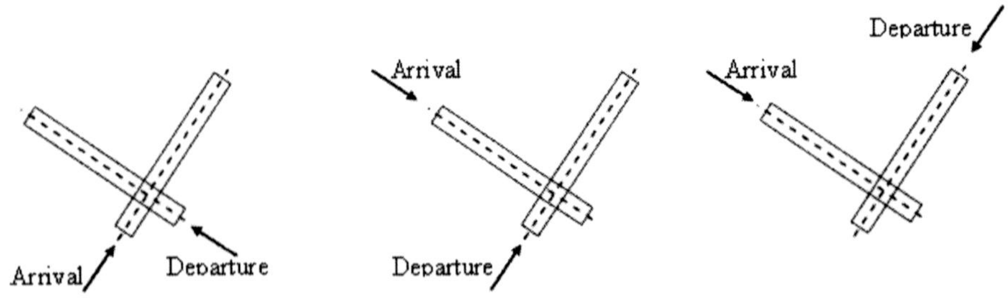

Airport Ground Safety Problems and Requirements of an A-SMGCS

Depending on the layout (and not necessarily on the size, i.e. larger vs. smaller airports) the airside runway and taxiway configuration could be complex in the sense that there could be conflict points between runway(s) and taxiway(s) (Figure 4). Conflict points arise when the trajectories of moving objects (as aircraft and land vehicles) converge towards a common point and then an accident could occur. As an example, two crossing runways have one conflict point reducing the potential safety level of both.

Apart from the airside configuration, aircraft and land vehicles moving in the airside can be potentially in conflict during operations such as passenger transferring to/from the aircraft, aircraft handling, and similar. Furthermore, during the movements on the airside, there could be potential conflicts between aircraft and aircraft or land vehicles and land vehicles, thus suggesting the need of suitable control systems in order to verify continuously the absence of critical situations.

The problem is relevant, because the consequences of ground collisions could be really catastrophic (see for example the disaster at Tenerife Los Rodeos airport, in 1977, or the more recent disaster at Milan Linate airport, in 2001).

To guarantee safety conditions during the ground movements of aircraft and operating vehicles, right separations among them should be assured.

Particularly, runway incursions should be avoided, where a runway incursion is defined by ICAO as: «Any occurrence at an aerodrome involving the incorrect presence of an aircraft, vehicle or person on the protected area of a surface designated for the landing and take off of aircraft» (ICAO, 2004). A protected area is an area including the runway in use, the runway strip and the several parts of the taxiways located between the runway holding positions and the runway in use.

A runway incursion may be caused by several factors, as: phraseology, level of knowledge of the used language, ATC procedures, equipment requirements, lighting and visual airport marks, knowledge of the airport layout and human factors. Situations potentially causing runways incursions (as low visibility) should be examined carefully and avoided by using suitable systems.

Risk severity should be examined by analysing the events leading to some "risk categories" (ENAC - Civil Aviation Italian Authority, 2008); as an example, five risk categories can be identified (from A to E), each one refers to space and time conditions preventing collisions (Figure 5):

A. **serious incident:** an accident could happen; in this case, there are very little separations among aircraft (or aircraft and vehicles) and/or very high speed; then, the collision probability is very high

Figure 5. Risk severity (from A to D) dependent on distances and reaction time

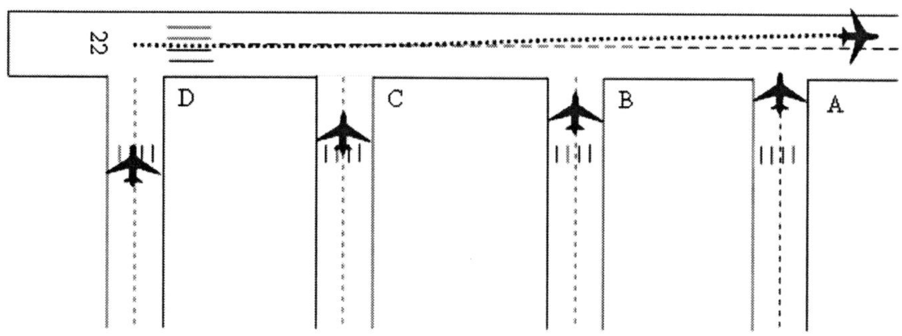

B. **major incident:** aircraft safety could decline; in this case, there are little separations among aircraft (or aircraft and vehicles), although not very little, and a potentially significant collision could occur
C. **significant incident:** threshold situation, as there could be a significant inconvenient if the risk would have not been controlled and/or another aircraft/vehicle/pedestrian would have been in the neighbourhood; there are little separations, but times and distances are sufficient to avoid a collision
D. **no safety effect:** any effect on safety; in this case, the situation is part of runway incursion definition, but the collision probability is negligible
E. **uncertain:** it is not possible to get to conclusions and/or conclusions seem uncertain

Events potentially causing runway incursions are associated to these most common types of actions:

- **pilot deviations:** any pilot action violating rules set by the Regulation Authority;
- **operational error:** any ATC action causing *i*) lack of minimum required distances among aircraft, or among aircraft and vehicles, or among aircraft and pedestrians; *ii*) aircraft landing or take-off on a closed runway;
- **vehicle/pedestrian deviations:** any action of vehicles or pedestrians moving on the airside without ATC permission and interfering with a taxiing aircraft.

Anyway, potential collisions do not concern only runways, but also other airside parts, as taxiways and aprons, although the most critical situations refer to runway incursions.

Finally, another important aspect concerning movements occurring in the movement area (consisting of runway(s), taxiways and aprons) refers to meteorological conditions. Generally, the airport location should be made by taking into account many factors as: distance from urban areas; land mechanical characteristics; absence of obstacles along the glide path (both buildings and natural obstacles as mountains or hills) and standard meteorological conditions in the area. This last aspect is important to guarantee that the airport is operative for the most part of time; for example, an area generally foggy or windy should not be used to build an airport as it will be inoperative many times.

Anyway, adverse meteorological conditions could happen at airports well located too, and in these cases safety conditions in terms of separations among moving objects should be assured by means of suitable devices.

If: *i*) visibility is good, *ii*) airside size is limited and *iii*) traffic is light, tower controllers can check on sight the movement of aircraft and vehicles along the taxiways and the airside as well. On the contrary, when visibility decreases (e.g., evening hours or adverse meteorological conditions), airside size is huge, and traffic is significant, controller's action must be supported by automatic monitoring systems.

By taking into account all the considerations above, ICAO (2004) defined the requirements for an Advanced Surface Movement Guidance and Control Systems (A-SMGCS) to assure safe operations in the airport area. In their most advanced version, they can be used not only to control the ground circulation, but also to optimal manage the ground movement (e.g., by reducing larger distances among vehicles on ground or by addressing them towards the best taxiway paths in order to reduce the runway occupancy time) so as to optimise the whole system in terms of capacity. Anyway, a right estimate of the capacity gain by using an A-SMGCS is difficult as there are many factors contributing to it. In this sense, simulation models (e.g., TAAM[5]) applied to specific contexts can be useful tools to estimate the advantages of an A-SMGCS in terms of capacity increase.

Following the ICAO recommendations, the functions carried out by an A-SMGCS should be:

- surveillance, to identify aircraft, land vehicles and objects within the movement area as well as their location;
- routing, to plan and assign paths to each aircraft or land vehicle moving within the movement area, so as to assure a safe, quick and efficient transfer from the current position to the final one;
- guidance, to give pilots and drivers suitable information, clearly and continuously, concerning the path to be followed and the relevant speed so as to continue safely
- control, to avoid any possible collision and runway incursion, by assuring safe, quick and efficient transfers within the movement area.

They depend on visibility conditions, traffic density and airport layout.

At this time, only a few airports have advanced ground traffic control systems, and most of them are not able either to locate and identify moving objects with the required precision or to convert signals acquired by detectors into digital data that can be used in automatic processing procedures. The current tendency is to implement surface movement ground control systems including more subsystems as: 1) mode S secondary radars (ICAO, 2005a, b) to locate aircraft equipped with suitable transponders ignoring interrogations not addressed with their unique identity code, thus reducing channel congestion; 2) several satellite systems, as GPS or Galileo European system in the next future (European Commission, 2001, 2009) to locate vehicles and aircraft suitably equipped; 3) primary radars (ICAO, 2005a,b) both to locate vehicles not equipped with transponders or GPS and to identify potential obstacles; 4) further sensors, as infrared video cameras. The subsystems should be linked by communication systems so as to give the most complete overview; in addition, they should send information, suitably processed by a control centre, to the several users.

Generally, greater airports require surface movement ground control systems mainly for safety issues, as they belong to the category of airports with complex layout and heavy traffic. Anyway, such systems carry out the capacity optimization function as well, as their employment can be very useful to reduce delays and improve the airport efficiency. Small airports too could need surface movement ground control systems, not only to optimize capacity, even if they generally have spare capacity, but above all to assure high safety standards in all conditions, both meteorological or peak hour traffic conditions. As said in the previous sections, when the positive relationship between airport and city leads to an increasing air traffic, the airport efficiency can decrease and, in turn, the city attractiveness as well; if attentive planning policies are undertaken to facilitate the development of city and airport together, then the potential problems due to the lack of airport efficiency should be considered a priori and ground traffic control systems described in the next section could then be implemented.

Satellite-Based/GIS General Architecture to Satisfy A-SMGCS Requirements

Satellite Systems are a source of positioning information that can be used in many applications. In the aviation field, Satellite Systems can be used to provide air traffic control, both along the routes and at the airside. According to Eurocontrol (2008), a gradual reliance on satellite navigation systems could be the best approach in terms of cost beneficial Air Traffic Management (ATM) solutions meeting safety and security requirements.

EU, together with the European Space Agency (ESA), started an important initiative concerning a satellite radio navigation system, called Galileo (European Commission, 2001, 2009); Galileo will work worldwide and will ensure complementarity with the existing GPS system, the US space-based radionavigation system currently providing positioning, navigation and timing services to civilian users on a continuous worldwide base. Galileo is based on a constellation of 30 satellites and ground stations providing information concerning the positioning of users. In the transport field, the main applications refer to vehicle location, route searching, speed control, guidance systems, and so on. Particularly, Galileo will support ITS, mainly to provide navigational functions, real time users information and optimised control to urban and interurban transportation systems. Expected positive impacts are above all: decreased travel times, increased capacity of transport elements (links, nodes) and increased safety.

Whatever system is used (i.e., GPS or Galileo), the satellite radio navigation is based on the satellites signal emission, processed so as to determine the location of any moving or motionless object equipped with an individual receiver. The precision of the object location measure is within one metre.

In terms of working principles, satellites in the constellation are provided with an atomic clock, able to measure time very precisely; they send out personalised signals indicating the precise time the signal itself leaves the satellite. Ground receivers, incorporated for example into a mobile phone or ground antennas or similar devices, have in their memories the precise details of the orbits of all the satellites in the constellation. By reading the incoming signal, a receiver can recognise the specific satellite, compute the time the signal has taken to arrive and then compute the distance from the satellite. Once the ground receiver gets the signals from at least four satellites simultaneously, it can calculate the exact position by using topographical triangulation methods (Hoffman-Wellenhof et al., 1997). The use of more satellites other than the three ones theoretically requested by the triangulation procedures needs because generally satellite measures are limited by several error sources (as irregular disposition of satellites on the sky at the moment of the survey, slackening of speed due to the ionosphere, and so on) that might generate location errors even significant. Furthermore, different operational survey modalities, as static and dynamic ones, help to overcome these problems (Hoffman-Wellenhof et al., 1997).

With reference to Geographical Information Systems (GIS), they are used to store and manage a large amount of information concerning an examined system, so as to provide user-friendly graphical representations. GIS applications include both geometrical and not-geometrical information of existing relationships (both topological and descriptive) among the different objects representing the system. For these reasons, many applications have been made (and some others may be thought) in the transport field, as GIS are particularly useful to represent geographical areas and their components, both in terms of transport infrastructures, transport services and land use.

Figure 6. General architecture of a satellite-based/GIS system for A-SMGCS implementations

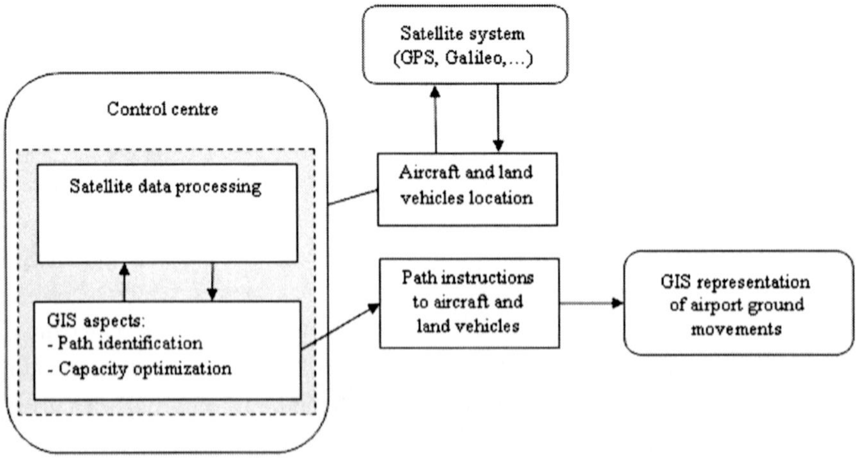

The combination of satellite advanced technologies and GIS user friendly performances can help to check at each instant the location of land vehicles and aircraft at the airport (Postorino *et al.*, 2006), as described below.

The basic architecture for control and management issues at an airport area, deriving from both location opportunities given by satellite systems and data processing/representation potentialities offered by GIS, is described in Figure 6.

Based on signals received by satellite systems, the equipped object (aircraft, land vehicle) transmits its position to the control centre, where information is processed in order to remove any positioning error due to different sources. Furthermore, the control centre uses the acquired location information to: *i*) verify the current situation so as to ensure there is no conflicts among moving vehicles; *ii*) select strategies to optimize capacity in accordance with the current situation; *iii*) assign specific paths to objects according to their requests (e.g., paths from apron to runway for take-off aircraft and from runway to apron for landing aircraft), congruently with the current movements and the capacity optimization strategy (see the previous section dealing with capacity features). The output of this step is obtained by a GIS tool, supporting both the strategy design and the graphical representation. In fact, GIS allows the geographic information and the position to be managed and visualised, by means of the implementation of suitable software functions to resolve the perception and information management problems and to update in real time the acquired data.

Then, each assigned, compulsory path is transmitted to aircraft and land vehicles that should conform to what is transmitted.

As result, the satellite-based/GIS architecture summarized in fig. 6 satisfy most of the A-SMGCS requirements; furthermore, the use of GPS Landing Systems (GLS) on aircraft (as the recent example of Air Berlin, that has received the German Federal Aviation Office approval to use GLS technology on its flights, Feuerle *et al.*, 2010) makes still more attractive the combined use of satellite and GIS systems to control the airside.

SUMMARY AND CONCLUSION

City competitiveness is an important aspect of modern society, as it can play a significant role to assure a suitable development and a better quality of life. Transport systems are important components of cities, not only because they guarantee accessibility among the several districts of a city, but also, and maybe above all, because they guarantee accessibility *among* cities.

For a given city, its attractiveness (and then its competitiveness) depends not only on how better its urban subsystems and services are (e.g., transport, sewage, electricity, water, education, health, and so on), but also how better its accessibility is in the wider, over regional perspective. In fact, decentralized cities, even well organized and efficient, unlikely will be chosen by external investors and visitors, because it is difficult to reach them. Then, all the other things being the same, they could be less competitive than cities more accessible. At the same time, also inhabitants can be dissatisfied, not because of the city subsystems but because it is difficult to reach external destinations.

From this point of view, airports are important nodes in the whole transport system as they represent the gateway to a city. Decentralized cities may gain a lot from airports in the surroundings, particularly when surface, long-distance transport networks are not developed and are not planned to be developed in short time. To take advantage of near airports, cities should improve the surface transport systems linking city and airport, particularly public transport systems in terms of coverage and reliability. In fact, incoming users will be attracted by the city best connected to the airport by public transport systems. Further, when integrated to the urban public transport modes, they also help the growth of the airport as "subsystem" of the city.

Policies able to facilitate the growth of the airport as part of the city will have positive effects on the city itself, but a positive demand trend at the airport should generate problems in terms of safety and capacity in the medium-long run. It is well known that a good planning activity should consider problems before they occur; then, an important goal is to maintain safety and capacity at the airport in order to not reverse the positive city-airport trend. In this light, when traffic demand begins to increase, the airport should be equipped with suitable apparatuses able to guarantee efficiency and safety.

In both cases, Information Science gives an useful contribution to develop tools/procedures able to manage, control and optimize the system. Particularly, Advanced Public Transport Systems and satellite-based/GIS architectures may give a great contribution to improve respectively reliability of public transport systems and safety and efficiency of airports, which in turn help city competitiveness.

REFERENCES

ACI-Europe. (2007). *Airport traffic statistics*. Retrieved from http://www.aci-europe.org

Akçelik, R. (1988). The highway capacity manual delay formula for signalized intersections. *ITE Journal*, *58*(3), 23–27.

ATAG. (2008). *The economic and social benefits of air transport*. Retrieved from http://www.atag.org/our-publications /latest/1/23-the-economic-and- social-benefits-of-air-transport.html

Button, K. J. (2010). Economic aspects of regional airport development. In Postorino, M. N. (Ed.), *Developments of regional airports* (pp. 11–27). Ashurst, UK: WIT Press. doi:10.2495/978-1-84564-143-6/01

Cascetta, E. (2001). *Transportation systems engineering: Theory and methods*. Dordrecht, The Netherlands: Kluwer Academic Publishers.

Chorianopoulos, I., Pagonis, T., Koukoulas, S., & Drymoniti, S. (2010). Planning, competitiveness and sprawl in the Mediterranean city: The case of Athens. *Cities (London, England)*, *27*(4), 249–259. doi:10.1016/j.cities.2009.12.011

Chorus, C. G., Molin, E. J. E., & Van Wee, B. (2006). Use and effects of Advanced Traveller Information Services (ATIS): A review of the literature. *Transport Reviews*, *26*(2), 127–149. doi:10.1080/01441640500333677

Crouch, G. I., & Ritchie, J. R. B. (1999). Tourism, competitiveness, and societal prosperity. *Journal of Business Research*, *44*(3), 137–152. doi:10.1016/S0148-2963(97)00196-3

de Neufville, R., & Odoni, A. (2003). *Airport systems: Planning, design and management*. New York, NY: McGraw-Hill.

ENAC. (2008). *Runway incursions – Avoidance and airport management. Circular APT-30*. Airport Series.

Erke, A., Sagberg, F., & Hagman, R. (2007). Effects of route guidance variable message signs (VMS) on driver behaviour. *Transportation Research Part F: Traffic Psychology and Behaviour*, *10*(6), 447–457. doi:10.1016/j.trf.2007.03.003

Eurocontrol. (2007). A place to stand: Airports in the European air network. *Eurocontrol Trends in Air Traffic* (vol. 3).

Eurocontrol. (2008). *Navigation application and NAVAID infrastructure strategy for the ECAC area up to 2020*. Retrieved May 2008 from http://www.ecacnav.com/downloads /NAV%20Application%20+%20NAVAID%20Infrastructure%20 Strategy%2015MAY08%20 Agreed%20at%20 SCG-8.pdf

European Commission. (2001, September 12). *European transport policy for 2010: Time to decide* [COM(2001) 370]. Bruxelles, EU: Commission of the European Communities.

European Commission. (2005). Community guidelines on financing of airports and start-up aid to airlines departing from regional airports. *Official Journal of the European Union, C 312*(1-C), 312-314.

European Commission. (2008, December 16). *Action plan for the deployment of intelligent transport systems in Europe* [COM(2008) 886]. Bruxelles, EU: Commission of the European Communities.

European Commission. (2009). *A sustainable future for transport: Towards an integrated, technology-led and user-friendly system*. Luxembourg: Publications Office of the European Union.

Federal Aviation Administration. (1983). Airport Capacity and Delay. *Advisory Circular*, *150*, 5060–5065.

Federal Transit Administration. (2000). *Benefit assessments of advanced public transportation system technologies*. Washington, DC: Office of Research, Demonstration and Innovation, Office of Mobility Innovation Advanced Public Transportation Systems Division.

Feuerle, T., Techt, T., & Altenscheidt, M. (2010, March). *GBAS - Pilots and airlines perspective*. Paper presented at the ICAO GBAS Workshop, Paris, France.

Golledge, R. G. (1997). *Dynamics and ITS: Behavioural responses to information available from ATIS*. Paper presented at the Conference Pre-Print IATBR '97, Austin, Texas.

Hansen, W. (1959). How accessibility shape land use. *Journal of the American Institute of Planners*, *25*(2), 73–76. doi:10.1080/01944365908978307

Hickman, M. D., & Wilson, N. H. M. (1995). Passenger travel time and path choice implications of real-time transit information. *Transportation Research*, *3*(4), 211–226. doi:10.1016/0968-090X(95)00007-6

Hoffman-Wellenhof, B., Lichtenegger, H., & Collins, J. (1997). *Global positioning system: Theory and practice*. New York, NY: Springer-Verlag.

ICAO. (2004). *Advanced surface movement guidance and control systems (A-SMGCS) manual (DOC 9830)*. Montreal, Canada: ICAO.

ICAO. (2005a). *Comparison for SSR Mode S system* (annex 10, amendment 73-77, ed. 4). Montreal, Canada: ICAO.

ICAO. (2005b). *Communications systems & Surveillance radar and collision avoidance systems* (annex 10, amendment 77, vol. 3-4). Montreal, Canada: ICAO.

Lee, W., Tseng, S., & Tsai, S. (2009). A knowledge based real-time travel time prediction system for urban network. *Expert Systems with Applications*, *36*(3), 4239–4247. doi:10.1016/j.eswa.2008.03.018

Murray, A., & Wu, X. (2003). Accessibility tradeoffs in public transit planning. *Journal of Geographical Systems*, *5*(1), 93–107. doi:10.1007/s101090300105

Ogden, K. (Ed.). (2003). World Congress on intelligent transport systems. *Transportation Research Part C, Emerging Technologies*, *11*(5), 329–342. doi:10.1016/S0968-090X(03)00028-7

Ortuzar, J. de D., & Willumsen, L. G. (2001). *Modelling transport* (3rd ed.). Hoboken, NJ: John Wiley and Sons.

Patriksson, M. (1994). *The traffic assignment problem: Models and methods*. Utrecht, The Netherlands: VSP.

Postorino, M. N. (Ed.). (2010). *Developments of regional airports: Theoretical Analyses and Case Studies* (pp. 29-53). Boston, MA: WIT Press. RITA. (2010-2014). *ITS Strategic Research Plan 2010-2014*. US Department of Transportation, Research and Innovative Technology Administration. Retrieved from www.its.dot.gov/strat_plan

Postorino, M. N., Barrile, V., & Cotroneo, F. (2006). Surface movement ground control by means of a GPS-GIS system. *Journal of Air Traffic Management*, *12*(6), 375–381. doi:10.1016/j.jairtraman.2006.09.003

Sheffi, Y. (1985). *Urban transportation networks*. Englewood Cliffs, NJ: Prentice Hall.

Transmodel. (2001). *Transmodel: The reference data model for public transport*. Retrieved from http://www.transmodel.org/ en/cadre1.html

Transportation Research Board. (2000). TRB publications index. *Highway Capacity Manual*, *193*, 22–23.

TRB. (2008). *AVL systems for bus transit: Update (TCRP synthesis 73)*. Washington, DC: Transportation Research Board.

Turok, I., & Bailey, N. (2004). Twin track cities? Competitiveness and cohesion in Glasgow and Edinburgh. *Progress in Planning*, *62*(3), 135–204. doi:10.1016/j.progress.2004.04.001

Wells, A. T. (2000). *Airport planning & management* (4th ed.). New York, NY: McGraw-Hill.

Williams, G. (1999). Metropolitan governance and strategic planning: A review of experience in Manchester, Melbourne and Toronto. *Progress in Planning*, *52*(1), 1–100. doi:10.1016/S0305-9006(99)90003-X

KEY TERMS AND DEFINITIONS

Accessibility: The intrinsic meaning is "ability to access", i.e. the ability to approach a place. As people want to participate in activities distributed over space and time, "accessibility" measures the combined performance of land use and transportation systems and determines how well that complex system serves its residents.

Land Use: In recent decades people realized that land is not a limitless commodity and then land use planning needs (e.g., land for agricultural, industrial, residential, recreational, or other purposes). The key to good land use planning is communication between the different land users to achieve a balance of land use and zoning that will satisfy as many people as possible.

Policy: A plan of action agreed or chosen by a political party, a municipality, the government on the basis of the benefits it could bring to the community.

Public Transport: The system of buses, trains, and so on, provided by the government or by companies, which people use to travel from one place to another.

Regional Airports: There can be more definitions, according to countries and transport situations as well as perspectives (e.g., transport demand vs. geographical or administration boundaries); the EU defines "regional" the airports that have from less then one million per year to five million per year passengers.

Transport System: The set of components and its relationships that determine the travel demand (both passengers and goods) among the several parts of a region and the transport supply (services and facilities) that needs to satisfy such demand.

Urban Planning: The control of the development of cities and their buildings, roads, and so on, so that they can be pleasant and convenient places for people to live in.

ENDNOTES

1 The definition of "Information Science" is not an easy task, because of regional differences and international terminology. Anyway, it can be said that Information Science focuses on systemic problems rather than single technological aspects within the system and mainly deals with collection, classification, manipulation, storing, retrieving and dissemination of any type of information. In the case of transport systems, Information Science is mainly associated to management aspects. Intelligent Transport Systems (ITS) are an example of the use of Information Science to manage and improve a transport system by providing user information.

2 Airline subsidies are the amount paid by a local or central authority to airlines that realize a public service for some specified routes which they would not normally operate on economic basis. Often, subsidies are provided as airport tax discounts. The European Community accepts the use of subsidies only in two cases: 1) when other transport systems cannot guarantee a suitable accessibility to some regions; 2) where subsidies have a social character to specific categories of passengers.

3 Flight frequencies and structure of the air network characterize the air service offered by an airlines at a given airport. Higher flight frequencies allow satisfying many travel exigencies in terms of desired departure and/or arrival times. Hub-and-spoke air networks make available more destinations to users with respect to point-to-point services, but require waiting times at the hub airport, that, in turn, depend on flight frequencies (the higher the frequencies the less, in general, the waiting time to board the connecting flight).

4 Automatic Vehicle Location (AVL) and monitoring stations receive data from two possible classes of detection systems: *i*) *on ground systems*, located in strategic points to detect and identify the vehicle (bus, carriage) transit; as their primary function is the vehicle identification, generally they belong to the following classes: a) optical detectors (automatic image processing is required in this case); b) radio detectors (wave identification is required in this case); *ii*) *on board systems*, located on the vehicles; they identify the vehicle location continuously and transmit such position to monitoring centres; generally they use GPS antennas and transmit data by using mobile phone networks or radio waves.

5 Total Airspace and Airport Modeller (TAAM): TAAM is a large-scale rules-based fast-time simulation package for modelling entire air traffic systems, developed by the Preston Group (TPG) in cooperation with the Australian Civil Aviation Authority (CAA). It is probably the most capable general purpose simulation for modelling airspace, runway, taxiway and apron operations. It has a high-fidelity graphics display and great flexibility for handling simulation options.

Chapter 5
From Intelligent to Smart Cities:
CoPs as Organizations for Developing Integrated Models of E-Government Services

Mark Deakin
Edinburgh Napier University, UK

ABSTRACT

This chapter develops the notion of the intelligent city as the smart provider of electronically-enhanced services. Set within the ongoing debate about competitive cities, it identifies how the growing interest in the notion of intelligent cities has led universities to explore the possibilities of using 'communities of practice' (CoPs) as a way of drawing upon the work-based learning such knowledge-based organizations offer to be smart in developing integrated models of e-government (eGov) services. It reports on the attempts made by a consortium of leading European cities to use the intelligence of CoPs as the organizational means to be smart in developing models of eGov services capable of integrating the e-learning needs, knowledge transfer requirements, and capacity building commitments of their socially-inclusive and participatory urban regeneration programmes.

DOI: 10.4018/978-1-61350-174-0.ch005

INTRODUCTION

Made up of researchers, computer engineers, informational managers and public sector service providers, the IntelCities CoP has worked to develop an integrated model of eGov services. This paper suggests the workings of the IntelCities CoP are worth studying because, as a network, it provides an example of a virtual organization set up to manage the learning needs and knowledge requirements of eGov services which are not only socially-inclusive, but that are also participatory. What follows proposes CoPs of this kind i.e. those geared towards meeting the electronically-enhanced governance needs of socially-inclusive and participatory urban regeneration programmes, offer the means required to:

- meet the learning needs, knowledge transfer requirements and capacity building commitments of the organization;
- co-design them as a set of services that are socially inclusive and participatory and which allow their users to learn about the availability of such services, how to access them and the opportunities they offer everyone to become engaged with and get involved in meeting the knowledge transfer requirements and capacity building commitments of their urban regeneration programmes;
- monitor and evaluate such service developments as part of a good governance test.

As the paper establishes, it is the e-learning platform that makes it possible for the online services under development to be integrated with the knowledge transfer and capacity-building technologies which are needed for this CoP to work as a shared enterprise. Something that in turn allows the citizens and businesses making up the organization in question to collaborate and build consensus on the competencies, skills, and training which is needed to service the required online developments. Together, the innovation and creativity of the partnerships responsible for organizing the development of these capacities make it possible to engage citizens and show how the active participation of communities is both intelligent and smart.

The paper also suggests, the overarching theoretical significance of such partnerships should not be overlooked either. Because as the following review of the on-going debate over competitive cities shall highlight, intelligent cities offer an index of the innovations CoPs have made to ensure the eGov services currently under development align competitiveness with the cohesive qualities of socially-inclusive policy agendas. It also goes on to suggest the overall significance of this realignment lies in the opportunity such a policy agenda offers to begin answering some of the big questions Harloe (2001:896) asks of the "*ongoing debate*". In particular, some of the big questions this policy agenda raises about *"competitiveness, exclusion, cohesion and social inclusion (ibid)"*.

In addressing this emerging policy agenda, the paper begins by studying the ongoing debate over competitive cities and goes on to examine CoPs as the institutional basis that have emerged to make competition less exclusionary and offer cities the organizational means by which to be cohesive and socially-inclusive. Having done this, it goes on to identify the two types of interaction that cultivate such qualities and which point towards the contribution on-line service developments make to such socially-inclusive and participatory practices. From here the paper goes on to study the defining features of the IntelCities CoP and integrated model of eGov services developed by this organization as part of their online service developments. This draws particular attention to the attempts made by this organization to realize the socially-inclusive and participatory virtues of such online service developments. That is by way of and through the technical platforms and semantically-interoperable content of the eGov services which have been developed as part of a 'real time' demonstration of what it means for intelligent cities to be the smart.

The Ongoing Debate over Competitive Cities

In reviewing the debate on the search for competitive cities, Malecki (2002, p. 930) states:

The competitiveness of places – localities, regions and nations – refers to the ability of the local economy and society to provide an increasing standard of living for its inhabitants.

As Malecki (2002) also states, the outcome of competition: growth in GDP, employment and levels of income, isn't always progressive, for while economically-efficient, it can just as easily be socially-exclusive and exhibit a tendency to close down, rather than open up access to increased standards of living. For while it is recognised competitiveness can raise standards of living, they in turn can just as easily increase the gap between the rich and poor and be socially divisive. That is fail to guarantee levels of income which generate wealth and reduce poverty, but instead run the risk of doing exactly the opposite. In short exclude some members of society access to the benefits rising standards of living offer others.

Given the potential such exclusion has to undermine the value cities place on competitiveness, governments have increasing come under pressure to develop policies whose competitiveness is not so much exclusive, as cohesive and socially-inclusive. Not exclusive, but cohesive and socially-inclusive in offering equal access to the benefits rising standards of living are expected to offer.

Reviewing the responses cities have made to the emerging policy agenda, Malecki (2002) draws attention to observations (Mitchell, 1999; Graham, 2000; Symonds, 2000; Warf, 2001; Selwyn, 2002) on what the development of electronically-enhanced services contribute to competitiveness. In particular the potential such developments have to make competition not just economically-efficient, but cohesive, and in that sense socially-inclusive in offering equal access to the opportunities and life chances which they in turn afford. Developing this line of reasoning, Malecki goes on to suggest:

Like firms, governments are also under pressure to meet rising expectations of service. Many people believe that a company without a Web site does not exist. The same might not be thought true of government – yet. However, cities and other customer- or citizen-responsive governments face the same expectations that firms face: provide information, facilitate links to the information the person wants, and anticipate needs. (Malecki, 2002, p. 940)

While the statement suggests this era may not yet be upon us, Malecki goes on to state:

As increasing numbers of consumers become used to the quality of service offered by the best web retailers and service providers, their willingness to accept slum standards in the public sector is coming under strain. If the same 24-hour, seven-days-a-week availability and convenience, fast delivery, customer focus and personalisation became the norm in the public sector, it would not just make life easier, it would fundamentally change the way that people view government itself. (Malecki, 2002, p. 940)

As Malecki (2002) goes on to stress, this means consumers shall have to begin viewing governments afresh. That is as the harbingers of a new 'reinvention' and perhaps more critically one which following Mitchell (1999) and Symonds (2000), leads to the development of intelligent cities. That is cities which are intelligent in the sense their competitiveness in not exclusive but cohesive and socially-inclusive in the way they use modern information and communication technologies (ICTs) as a means to bridge the 'digital divide'.

What follows shall suggest Malecki's (2002) account of the e-government underlying the emergence of intelligent cities is instructive in the manner it foreshadows many of the innovations supporting the type of electronically-enhanced service developments referred to in his paper. That is not only in the sense the innovations in question tend to foreshadow such developments, but the way in which they also call for a constructive alignment of competition with the cohesive qualities of a socially-inclusive driven policy agenda.

In view of the considerable potential such a policy realignment has to reconcile competition with the cohesive qualities of socially-inclusion, this paper suggests any attempts to study the 'e-government of intelligent cities', calls for a new bottom-up and institutionally grounded method of enquiry. That is a method of enquiry grounded in what is known about CoPs and their institutionalization as organizations associated with the development of electronically enhanced services.

CoPs

Surveys of CoPs reveals many different kinds of situated practices, all of them displaying quite varied processes of learning and knowledge gathered around distinct forms of social interaction. In this respect, Wenger's (1998, 2000) studies of CoPs is of the ways that insurance claim processors and other such occupational groups learn to be effective in their job. Orr (1996) also studies the importance of CoPs amongst photocopier repair technicians. Osterlund (1996) studies are of CoPs as learning organizations that cut across craft, occupational and professional divisions and which transfer knowledge between them. The collective representation of CoPs in the literature suggests such organizations have the characteristics outlined in Table 1.

Table 1. Key characteristics of a community of practice

• sustained mutual relationships • shared ways of engaging in doing things together • the rapid flow of information and propagation of innovation • absence of introductory preambles, as if conversations and interactions were merely the continuation of an ongoing process • very quick setup of a problem to be discussed • substantial overlap in participants' descriptions of who belongs • knowing what others know, what they can do, and how they can contribute to an enterprise • mutually defining identities • the ability to assess the appropriateness of actions and products • specific tools, representations, and other artefacts • local lore, shared stories, inside jokes, knowing laughter • jargon and shortcuts to communication as well as the ease of producing new ones • certain styles recognised as displaying membership • a shared discourse reflecting a certain perspective on the world

Source: Compiled from Wenger (1998)

Taking this representation of CoPs as a starting point for their examination, Amin and Roberts (2008) suggest there are four distinct types of inter-organizational learning and knowledge transfer. These being: craft, professional, creative and virtual.[1] As Amin and Roberts (2008) point out, until recently it has been assumed that virtual organizations are not capable of promoting learning and transferring knowledge. Although, as it becomes easier to communicate with 'distant others' in real time and in increasingly rich ways, the resulting proliferation of online learning means interest is now centering on how the knowledge dynamics of such organizations differ from CoPs dependant on social familiarity and direct engagement (Ellis and Vasconelos, 2004; Johnson, 2001).

TWO TYPES OF ONLINE INTERACTION

As Amin and Roberts (2008) acknowledge, there are now two types of online interaction that merit close attention as spaces where CoPs engage in learning and get involved in knowledge generation. Firstly, innovation-seeking projects that can involve a large number of participants and secondly, relatively closed interest groups which face specific problems and are consciously organized as platforms needed for learning about and gaining a knowledge of, how to build the capacity required to include 'distant others' as participants in such projects.

As they say: open source software groups provide a good example of the first CoP. Typically, they involve short-lived projects that make source code freely available to technical experts who are motivated by the challenge of solving a difficult programming problem. Successful projects of this kind are those guided by shared notions of the problem and by a core group of highly motivated experts who associate with one another to learn about the subject and transfer the knowledge generated to distant others.

More recently, however, we have seen a rapid rise in the development of the second type of CoP. These are established explicitly by professionals, experts, or lay people to advance knowledge. Typically, they involve experts interested in developing and exchanging best practice, or lay people wishing to not only learn about, or transfer knowledge about a given subject, but build the capacity for this to take place via electronically-mediated communication.

Here a CoP is seen to emerge once the technologies for the virtual organization are available and success is seen to emerge from the ability such platforms have to transfer knowledge. Furthermore, it is also stressed that with this type of CoP the technology available to support the development of such virtual learning organizations is something which has to be managed. That is and as Josefsson (2005) points out, managed in accordance with a 'netiquette', where semantically-rich language is used to develop a culture of engagement, replete with humor, empathy, kindness, tact, and support. This way virtual learning organizations are seen to replicate the rich texture of social interaction normally associated with CoPs marked by high levels of inter-personal trust and reciprocity, or collaborations built around strong professional or occupational ties.

Defining Features of the IntelCities CoP

Made up of both open source software groups, experts and lay people, the IntelCities CoP is unique in the sense its network provides an example of a virtual organization set up to manage the learning needs and knowledge requirements of a technological platform. Set up under the name of the IntelCities Project, this CoP:

- offers the means to meet the learning needs, knowledge transfer requirements and capacity building commitments of the organization;
- co-design them as a set of services that are socially inclusive and participatory and which allow users to learn about the availability of such services, how to access them and the opportunities they offer everyone to become engaged with and get involved in meeting the knowledge transfer requirements and capacity building commitments of their urban regeneration programmes;
- also allows for the monitoring and evaluation of such actions.

These are the three features that define the IntelCities CoP and which give it meaning and a sense of purpose. These can be paraphrased as: building the capacity for a shared enterprise, the co-design of their online services and both their monitoring and evaluation. The working definitions of capacity-building, co-design and monitoring and evaluation adopted by the CoP can be found in the keyword definition section located at the back of this paper.

Building the Capacity for a Shared Enterprise

It is the CoP's e-learning platform's that makes it possible for the online services under development to be integrated with the knowledge transfer and capacity-building technologies which are needed for this to work as a shared enterprise. This is because this platform alone makes it possible for the citizens, communities and organizations in question to collaborate with one another and build consensus on the competencies, skills and training needed for the development of the required online services.

Such a shared enterprise is made possible because:

- the ICT-enabled networks underpinning all of this are innovative in developing an e-learning platform based on open-source technologies interoperable across online services;
- in satisfying the need for a formal learning community, this high-tech, digitally enabled network allows for the planning, development and design of the online service requirements;
- such services in turn, allow these applications to be integrated with the e-learning, knowledge transfer and capacity-building technologies supporting the regeneration programmes under review.
- such technologies allow citizens and communities to collaborate and build consensus on the competencies, skills and training needed for the online services under development to support the quintessentially civic values of this regeneration programme.
- together these networks, innovations and partnerships create the trust needed to engage citizens and show how the active participation of communities in digitally-inclusive regeneration is both intelligent and smart in developing the social capital -norms, rules, and civic values- of the ecological integrity and equity underlying this modernization.
- here the ecological integrity and equity of the democratic renewal take the form of consultations and deliberations in government and citizen-led decision making that engages citizens as members of a community which participate in the governance of this ongoing modernization.

The resulting platform supports the distribution, storage, retrieval of learning material, skill packages and training materials needed for such engagement and participation to bridge the digital divides that currently exist, build the capacity which exists for inclusive decision making and transfer the knowledge required for citizens to bond with one another as members of a community.

The Co-Design of On-Line Services

The IntelCities CoP has sought to co-design such eGov services by overcoming the limitations of a customer focussed approach and by supplementing this with a more user-centric strategy. (Lombardi, et.al, 2009). Here collaboration is not only based on notions of either a sovereign consumer, or customer, but on the consensus built between those citizens who participate in such a process of co-design (Berger, 2002; Bergen, 2005). Here the informational and transactional-logic of mass customisation is seen as being supplemented with a process of participatory co-design that is more democratic in the way it goes about meeting personal preferences. Such a strategy advocates that citizens participate in the co-design of products not as customers, but as users of the services and through their involvement in workshops which promote the authoring, self-documentation and recording of their creative experiences. Here the objective of the co-design strategy is not the mass customisation of products, or personalisation of service provision, but collectivisation of the process in ways that allow citizens to collaborate with one another as a community of subjects. As a community of subjects sufficiently empowered to govern such developments (Nicklaus et.al., 2008, Binder, et.al., 2008).

Monitoring and Evaluation

For those co-designing the eGov services, it is not so much their customisation, or the multi-channel access opportunities which this offers, but outcomes of the co-design process that ought to be subject to monitoring and evaluation. This is because for this group of stakeholders, co-design provides a basic measure of value and demonstrates, not whether a service can be trans-acted, but if this is useful. Not in that sense whether services can be linked to the transactional-based logic of customisation, but if in doing so they can also be connected to the user-centric reasoning of social need which eGov services aim to meet the material requirements of.

As the defining features of the IntelCities CoP, the aforementioned align with the characteristics previously highlighted by Amin and Roberts (2008) and set out in Table 1.

Table 2 underlines the importance of these as characteristics and adds another seven that have been exploited by the network to develop a virtual learning organization capable of bridging the gap which exists between the Type 1 and 2 (innovation-seeking and knowledge generating) classifications of virtual CoPs offered by Amin and Roberts (2008). What follows should like to suggest the extra characteristics are those needed to span the divide between the transactional-based logic ofType 1 and the user-centric reasoning Type 2 adopts towards the co-design of eGov services.

In line with current definitions of CoPs as shared enterprises, the additional features clearly highlight these particular qualities and reflect their importance, but in addition to this they underline the technical, social and material purpose of the virtual organization in question. This suggests

Table 2. Defining characteristics of the IntelCities CoP

• sustained mutual relationships • shared ways of engaging and in doing things together • the rapid flow of information and propagation of innovation • absence of introductory preambles, as if conversations and interactions were merely the continuation of an ongoing process • very quick setup of a problem to be discussed • substantial overlap in participants' descriptions of who belongs • knowing what others know, what they can do, and how they can contribute to an enterprise • a shared discourse reflecting a certain perspective on the world • shared enterprise between research institutes, ICT companies and cities joint venture commitment to product development • building the capacity for ICTs to be used as a means of bridging the digital divide • the co-design of services • shared commitment to social-inclusion and participatory urban regeneration programmes as a means to close the gap between the information-rich and poor • support for the modernisation of local government service provision using technological platforms • consensus-based decision making, consultative and deliberative in nature • monitoring and evaluation

that in building the capacity to co-design eGov services, it is not possible for intelligent cities to develop as either Type 1 or 2 CoPs and this is because they have to be technical and social in equal measures. In other words, rest on the transactional-based logic of (innovation seeking) customisation and the user-centric reasoning of (knowledge-generating) eGov services in equal measures.

The following examination of the IntelCities CoP shall to a large extent reflect this position. It shall begin by examining the capacity that the CoP has built to co-design an integrated model of eGov services and IT underlying the eCity platform developed as an intelligent solution for the virtual organization's learning needs and knowledge transfer requirements. The examination shall then reflect on the CoPs search for an intelligent city solution and the step-wise logic which has been adopted to meet the challenge the learning needs and knowledge transfer requirements of such virtual organizations pose. From here the e-learning platform, knowledge management system and digital library developed for such purposes shall be outlined. Having done this, attention shall turn to the innovative features of this platform, management system and library and the semantically-interoperable qualities of the learning, knowledge and repository services this offers shall be reviewed. From here the examination turns attention towards a review of how the learning, knowledge management and digital library services now available as eGov services are integrated into the eCity platform and made available over the web.

This turns attention to what is termed the eTopia demonstrator developed to illustrate the functionality of the semantically-rich eGov services in question. This term is borrowed from Mitchell's (2000) account of intelligent cities as e-topias and as organizations that are: 'SMART', lean, mean, green software systems, driven by networked communities which are virtual (see, Deakin and Allwinkle, 2007; Deakin, 2007). Those organizational characteristics which the author would add are built on the learning needs, knowledge management requirements and digital libraries of electronically-enhanced government services and that are available on the eCity platform as a pool of integrated eGov services.

All of the material supporting this examination of the Intelcities CoP is drawn from the findings of the IntelCites integrating project, funded under the 6th Framework of the EC's IST Research and Technical Development Programme. Further information on this project can be found at: http://en.wikipedia.org/wiki/Intelcities.

The Integrated Model of eGov Services

Figure 1 outlines the integrated eGov services model developed by the CoP. At the front-end there are a range of eGov services under development, highlighted as social inclusion, participation and regeneration and shown in terms of the middleware integrating them between the front-end presentation tier and back-office core interoperability and infrastructure service layer of the eCity platform. This also illustrates the services located in the back-office and the relationship this develops between the organisation's e-learning platform, KMS and digital library.

This shows that it is the middleware of the eCity platform which integrates the front-end delivery of government services to citizens with the back-office business functions. Figure 1 also shows that it is the middleware which in turn provides the opportunity for the e-learning platform, KMS and digital library making up the back office functions, to do the same and become an integral part of the eCity platform, supporting the pool of eGov services which are available for citizens to access at the front-end. This integration of the e-learning platform, KMS and digital library into the middleware and use of it as the eCity platform supporting the presentation of eGov services to citizens at the front-end, is the challenge the IntelCities CoP has set out to meet and customisation it has sought to co-design a solution for.

THE IT UNDERLYING THE INTELLIGENT CITY SOLUTION

The main challenge for the IntelCities CoP has been that of finding a solution offering the intelligence which cities need for the information technology (IT) underpinning the presentation of eGov services to be extensible, flexible and also have the capacity to carry existing local government

Figure 1. The eGov services model

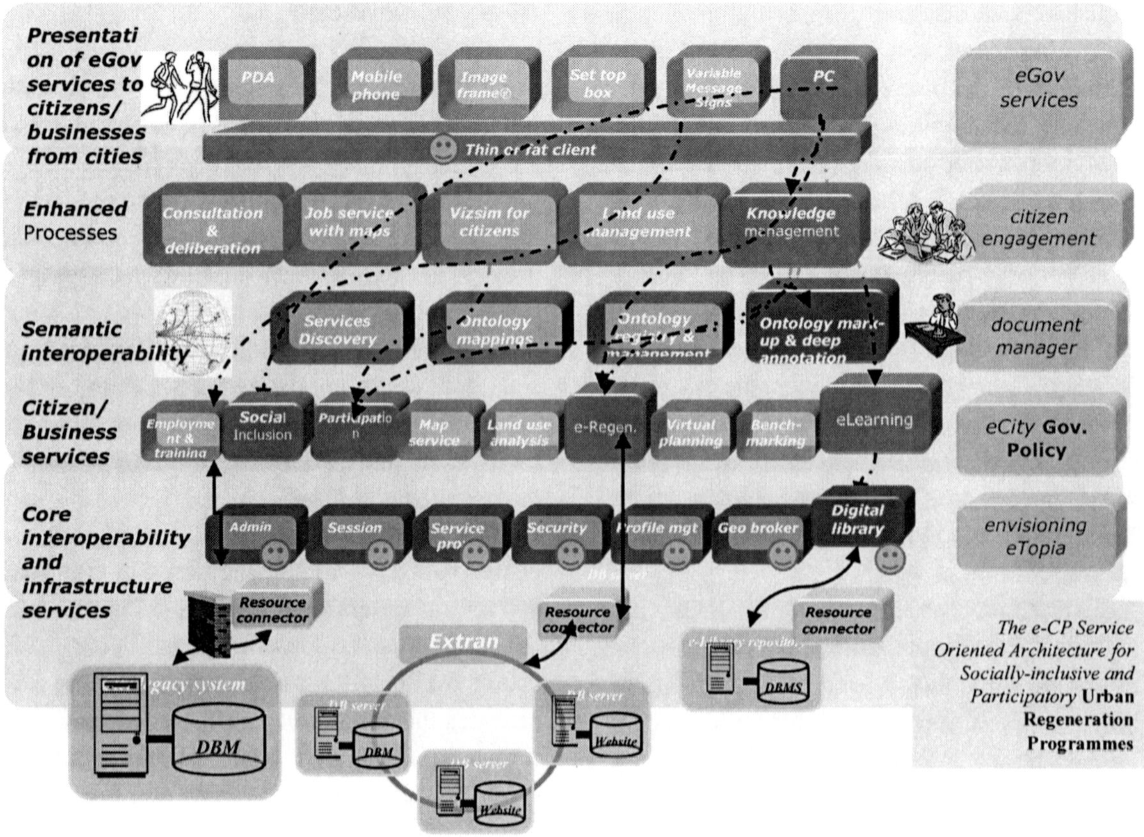

legacy systems. The Services Oriented Architecture (SOA) of the enterprise-wide business model adopted as the joint venture vehicle for such an 'intelligent solution' meets this challenge by offering the IntelCities CoP a distributed, web-based and extendable access system. This intelligence in turn offers cities the opportunity to build a web services enabled platform of eGov services, with XML IT utilisation and SOAP communication.

An important element in the initial system design relates to the use of the Unique Modelling Language (UML) and Rational Unified Process (RUP) methodology used for developing the integrated model of electronically-enhanced government (eGov) services. This allows for the development of complex 'N-tiered' systems and the possibility of cities hosting eGov services on e-learning platforms, KM systems and digital libraries utilising the intelligence such IT offers. This has the advantage of offering a homogenous platform solution supporting the development of specific service applications meeting the e-learning needs, knowledge transfer requirements and capacity building commitments of the IntelCities CoP. It also manages to do this while leaving open the possibility for this customisation of services to be co-designed by other organizations not yet integrated into the eGov services model and eCity platform supporting this particular organization's e-learning needs, knowledge transfer requirements and capacity building commitments.

THE SEARCH FOR AN INTELLIGENT SOLUTION

The search for an intelligent solution to the e-learning needs and KM requirements of the eCity platform has progressed by applying a stepwise logic to the challenge it poses the IntelCities CoP. This has taken the following form:

- a survey of user learning needs;
- analysis of the knowledge requirements;
- review of learning and knowledge services leading city portals provide;
- benchmarking of existing e-learning platforms against the user's knowledge transfer and capacity building requirements;
- selecting the e-learning platform able to meet these requirements and develop as a KM system supported by a digital library;
- integrating the aforesaid into the IntelCities middleware as a platform of eGov services delivered to citizens at the front-end.

Following this step-wise logic has meant focusing attention on the underlying pedagogical issues, the competencies, skills and training requirements of IntelCities. The next step involved a review of the learning services leading city portals offer as legacy systems and benchmarking of the e-learning platforms these systems are based upon against the knowledge transfer and capacity building requirements of the IntelCities CoP. Here the learning services of five leading city portals were reviewed. These included the learning services provided on the city portals of: Edinburgh, Dublin, Glasgow (Drumchapel), Helsinki (Arabianranta and Munala) and Reykjavic (Garoabaer). The review found:

- the said city portals provide learning services for citizens;
- these portals provide citizens with a community grid for learning;
- much of the data available to the community is informative, telling citizens about learning opportunities in their neighborhoods' and providing links to the service providers;
- while being used by up to 10% of the population and offering free email and storage, most of the services provided by the city portals are insufficiently engaging for citizens to use them as grids for communities to base the development of learning partnerships with cities.

As legacy systems, the review found these e-learning platforms were insufficient to meet the knowledge transfer requirements of the IntelCities and needed to represent the point of departure for the CoP. However, on a more positive note, the review made clear the focus of the IntelCities e-learning platform should be the needs of the citizen, their knowledge requirements and the technology adopted to deliver this ought to break with the tradition of existing city portals, be more socially-inclusive and offer greater opportunity for communities to participate in their development. With this in mind, the examination went on to benchmark the e-learning systems which existing portals are based on and examine them against the knowledge transfer and capacity building requirements they set.

The e-Learning Platform

Table 3 illustrates the results of this benchmarking exercise, presenting the average percentage scores of tools provided by 67 commercial e-learning platforms and compares these against the industry standard (Web CT) & European Dynamics' OSS eOWL system. This benchmarking exercise has in turn produced an OSS (Open Source Standards) approach to e-learning. This has opened up the opportunity to get beyond the tendency for city

Table 3. Results of the e-learning Platform Benchmarking Exercise

Learner tools	Commercial platforms[1]	WebCT[2]	IntelCities platform[2]
Communication Tools	57%	71%	86%
Learning Tools	62%	60%	60%
Learner Involvement Tools	64%	75%	100%
Administration Tools	79%	75%	100%
Course Delivery Tools	72%	80%	100%
Course Design	56%	83%	83%
Hardware/Software	70%	80%	63%
Pricing/Licensing	80%	40%	100%

Source: Deakin et.al (2004)

[1] This column illustrates the average percentage of learner tools covered by the 67 commercial e-Learning platforms surveyed.

[2] This highlights the percentage of functionality of individual learning tool covered by services available on WebCT and European Dynamics' OSS (e-OWL) platform.

learning portals to merely provide links to resources held elsewhere and provided the means to customise an e-learning platform capable of meeting the particular knowledge transfer requirements of the IntelCities CoP.

The Learning Management System

The Learning Management System (LMS) developed for such purposes lies at the centre of the platform. This management system provides the common ground between course tutors, trainers and learners, a virtual space where they can co-operate with one another by sharing experiences and offering personal and confidential advice on the available courses, content and communication tools. It is designed as a set of modules where tutors can create content, administer the resulting course and create assessments for learners, while learners are able to work with the related material. The services offered by the LMS are underpinned by a set of repositories holding information on the personal data of registered members, learner's profiles, material available to support the structured course of studies and other unstructured data also available to learners.

The system architecture rests on three levels, each supported by dedicated administrators, course coordinators, tutors and trainers. Here the administrator is responsible for managing the directory of members registered to a course (this provides the interface between the course provider and the learner), whilst the tutor/trainer will be the course content creator, and the coordinator is responsible for distributing the course(s) to the learner and the services supporting the relating studies. This is supported by core services that provide the learning content, communication, collaboration, assessment and administration of the IntelCities courses (i.e. the learning materials, skill packages and training exercises used for developing socially-inclusive and participatory urban regeneration programmes) available to the CoP.

The e-Learning Materials and Courses

The e-learning materials are made up of three IntelCities courses. The first short course is aimed at members of the public with an interest in becoming more involved in civic life via the use of new technologies. The second course targets administrators within the public sector: those responsible for meeting citizens' expectations, in terms of access to electronically-enhanced eGov services. The third is aimed at policy-makers and strategists within city administrations who want to make their cities leading examples of the digitally-inclusive knowledge society. Together, these three courses make up the CoPs eCitizenship module.

Under this heading, the course materials tackle the same core concepts: digital inclusion; citizens' expectations and the means available for cities to meet the needs of their e-ready citizens, whilst enabling access for those currently excluded from the knowledge management systems and digital technologies underlying the community's use of online services. The pitch and tone vary accordingly across the suite of materials, yet each progresses the learner towards an understanding of the tools and methods currently available for cities to use to engage citizens as members of an online community.

Whilst the short course on digital inclusion provides as set of 'taster' sessions on citizens' engagement with digital technologies, no prior experience of ICTs is needed as a prerequisite for the learning. It is designed to be open to everyone and provide universal access as a bottom line for the learning experiences to follow. Level 2 is targeted at citizens with different levels of experiential learning and therefore, abilities. Those collaborating on the development of learning materials for Levels 2 have developed three representative e-service users, each with different levels of familiarity with ICTs. The novice user is characterised as a citizen with little experience in using computers or the internet, but an interest in learning how to find information and pay bills online. The semi-skilled or intermediate level user is a citizen with regular access to a computer and average-to-good ICT skills. At this level of ICT ability, the citizen is interested in locating detailed, up to- date information online and in submitting comments and feedback to the City. The advanced user has frequent access to ICTs and is highly skilled and confident in their ability to interact using the internet. This user wants maximum benefit from new technologies and is keen to interact with the City via services such as online debates and e-petitions.

These three characterisations serve to elicit the relationship between citizens' ICT skills and competencies and the e-services they expect their cities to provide. Figure 3 summarises this relationship. The left-hand column details the expectations of novice ICT users, the challenges these represent and action cities can take in response to them. With little access to ICTs, such as home PCs or 3G mobile phones, the novice ICT user has little confidence in the e-services under development and the potential benefits they offer. In terms of their priorities, citizens at this level prioritise the accessibility of new online services. The challenge cities are faced with is that of meeting these very basic requirements without alienating those which have higher skill levels.

As Figure 2 indicates, citizens with minimal ICT skills are unable to make use of cutting-edge, interactive technologies. The digitally-excluded, often amongst the most socially deprived, risk being further alienated for the reason they are not members of the online communities cities can use as a means to consult with citizens and deliberate on issues of public concern. By investing in community-based training initiatives and online user support, cities ensure that citizens with little-or-no-ICT experience are offered the chance to develop their skills, be included and participate in the customisation and co-design of more complex inter-active online activities.

Figure 2 identifies citizens at the lowest skill level as seeking engagement at an informational level. Citizens who have progressed beyond basic ICT skills are referred to as seeking engagement at a transactional level. The semi-skilled, or intermediate level, user has better access to ICTs than the novice and is already comfortable accessing basic information and making bill payments online. It also identifies the intermediate user's expectations: for up to-date information of a high quality and the seamless transition between different online services and websites.

As with the novice, the user at this level requires services pitched at the appropriate skill

Figure 2. Citizens' skills and competencies. Source:Campbell and Deakin (2005).

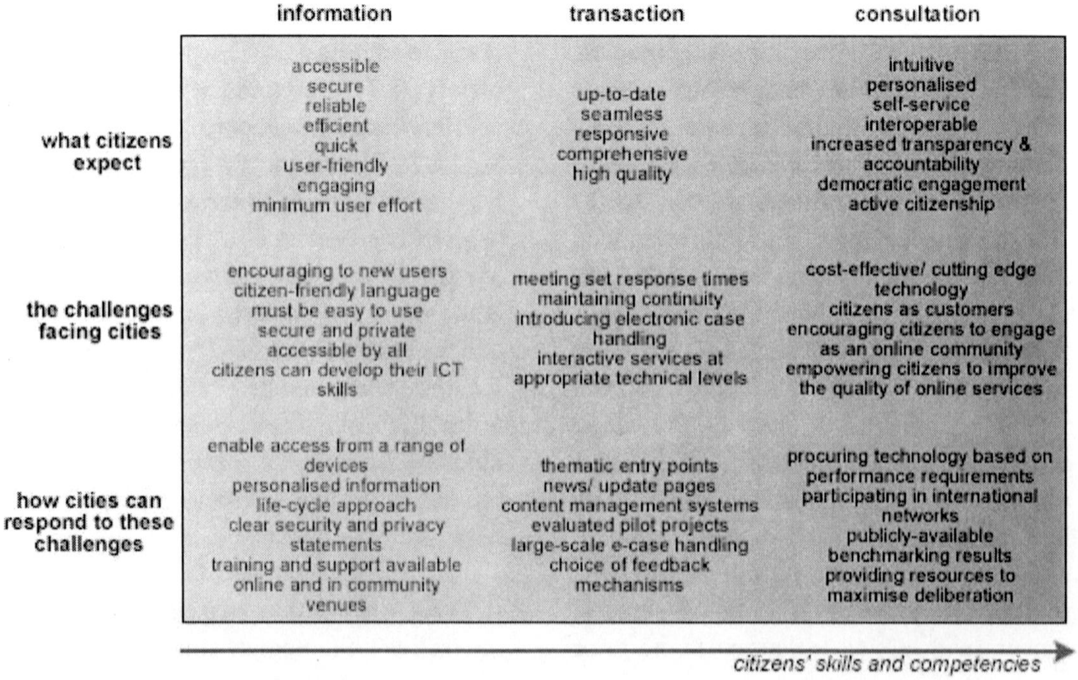

level, again presenting the city with the challenge of meeting the needs of a diverse society. At this 'transactional' level, citizens are interested in establishing online communication with the City and in order to engage these users and encourage repeated use of such services, they in turn are required to respond within set times. Electronic case handling is listed in Figure 3 as one method of managing the information flow and building citizens' trust in e-services, as are content management systems to ensure continuity across a range of Web pages and services.

Citizens with advanced ICT skills and regular access pose an addition set of challenges to their cities, given their expectations of personalised and intuitive services like those offered in e-commerce. However, citizens at this level of ability are also able to make use of the more complex technologies cities can offer to encourage online consultative and deliberative participation. By engaging increasing numbers of citizens in online dialogue, city administrations harness the knowledge and experiences of local people as a means to improve the quality of services they customise by way of and through their co-design.

The Level 3 set of lessons examines the skill bases and competencies of a user who has just such abilities, who expects their city to provide personalised and intuitive services and to make use of the more complex technologies cities can offer to encourage online consultation and deliberative participation. Level 3 provides a set of lessons on how cities can use the skills and competencies their citizens have to make use of these complex technologies and become leading examples of the IntelCities CoP. Two inter-active video lessons have also been produced to support this set of lessons.

Figure 3. Integration into the eCity platform

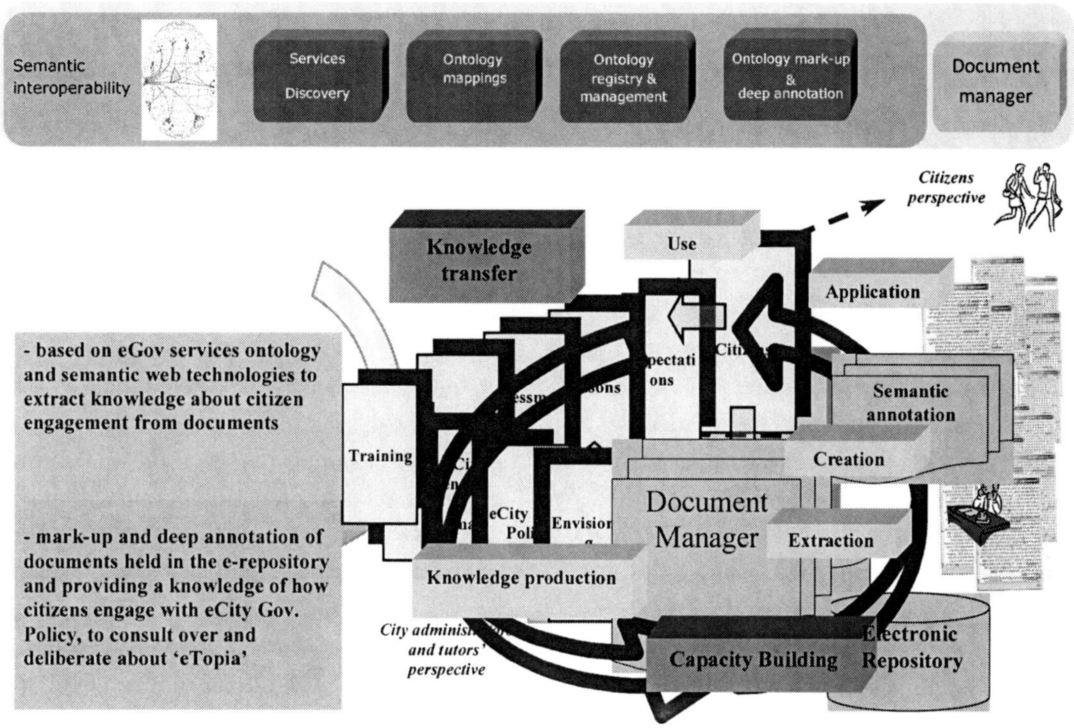

THE KNOWLEDGE MANAGEMENT SYSTEM AND DIGITAL LIBRARY

The KMS is organised and grouped according to the requirements of a pre-specified, but evolving, eGov services ontology. The overriding objective of the IntelCities CoP is to provide an e-learning platform that allows access to a KMS and which is both accessible and usable. This objective has been met by developing the KMS's document manager (DM). The DM developed has built the capacity to perform Ontology-based Annotation in Semantic Web for the easy creation, application and use of semantic data. This is particularly important where learners require the KMS to perform a deep and semantically-rich annotation of materials.

The digital library is the electronic repository storing the information available for extraction by the KMS. The rationale for developing the digital library as part of the KM system lies with the potential the DM has to function as a service capable of:

- capturing, storing, indexing and (re)distributing the learning materials, skill packages and training manuals;
- extending this to include the formal semantics (metadata, knowledge) for the retrieval and extraction of the said materials, packages and manuals available to support the integrated modelling of eGov services;
- offering access to the extensive range of products stored as knowledge objects in the digital library and available for extraction by those managing the development of the middleware as a platform for pooling the said eGov services together and extending delivery of them to citizens as front-end users.

SEMANTICALLY-INTEROPERABLE EGOV. SERVICES

Utilising the Semantic Web paradigm, the e-learning platform is capable of delivering data to its users in a way that enables a more effective 'query-minded' discovery, integration and reuse of the knowledge which can be accessed from the digital library. Through the platform's utilisation of Semantic Web technologies, data uploaded by the KMS (as information available from the system's DM) presents knowledge products codified in ways that not only correspond to documents (web pages, images, audio clips, etc. as the internet currently does), but more pre-defined objects, such as people, places, organisations and events which are deposited in the digital library. Using a pre-defined ontology of this type, the DM allows multiple relations between objects to be created.

It is here the IntelCities project achieves its goal of integrating Type 1 and 2 CoPs and builds the capacity for the organization to be both innovation seeking and knowledge generating. Both innovation seeking and knowledge generating in the sense which the organization is able to co-design eGov services as semantically-interoperable codifications of the objects they relate to. Currently none of the e-learning platforms forming the basis of the CoP's S.W.O.T. analysis offer such services. Until now it has only been common to see references to the possible convergence of e-learning platforms, KMS and digital libraries. This platform and system gets beyond the call for the convergence of such technologies and begins to integrate eGov services with the ICTs available to achieve this. Perhaps most importantly of all, the outcome of all this is an e-learning platform, KMS and digital library with the embedded intelligence cities need to deliver semantically interoperable eGov services and meet this requirement as a standard measure of the socially-inclusive and participatory urban regeneration programmes the IntelCities CoP has a particular interest in.

INTEGRATION INTO THE ECITY PLATFORM

Figure 3 illustrates how the IntelCities CoP proposes the integration of the e-learning platform, KMS and digital library should take place and shows the workflow supporting this. This shows the workflow as having its basis in the digital library and KMS of the e-learning platform. It also shows the workflow between the courses held on the said platform and KMS. Here the system's DM is shown to semantically annotate the learning materials, skill packages and training manuals supporting the courses held on the platform and mark them up in line with the index and classification of the eGov services ontology evolving to manage the knowledge drawn from the digital library.

These back-office functions in turn lead to the creation of the citizen engagement matrix, designed as a semantically-rich grid, allowing communities to be inclusive and actively participate (via consultative and deliberative operations) in the development of the middleware as applications which this platform of eGov services delivers to the front-end. These developments provide the knowledge management toolkit. This term is preferred to "e-learning platform" because this best captures the contribution the tools: the electronic repository, document manager, semantic annotation and mark-up system, make to the type of knowledge management and digital library services currently found on city portals.

While it is recognised this journey from the front-end eGov services to the middleware and towards the back-office functions, represents a significant detour, it is undertaken because the path taken does mark a significant step forward. Not only in terms of the additional learning services that existing city portals are now able to offer, but in 'squaring of the circle' and providing a platform with the intelligence - KMS and digital library - to integrate the front, middle and back-office sections of their organization as a virtual

Table 4. Extract of the consultation and deliberation services

Here the ecological integrity, equity and participatory nature of the democratic renewal take the form of *consultations and deliberations* in government and citizen-led decision making as members of an on-line community. The resulting platform supports the distribution, storage and retrieval of learning material, skill packages and training materials needed for such engagement and participation to bridge the digital divides that currently exist, build the capacity for inclusive decision making and transfer the knowledge required for the cohesiveness qualities of this modernization process to work. Here the building of the capacity for ecological integrity and equity takes the form of decisions over the urban regeneration's footprint, bio-diversity and environmental loading. The democratic renewal is content to promote the shift from government to citizen-led decision making. This involves the use of *advisory panels, discussion boards, opinion polls, focus groups, petitions, citizens juries, ballots and online voting* as part of the visioning and scenario building exercises gaining consensus on the norms of energy consumption, waste and emissions, as a set of rules underlying the ecological integrity and equity of the democratic renewal governing this modernization.

CoP. That is a CoP which in this instance is based on standards that are interoperable across a growing pool of extensible eGov services and which have the capacity, customisation qualities and co-design features to support socially-inclusive and participatory urban regeneration programmes.

The 'eTopia' Demonstrator

At present this integration is mainly technical, concerning the software developments needed to host such services and meet with the semantics of the platform's e-learning needs, knowledge transfer requirements and capacity building commitments. This currently takes the form of an 'eTopia' demonstrator, showing in 'session-managed logic' how the eCity platform accesses the extensive pool of eGov services located in the back-office and uses the intelligence embedded in the middleware to deliver Level 3 (advanced e-Citizenship) courses on the consultative needs and deliberative requirements of such developments. This provides a 'real time' demonstration' of the platform's capacity to be 'SMART' in developing both the technical and semantically-rich content required for the middleware to begin supporting the socially-inclusive consultations and participatory deliberations of urban regeneration programmes. These enhanced processes of consultation and deliberation also have the advantage of offering citizens multi-channel access to such eGov services, customised, co-designed and bundled together as socially-inclusive and participatory urban regeneration programmes for bringing about improvements in the quality of life (Deakin and Allwinkle, 2007; Deakin, 2009; Lombardi, 2009). This goes a long way to:

- uncover the business logic needed to base the intelligence-driven (re)organisation of cities on and standards required to benchmark the performance of the platform against;
- provide the performance-based measures needed to assess whether any plans cities posses to develop eGov services (over the platform) have the embedded intelligence (the learning, knowledge-based competencies and skills) required to support such actions;
- also provide the means to evaluate if such planned developments build the (intellectual) capacity – learning, knowledge-based competencies and skills – society needs to support such actions.

These service developments have more recently been bundled together and augmented under the SmartCities project. Under this fourth phase i.e. the democratization of city portals, the all too often 'top-to-bottom' logic overhanging some intelligent city thinking has firmly been turned on its head. Here Web2.0 technologies are used as the social basis to build the capacity for

CoPs to customize their co-design of post-transactional services and capitalize upon these developments via a platform of multi-channeled access smart enough for the environments this generates to be matched with given user-profiles (Deakin, 2010a, 2010b).

With this platform the top level citizen engagement processes previously outlined in Figure 1, now become the baseline issues for SmartCities and their deployment as a means to monitor and evaluate how and in what way, the socially-inclusive and participatory service developments of the Level 3 Citizenship courses, work as programmes of urban regeneration capable of sustaining such a process of democratic renewal.[2]

The terms of reference for the SmartCities project are noteworthy for a number of reasons: firstly, because, unlike the IntelCities project, they do not focus so much on the problem i.e. the development of an intelligent system with service platforms able to support cities, but in exploiting the solutions they offer. That is the solutions they offer cities as business-to citizen applications which are capable of being accessed by members of the public as front-end services. Secondly, while IntelCities tackled the infrastructure problems experienced by large metropolitan cities, the solution corridor SmartCities opens up rests upon their application. Their application that takes place by way of a customization which can be accessed by members of the public through a process of co-design, targeting the front-end services of medium-to-small towns and cities.

While this does much to verify the 'scalability' of the eGov services model outlined, it also demonstrates how, with the development of Web2.0 technologies, the possibility now exists to short-circuit many of the capacity-building exercises previously embarked upon and focus in on the co-design stage, monitoring and evaluation of the developments in service provision. This having the advantage of building upon the capacities founded by the intelligent cities movement and using the skills, competencies and training exercises this lays down as a platform to be smart. That is smart in not only reducing the initial start up costs associated with their construction, but subsequent ongoing maintenance charges which relate to their front-end service developments.

CONCLUSION

Set within the ongoing debate over competitive cities, this paper has sought to examine CoPs as the institutional basis that has emerged over the past decade to make competition less exclusionary. Make competition less exclusionary and in turn provide the organizational means to be cohesive and socially-inclusive. Having done this, the paper has gone on to identify the two types of interaction that cultivate such non-exclusionary and cohesive qualities and which point towards the contribution on-line service developments make to such socially-inclusive and participatory practices.

Set within this ongoing policy debate, the paper has also sought to build on the critical insights Malecki (2002) offers into the 'potential of intelligent cities'. What-is-more realise this by fostering the development of CoPs as a platform of eGov services capable of tackling social exclusion and combating the poverty of area-based deprivation. That is as a platform of eGov service developments capable of bridging the digital divide and spanning the gap which has opened up between the information rich and poor by capitalizing on the potential such organizations have to realign competition with the cohesive qualities of a socially-inclusive policy agenda. In particular, with the agenda that has surfaced about the learning needs, knowledge transfer requirements and capacity-building commitments which surround socially-inclusive and participatory urban regeneration programmes.

In view of the considerable potential which such a policy realignment holds out to reconcile competition with the anti-exclusionary sentiments and cohesive qualities of such socially-inclusion agendas, the paper has also suggested that any attempts to study intelligent cities calls for a new, bottom-up and institutionally grounded method of enquiry. For a new, bottom-up and institutionally grounded method of enquiry that doesn't base competitiveness on the natural advantage of given resource endowments, human capital, or technological advances, so much as the social capital which underlies them. That social capital which underlies their institutionalisation and that in turn makes it possible for CoPs to organize their development of eGov services in a way which is smart.

In a way which is smart, it perhaps, might well be added, in out-rivalling the intelligence of their competitors in other cities and deploying CoPs as organizations that are able to exploit the opportunities eGov service developments offer to be constructive. That is constructive in building the capacity which such organizations have to align their service developments with the cohesive qualities of a socially-inclusive policy agenda. Not only because their participation in such a realignment helps build the capacity to co-design, monitor and evaluate what service developments of this kind offer members of the public, but for the reason they also provide the means to measure their performance. In particular, as services that improve the quality of life, which they, alongside their rivals, compete with one another to effect.

Adopting this bottom-up method of enquiry, the paper has gone on to suggest there are three features that define the IntelCities CoP as an organization which exemplifies the constructive nature of such a realignment. These are: building the capacity for the shared enterprise underlying the venture, the co-design of the online services supporting it and both the monitoring and evaluation of the very improvements in the quality of life such a policy agenda offers the prospect of. It has also gone on to underline the importance of these characteristics by adding another eight qualities that have been exploited by this particular CoP to develop a virtual learning organization which is capable of bridging the gap between the type 1 and 2 classifications offered by Amin and Roberts (2008). This has been done in the interests of illustrating how the CoP spans the divide between the transactional-based logic (competitive) and user-centric (in this instance, cohesive, socially-inclusive and participatory) reasoning behind the development of eGov service provision.

The additional features referred to clearly highlight these qualities and reflect their importance, but in addition to this, they also serve to underscore the technical rational and social purpose of the virtual organization responsible for the capacity-building and co-design components of the service developments in question. This suggests that in developing integrated eGov service models, it is not possible for intelligent cities to develop as either type 1 or 2 CoPs because the shared enterprise and joint ventures which have come to charactrerise such virtual learning organizations, means their customisation has to be co-designed in such a way they are smart enough to be both innovation-seeking and knowledge-generating.

Having gone on to discuss the technical solutions adopted to integrate the eGov services model advanced with the legacy systems currently operated by cities involved in this enterprise, attention has turned to the SOA adopted as the business case for such an eCity platform. As has been shown, these developments are valuable because they provide the means to address the criticisms of the learning services currently available on city portals and offer the opportunity for the emerging technologies of the e-learning platform, KMS and digital libraries, to meet the learning needs, knowledge transfer requirements and capacity building commitments of the CoP. This it has been suggested, marks a significant step forward in the development of learning services and offers the opportunity for platforms of this type to develop as a KMS supported by digital libraries.

In view of this, the chapter has suggested that if the full significance of these technical innovations is to be realised, then this integration needs to progress and requires the e-learning platform,

KM system and digital library developed for such purposes, to not only be interoperable across the middleware, but all the eGov services which are available for citizens to access at the front-end. The way in which the CoP proposes to achieve this is particularly innovative. For as an organization it offers a strategy to consolidate the underlying learning aspirations of city portals, but as particular types of eGov services that have previously remained beyond the reach of the platforms developed for such purposes. That is to say, beyond the reach and out with the grasp of previous attempts which have been made by such organizations to develop a knowledge-base capable of delivering on-line consultation and deliberation services. On-line consultation and deliberation services smart enough for their socially-inclusive and participatory urban regeneration programmes to democratize all of this.

REFERENCES

Amin, A., & Roberts, J. (2008). Knowing in action: Beyond communities of practice. *Research Policy*, *37*(2), 353–369. doi:10.1016/j.respol.2007.11.003

Berger, C., Möslein, K., Piller, F., & Reichwald, R. (2005). Cooperation between manufacturers, retailers, and customers for user co-design: Learning from exploratory research. *European Management Review*, *2*(1), 70–87. doi:10.1057/palgrave.emr.1500030

Binder, T., Brandt, E., & Gregory, J. (2008). Design participation(-s): A creative commons for ongoing change. *CoDesign*, *4*(2), 79–83. doi:10.1080/15710880802114458

Campbell, F., & Deakin, M. (2005). Cities as leading examples of digitally-inclusive knowledge societies: The e-citizenship course, representative users, pedagogy and engagement matrix. In Osborne, M., & Wilson, B. (Eds.), *Making knowledge work. Stirling, Scotland.* UK: Stirling University.

Curwell, S., Deakin, M., Cooper, I., Paskaleva-Shapira, K., Ravetz, J., & Babicki, D. (2005). Citizens expectations of information cities: Implications for urban planning and design. *Building Research and Information*, *22*(1), 55–66. doi:10.1080/0961321042000329422

Deakin, M. (2007). E-Topia, SUD and ICTs: Taking the digitally-inclusive urban regeneration thesis full circle. *Journal of Urban Technology*, *14*(3), 131-139.

Deakin, M. (2009). The IntelCities community of practice: The eGov services model for socially inclusive and participatory urban regeneration programs. In Karacapilidis, N. (Ed.), *Web-based learning solutions for communities of practice: Developing virtual environments for social and pedagogical advancement*. Hershey, PA: Information Science Reference. doi:10.4018/978-1-60566-711-9.ch018

Deakin, M. (2010a). A review of city portals: The transformation of service provision under the democratization of the fourth phase. In Reddick, C. (Ed.), *Politics, democracy and e-government: Participation and service delivery*. Hershey, PA: IGI Global. doi:10.4018/978-1-61520-933-0.ch013

Deakin, M. (2010b). SCRAN: The SmartCities (inter) regional academic network supporting the development of a trans-national comparator for the standardisation of egovernment services. In Reddick, C. (Ed.), *Comparative e-government: An examination of e-government across countries*. Berlin, Germany: Springer Press.

Deakin, M., & Allwinkle, S. (2006). The IntelCities community of practice: The e-learning platform, knowledge management system and digital library for semantically-interoperable e-governance services. *International Journal of Knowledge. Culture and Change Management*, *6*(3), 155–162.

Deakin, M., & Allwinkle, S. (2007). E-Topia, SUD and ICTs: The post-human nature, embedded intelligence, cyborg-self and agency of digitally-inclusive regeneration platforms. *International Journal of the Humanities*, *5*(2), 199–208.

Deakin, M., & Allwinkle, S. (2007). Urban regeneration and sustainable communities: The role of networks, innovation and creativity in building successful partnerships. *Journal of Urban Technology*, *14*(1), 77–91. doi:10.1080/10630730701260118

Deakin, M., Van Isacker, K., & Wong, A. (2004). *Review of the IntelCities knowledge capture requirements using a S.W.O.T. analysis*. Edinburgh, Scotland: Napier University.

Ellis, D., Oldridge, R., & Vasconcelos, A. (2004). Community and Virtual Community. *Annual Review of Information Science & Technology*, *38*, 146–186.

Graham, S. (2000). Constructing premium network spaces: Reflections on infrastructure networks and con-temporary urban development. *International Journal of Urban and Regional Research*, *24*(1), 183–200. doi:10.1111/1468-2427.00242

Graham, S., & Marvin, S. (1996). *Telecommunications and the city*. London, UK: Routledge. doi:10.4324/9780203430453

Harloe, M. (2001). Social justice and the city. *International Journal of Urban and Regional Research*, *25*(4), 799–797. doi:10.1111/1468-2427.00350

Johnson, C. (2001). A survey of current research on online communities of practice. *The Internet and Higher Education*, *4*(1), 45–60. doi:10.1016/S1096-7516(01)00047-1

Josefsson, U. (2005). Coping with illness online: The case of patients online communities. *The Information Society*, *21*(2), 143–153. doi:10.1080/01972240590925357

Lombardi, P., Cooper, I., Paskaleva, K., & Deakin, M. (2009). The challenge of designing user-centric e-services: European dimensions. In Riddeck, C. (Ed.), *Research strategies for e-government service adoption*. Hershey, PA: Idea Group Publishing.

Lombardi, P., & Curwell, S. (2005). INTELCITY Scenarios for the city of the future. In Miller, D., & Patassini, D. (Eds.), *Beyond benefit cost analysis*. Aldershot, UK: Ashgate.

Malecki, E. (2002). Hard and soft networks of urban competitiveness. *Urban Studies (Edinburgh, Scotland)*, *29*(5/6), 929–946. doi:10.1080/00420980220128381

Mitchell, W. (1995). *City of bits: Space, place, and the Infobahn*. Cambridge, MA: MIT Press.

Mitchell, W. (2000). *E-Topia: Urban life, Jim-But not as you know it*. Cambridge, MA: MIT Press.

Nikolaus, F., Keinz, P., & Schreier, M. (2008). Complementing mass customization toolkits with user communities: How peer input improves customer self-design. *Journal of Product Innovation Management*, *25*(6), 546–559. doi:10.1111/j.1540-5885.2008.00321.x

Orr, J. (1996). *Talking about machines: An ethnography of a modern job*. New York, NY: IRL Press.

Selwyn, N. (2002). E-stablishing an inclusive society: Technology, social inclusion and UK government policy. *Journal of Social Policy*, *21*(1), 1 20.

Symonds, M. (2000). The next revolution: A survey of government and the Internet. *The Economist*. Retrieved from http://www.economist.com/ surveys/ showsurvey.cfm?issue= 20000624

Warf, B. (2001). Segueways into cyberspace: Multiple geographies of the digital divide. *Environment and Planning. B, Planning & Design*, *28*(1), 3–19. doi:10.1068/b2691

Wenger, E. (1998). *Communities of practice: Learning, meaning, and identity*. Cambridge, MA: Cambridge University Press.

Wenger, E. (2000). Communities of practice and social learning systems. *Organization*, *7*(2), 225–246. doi:10.1177/135050840072002

Wind, J., & Rangaswamy, A. (2001). Customerization: The next revolution in mass customization. *Journal of Interactive Marketing*, *15*(1), 13–32. doi:10.1002/1520-6653(200124)15:1<13::AID-DIR1001>3.0.CO;2-#

KEY TERMS AND DEFINITIONS

Capacity Building: Refers to assistance which is provided to organisations which have a need to develop a certain skill or competence, or a general upgrading of performance ability. Most capacity is built by societies themselves, sometimes in the public, sometimes in the non-governmental and sometimes in the private sector. They are activities which strengthen the knowledge, abilities, skills and behaviour of individuals and improve institutional structures and processes such that an organization can efficiently meet its mission and goals in a sustainable way.

Citizenship: represents "a bundle of rights" - primarily, political participation in the life of the community, the right to vote, and the right to receive certain protection from the community, as well as obligations." As a state of being citizenship carries with it both rights and responsibilities, in particular the notion of "active citizenship" and idea that citizens should participate with one another to better their community.

Co-Design: here there is an understanding that all human artifacts are designed and with a purpose. In co-design one tries to include those perspectives that are related to the design in the process. It is generally recognized that the quality of design increases if the stakeholders interests are considered in the design process.

Community of Practice: Groups of people who share a concern or a passion for something they do and learn how to do it better as they regularly interact with one another as knowing subjects.

Community: The term is derived from the word *communité* which is derived from the Latin *communitas* (*cum*, "with/together" + *munus*, "gift"), a broad term for fellowship or organized society. In biological terms, a community is a group of interacting species sharing an environment. In human communities, intent, belief, resources, preferences, needs, risks, and a number of other conditions may be present and common, affecting the identity of the participants and their degree of cohesiveness. In sociology, the concept of community has led to significant debate, and sociologists are yet to reach agreement on a definition of the term. Traditionally a "community" has been defined as a group of interacting people living in a common location. The word is often used to refer to a group that is organized around common values and is attributed with social cohesion within a shared geographical location, generally in social units larger than a household. The word can also refer to the national community or global community. Since the advent of the Internet, the concept of community no longer has geographical limitations, as people can now virtually gather in an online community and share common interests as citizens regardless of physical location.

Digital Inclusion: is the term used to encompass activities related to the achievement of an inclusive information society. In this vein, new developments in technology turns the risk of a digital divide into "digital cohesion" and opportunity, bringing the benefit of the internet and related technology into all segments of the population, including people who are disadvantaged due to education, disabilities, gender, ethnicity, or those living in remote regions. Digital inclusion covers mainly the development of appropriate policies, maintenance of a knowledge base, research and technology development and deployment and best practices dissemination.

Digital Library: A library in which collections are stored in digital formats (as opposed to print, microform, or other media) and is accessible by computers. The digital content may be stored locally, or accessed remotely via computer networks. The terms is diffuse enough to be applied to a wide range of collections and organizations, but, to be considered a digital library, an online collection of information must be managed by and made accessible to a community of users. Some web sites can be considered digital libraries, but they may not offer such functionality.

e-Government Services: are internet technologies that act as a platform for exchanging information, providing services and transacting with citizens, businesses, and other arms of government. Such e-Government services include: a) Pushing information over the Internet, e.g: regulatory services, general holidays, public hearing schedules, issue briefs, notifications, etc. b) Two-way communications between the agency and the citizen, a business, or another government agency. In this model, users can engage in dialogue with agencies and post problems, comments, or requests to the agency. c) Conducting transactions, e.g: lodging tax returns, applying for services and grants. d) Governance, e.g: online polling, voting, and campaigning. The most important anticipated benefits of e-government include improved efficiency, convenience, and better accessibility of public services.

e-Learning: A general term used to refer to a form of learning in which the instructor and student are separated by space or time where the gap between the two is bridged through the use of online technologies. The term is used interchangeably in a wide variety of contexts and can be used to define a specific mode to attend a course or programmes of study where learners rarely, if ever, attend face-to-face contact, or rely upon such direct support.

Governance: is the activity of governing. It relates to decisions that define expectations, grant power, or verify performance. In terms of distinguishing the term governance from government - "governance" is what a "government" does. It might be a geo-political government (nation-state), a corporate government (business entity), a socio-political government (tribe, family, etc.), or any number of different kinds of government. But, irrespective of this, governance is the kinetic exercise of managing power and administering policy.

Knowledge Transfer: The practical problem of transferring knowledge from one part of the organization to another (or all other) parts of the organization. It seeks to organize, create, capture or distribute knowledge and ensure its availability for future users. It is considered to be more than just a communication problem and more complex because: a) Knowledge resides in organizational members, tools, tasks, and their sub-networks. b) Much of the knowledge organizations have is tacit or hard to articulate in direct communication.

Middleware: The enabling technology. It functions as a piece of software that connects two or more applications, allowing them to exchange data. It is computer software that connects software components or applications. The software consists of a set of enabling services that allow multiple processes running on one or more machines to interact across a network. This technology evolved to provide for interoperability in support of the move to coherent distributed architectures, which are used most often to support and simplify complex, distributed applications. It is especially integral to modern information technology based on XML, SOAP, Web services, and service-oriented architecture.

Open Source Software: Computer software for which the human-readable source code is made available under a copyright license, or arrangement This permits users to use, change, and improve the software, and to redistribute it in modified or unmodified form. It is often developed in a public, collaborative manner.

Semantic-Interoperability: The ability of two or more systems or components to exchange or harmonize cognate subject vocabularies and/or knowledge organization schemes to be used for the purpose of effective and efficient resource discovery without significant loss of lexical or connotative meaning and without special effort by the user.

ENDNOTES

1 While the title of the article by Amin and Roberts (2008) goes under the curious name of "beyond communities of practice", they use the phrase to suggest the need to "get beyond" the "undifferentiated" use of the term and requirement for more "contextualised" studies of the type set out in this paper.

2 http://www.smartcities.info/

Chapter 6
Developing a Competitive City through Healthy Decision-Making

Ori Gudes
Queensland University of Technology, Australia & Griffith University, Australia

Elizabeth Kendall
Griffith University, Australia

Tan Yigitcanlar
Queensland University of Technology, Australia

Jung Hoon Han
The University of New South Wales, Australia

Virendra Pathak
Queensland University of Technology, Australia

ABSTRACT

This chapter investigates the challenges and opportunities associated with planning for a competitive city. The chapter is based on the assumption that a healthy city is a fundamental prerequisite for a competitive city. Thus, it is critical to examine the local determinants of health and factor these into any planning efforts. The main focus of the chapter is on e-health planning by utilising Web-based geographic decision support systems. The proposed novel decision support system would provide a powerful and effective platform for stakeholders to access essential data for decision-making purposes. The chapter also highlights the need for a comprehensive information framework to guide the process of planning for healthy cities. Additionally, it discusses the prospects and constraints of such an approach. In summary, this chapter outlines the potential insights of using an information science-based framework and suggests practical planning methods as part of a broader e-health approach for improving the health characteristics of competitive cities.

DOI: 10.4018/978-1-61350-174-0.ch006

BACKGROUND

For the past few decades, there has been emerging evidence of a close relationship between health and place. The Healthy Cities movement (WHO, 1999) is a good example of the effort that is now being put into the creation of places that can promote health and wellbeing. This initiative was officially introduced in 1986 by Ilona Kickbusch at a conference of the World Health Organisation (WHO) in Copenhagen, Denmark. To date, according to Health Cities Illawarra (2010), *"since 1985 over 3000 healthy cities, towns, villages and islands have been established throughout the world"*

A healthy community is a critical component of a competitive city (Johnson, 2002). The qualities that reflect health can make a 'location' more attractive and, therefore, competitive in comparison to global standards. The term 'competitiveness' has been well defined by Webster and Muller (2000, p. 1): *"The ability of an urban region to produce and market a set of products (goods and services) that represent good value (not necessarily lowest price) in relation to comparable products of other urban regions. Non-tradeables, e.g., local services, are part of the competitiveness equation. An urban economy that produces goods and services for local people of high value relative to price, supports the export economy of the city, making it more competitive, as well as directly raising the quality of life and standard of living for people living in the urban region"*. Thus, like the healthy cities movement, the call for competitive cities is based on the notion of better city lifestyles. The economic competitiveness of a city is defined by the prosperity of its citizens. A competitive city is known as 'a city of opportunities for families, lifestyle and business' (See Figure 1 as an example). Similarly, a healthy city is *"one that is continually creating and improving [its] physical and social environments and strengthening the community resources which enable people to mutually support each other in performing all the functions of life and achieving their maximum potential"* (Flynn, 1996, p. 300).

Figure 1. A promotional tool of Logan, Australia

It is not surprising, then, that the task of applying knowledge to the process of building a healthy and competitive city has become an important focus for urban planners. However, there is evidence that planners require a new approach to enable them to respond to this agenda. Specifically, they need timely access to local information, collaborative planning processes and mechanisms for engaging the public in decision-making. The WHO has concluded that e-Health systems hold great promise for both low- and high-income countries. The benefits of an online approach to health planning apply not only to effective and efficient health-care delivery, but also to public health governance, finance, education, research, and health-related economic activities (WHO, 2008). The purpose of this chapter is to examine an online approach to planning that can respond to the call for healthy and competitive cities. In this chapter, we will first review competitive cities in the context of health. Then, we provide discussion on e-health decision support systems in the context of collaborative health planning practice. Next, we present our suggested conceptual framework for planning healthy and competitive cities. Subsequently, and to apply this framework in practice, we introduce a participatory model for implementing collaborative health planning. In summary, we conclude that e-health Geographical Information System (GIS) based Decision Support System (DSS) can contribute to the development of healthy and competitive cities, particularly if the challenges presented by an online environment can be addressed.

COMPETITIVE CITIES IN THE CONTEXT OF HEALTH

Recent literature points to the importance of public health and community well-being for the development of a competitive city. Indeed, according to Kipfer and Keil (2002), public health policies are amongst the most important components of a competitive city agenda.

Jonhnson (2002) argued that there is a direct link between enhancing community health and promoting the competitiveness of a city. As he stated: *"the model for enhancing community health and competitiveness is predicated on the ability of the leadership of city government to leverage its polity capital assets to forge win-win and/or reciprocal relationships with for-profit (business) and nonprofit institutions"* (Jonhnson, 2002, p. 764). Improving the local health and socio-economic state of the community may improve the overall cities' competitiveness in the global marketplace. For example, the city of Charlotte, North Carolina – United States, has worked jointly with some of its private sector companies (i.e., banks) over the years to improve the community health and well-being of the city. As a result of these collaborative efforts, the city of Charlotte has experienced population economic growth and greater prosperity (Perlmutt, 2001; Jonhnson 2002). The 'Livable city' Toronto hinged on the high degree of environmental-health enjoyed by its residents (Porter, 1995). In his seminal work, Porter (1995) focused on the world class health facilities in the inner city of Boston as the most important contributor to the city's competitiveness. By harnessing the private sector and improving their commitment to the local community health, a significant positive impact on cities' competitiveness was generated in the global market arena. Thus, well-being, health and lifestyle have become essential promotional tools in the quest to make a location more competitive (See Figure 1).

However, in discussing the creation of competitive and healthy cities, Moutinho (2006) emphasised the need for more attention to be given to the digital infrastructure including local E-government, E-health, E-business and entertainment. Further, in defining successful 'Digital Cities', Moutinho concluded that they require consistent public policies, evaluation frameworks, innovative systems, openness to the society and large investments if they are to create the conditions for enhanced competitiveness and most importantly, competitive cities need to foster a 'culture of knowledge'. Drawing on all these arguments, it is clear that a healthy city depends on the creation of wellbeing through excellent health services and health promotion. These qualities are also precursors to, and features of, a competitive city. Both healthy and competitive cities depend on the efficient use of digital infrastructure, which in turn depends on a culture of knowledge, collaborative planning and timely access to local place-based knowledge.

To plan effectively for healthy and competitive cities, it is necessary to revive the historic collaboration between urban planning and public health professionals, who together can conduct informed evidence-based decision-making (Northridge et al., 2003). Further, planning efforts must focus on the creation of structures and processes that actively work to dismantle existing health inequalities and create economic, political, and social equality (Schulz & Northridge, 2004). To date, urban planning has not focused on the creation of cities that can promote health and prevent disease. Although health is a consequence of the activity planners undertake (i.e., developing effective transport systems, planning community facilities), it has not yet been their direct focus. One of the reasons why planning has not yet been able to effectively contribute to the healthy cities movement is that there are no models to define the type of information that must be considered by planners and there is no method for sharing this information in a meaningful form (Gudes et al., 2010). As Flynn (1996) concluded, every community is unique, with different physical, social, political and cultural contexts that must be understood in the planning process. As these contexts are likely to impact heavily on health, it is necessary for planners to develop a thorough understanding of each individual community health profile and the features that influence health.

To promote health and, therefore, competitiveness, two critical features must be present in a city, namely the ability to organise, access and share local information and the ability to plan collaboratively across multiple sectors and systems. One of the ways to increase accessibility to health information and promote collaborative use of that information is to adopt an e-health approach to decision-making.

E-HEALTH DECISION SUPPORT SYSTEMS AND COLLABORATIVE HEALTH PLANNING PRACTICE

Decision support systems (DSS) are information and communication technology (ICT) tools that provide the mechanism to help decision-makers and related stakeholders assess complex problems and solve these in a meaningful way (Shim et al., 2002). The overall aim of DSS, without substituting decision-makers, is to improve the efficiency of the decisions made by stakeholders, optimising their overall performance and minimising judgemental biases (Turban, 1993). Technologies such DSS are reported in the literature as being the best means yet of "*circulating*" information and improving its 'interoperability'.

DSS incorporates two main domains: policy-making, which entails the process of making decisions to solve problems, and technology, which uses computational problem solving tools to support those decisions. Given that cities represent distinct geographical locations, one of the most relevant methods for this type of planning is the spatially-based Geographical Information Systems (GIS) technology. GIS has become a central spatial component of DSS because it embraces computational, analytical, problem solving and visualisation capabilities (Gudes et al., 2009). As Conway et al. (2008) indicated, spatial maps are likely to be particularly helpful to stakeholders, improving decision-making by visualising and simulating spatial phenomena.

However, it is essential to expand the use of DSS through online ICT platforms to capitalise on broader e-health initiatives. A collaborative online platform can improve accessibility to information, conceivably to a point that even the public could engage in the process of decision-making and health policy-making. The capacity to share information in a variety of forms, improves stakeholders' involvement in decision-making, creates horizontal knowledge sharing and simplicity of the decision process (Dur et al., 2009). For example, the West Australia Department of Health has developed an online GIS-based DSS application named "Healthtracks" which improves their health planning capacity (Spatial Vision, 2010). Another example is the Social Health Atlas online (www.publichealth.gov.au) that allows user to map specific health outcomes and associated socio-economic status for administratively defined boundaries (Baum et. al., 2010). Although there are several good examples of GIS-based DSS models in Australia (mostly facilitated by Victoria, West Australia and South Australia health government initiatives), the majority of the applications are focused on specific health issues (e.g. National Diabetes Service Scheme and the Social Health Atlas of Adelaide Health Service website) (Central Northern Adelaide Health Service, 2008) rather than the promotion of health in the community in its broad sense. Baum et. al. (2010, p.30) concluded that: "*In Australia there appears to be a lack of examples illustrating the use of these online tools*". Although, these examples demonstrate the potential utility of GIS-based DSS models, yet, all these models lack the communication channels and the collaboration mechanism which are so imperative for developing healthy cities. Thus, online GIS-based DSS are likely to provide a 'place' in which collaborative planning can occur, it is also necessary to adopt a systematic way of organising local information to increase its usability. Specifically, if planners aim to develop healthy and competitive cities, then the information they need to access should address the determinants of these outcomes.

A CONCEPTUAL FRAMEWORK FOR PLANNING HEALTHY AND COMPETITIVE CITIES

Mooney and Fohtung (2008) stated that thorough understanding of the complex relationships between the broad determinants of health and actual health outcomes may lead to more knowledgeable decisions about policies that can influence the health of the city. It is therefore imperative that the DSS be based on a broad information framework that will allow planners to develop powerful and effective spatial analyses of complex questions. In this chapter, we propose a framework for the organisation of local information to support collaborative health planning. Given that the elements of this framework have been found to determine health outcomes, they are also likely to contribute to the competitiveness of a city. Indeed, some elements of the framework have been directly linked to competitive cities by having the potential to improve health and wellbeing (Johnson, 2002)..

In this regard, Schulz and Northridge (2004) developed a public health framework for health impact assessments (Figure 2). This framework summarises the different levels of factors that impact upon health and, therefore, should be considered in health planning. According to Northridge et al. (2003), factors that contribute to health can be divided into four levels, namely: Macro, Meso, Micro and Individual, which combine to a health profile of a particular community. According to the model, these factors interact to contribute to health in the community. For instance, the natural environment, macro social factors, and inequalities (i.e., fundamental factors) influence health outcomes and well-being (i.e. individual level factors) via multiple pathways through differential access to power, information, and resources. These fundamental factors, in turn, influence intermediate factors (i.e. the built environment and the social context). Intermedi-

Figure 2. Public health framework for health impact assessment and health profiling (derived from Schulz & Northridge 2004)

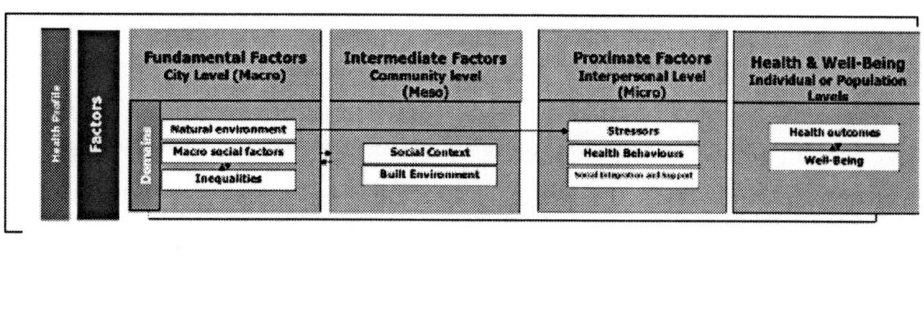

Figure 3. A conceptual framework for planning a healthy city (derived from WHO 1997b; Schulz & Northridge 2004)

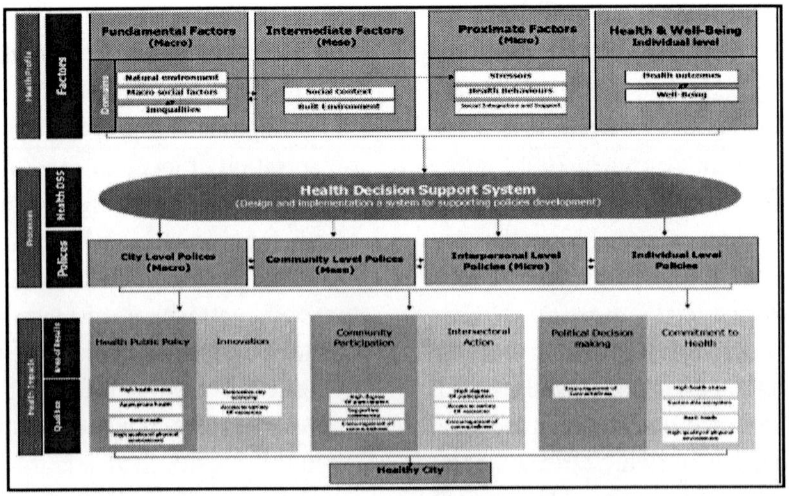

ate factors include the development of land use policies. Consequently, it is here that the impact of the built environment is especially subject to policy management by planners.

Moving to thc proximate factors (public health practitioner territory), it can be observed by three domains: stressors, social integration and social support, and health behaviours. The literature argues that proximate factors have been given greater scientific attention in recent years (Northridge et al., 2003). The last column in Figure 2 contains two key domains: health outcomes and well-being, as these in turn influence the individual habits. Figure 2 also illustrates the interactive and dynamic relationships among the various domains, between the fundamental and intermediate factors as well as between the intermediate and proximate factors, and their impacts on health outcomes across an individual's life course. Thus, it is suggested that by adopting such a framework, we can improve our understanding of health in place, increase accessibility to health information and the build knowledge about the role of the social determinants of health.

A collaborative health-planning framework that has been developed for this study illustrates the overall place of DSS within a healthy cities' planning initiative (See Figure 3). However, the literature emphasises the importance of grounding a DSS in a broad information framework. Specifically, it is suggested that the Information Management Framework as described by Schulz and Northridge's (2004) should guide the development of a community health profile, with information derived from multiple sources. In this regard, Duhl and Sanchez (1999) defined a list of six fundamental characteristics (health public policy, innovation, community participation, intersectional action, policy decision making and commitment to health) that are needed to create a healthy city. If these characteristics are adopted, it is likely that a healthy city will emerge. Thus, this framework suggests that by utilising a DSS as part of a broader city health- planning process, it is more likely that a healthy community will form, and this in turn can improve the city's competitiveness.

However, increasing accessibility to effectively organised information through e-health DSS initiatives is not sufficient to ensure effective planning for a healthy and competitive city. Unless planning is being practiced in a collaborative manner, the solutions may only suit a small minority of the residents. Although the online environment enhances the likelihood of simultaneous access by multiple parties and shared knowledge to underpin decisions, it does not imply a collaborative and participatory process. Thus, a participatory planning process has also been proposed.

PARTICIPATORY ACTION RESEARCH: A METHOD FOR IMPLEMENTING COLLABORATIVE HEALTH PLANNING

Involvement of a wide range of stakeholders and the affected communities in decision-making is likely to improve the ability of cities to address health inequalities (Beyer & Rushton, 2009). Indeed, collaborative health planning procedures are essential to the policy-making process. Ridley and Jones (2001) argued that most significant forms of collaboration are those that become part of the day-to-day practice of health care and planning. In collaborative health planning, it is necessary to have a clear understanding and recognition of the stakeholder relationships, behaviour patterns and potential outcomes. This level of clarity, in turn, may empower users and enable them to influence the collaborative policy-making dialogue. The broad involvement of different segments of the community remains a promising approach for dealing with urban problems and promoting positive conditions within cities (Flynn, 1996).

One of the methods that can be used to implement collaborative health planning is Participatory Action Research (PAR). PAR has increasingly been used as an overarching name for an orientation to practice that places all stakeholders in equal positions of co-learners. PAR puts a heavy accent on community participation and the translation of knowledge into action (Minkler, 2000). Recently, the method is gaining increased attention in health, particularly in the public health context (Minkler & Wallerstein, 2003). One of the most important characteristics of PAR is the fact that participants whose lives are affected by the knowledge or decisions take an active role of its design. Israel et al. (2001) defined PAR by the following principles:

- Participation of community members in planning;
- Engaging all stakeholders in a joint process where they contribute equally;
- A co-learning process for all stakeholders;
- A method for systems development and local community capacity building;
- An empowering process through which participants can increase control over their lives nurturing community strengths and problem-solving abilities; and
- A way to balance knowledge and action.

Amongst the advantages which inherently co-exist in PAR approach, we can observe the following: it is a ground-up approach rather than top-down by expert, it strengthens the degree of community input into decision-making, it uses a democratic participatory process, driven by community priorities, based on community input to create a health community. Duhl (1996) argued that PAR is a community-driven approach oriented toward social change. Thus, stakeholders involved in a PAR initiative come together as co-learners rather than teachers or students (Minkler, 2000).

However, a few important limitations of PAR need to be noted. Implementing PAR within a health planning initiative is a time consuming task, and requires significant attention to issues of power, trust, rigor of information and conflicting interests or agendas among the stakeholders (i.e., scientists, policy-makers, providers and citizens). However, the literature reveals that through effective community consultation processes, many healthy cities have effectively incorporated a high level of participation from all stakeholders (Minkler, 2000; Stern et al., 2009). As Minkler (2000) emphasised, PAR offers a promising ap-

Figure 4. A practical PAR model for designing and evaluating an e-health-based DSS

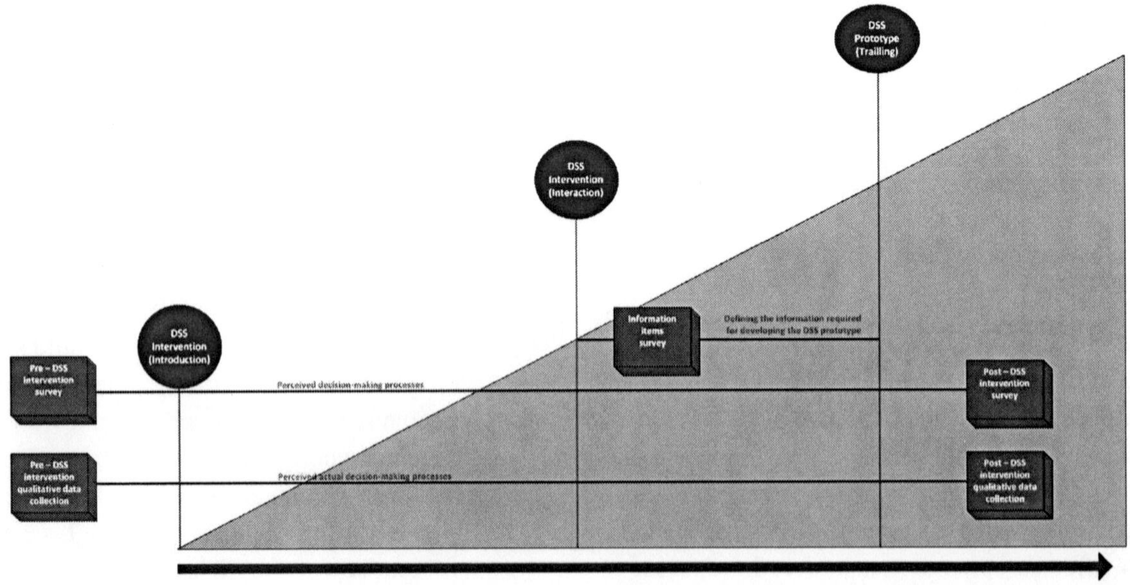

proach for realising community participation in planning, thus articulating the vision of the healthy cities movement.

PAR could easily be applied to the process of planning using an e-health GIS-based DSS. For instance, spatial information contained in the DSS can be shared and discussed through iterative focus groups, consultation meetings and surveys, as suggested by Rowe and Frewer (2009). All stakeholders (e.g. health planners, professionals, researchers, community representatives etc.) could be consulted in iterative cycles, with each consultation feeding into the further refinement and development of both the DSS itself and the decisions or policies that are made. To exemplify these statements, and to address the collaborative challenge of developing and evaluating e-health based DSS, this study suggests a practical PAR model (illustrated in Figure 4) to collect quantitative and qualitative data for evaluating the application of such a system. Further, the model also embraces the practical procedures required for designing such systems (e.g., information items etc).

In broad terms, two phases of study data collection are suggested: Once before the implementation of DSS (pre-DSS intervention) and then after the implementation of DSS (post-DSS intervention). The rationale for this method is to enable to identify the ways in which decisions (i.e. health planning) were being made before and after DSS intervention occurred, and draw the differences accordingly. The post-DSS study data collection should be conducted after trialling the DSS interface for several months. Consequently, it may give decision-makers sufficient time to familiarise themselves with the DSS interface and to incorporate it into their daily routine. This is represented in Figure 4 by the rubrics placed subsequent to the DSS intervention (i.e., post-DSS). In line with the recommendation of Meang and Nedovic-Budic (2010), it is suggested that qualitative data should also be collected in order to support the quantitative data collection.

As for the DSS design, it is suggested that a series of consultation meetings should be conducted to obtain input from end-users about the prospective GIS information items, workflows and functionality to be included in the DSS interface.

For instance, the DSS intervention phase may commence with an introductory presentation about GIS, followed by a series of DSS consultation meetings (i.e., Interaction Phase), whereas additional data obtained from prospective participants and end-users (e.g., information items) should be used for developing the DSS interface. This process may conclude with trialling the DSS prototype (i.e., Trialling Phase) for several months. Thus, the PAR approach could be utilised to develop a practical and efficient model to design an e-health DSS for health planning.

However, the use of a PAR approach to planning raises a few key barriers to the implementation of an online e-health sharing platform, including accessibility to secure and private data, infrastructures, regulations and procedures. The following section provides more details about these challenges in practice.

CHALLENGES FOR AN E-HEALTH APPROCH TO DECISION-MAKING

Within the plethora of studies dealing with e-health, there are two major lines of interest: those focused on clinical informatics, such as electronic patient or health records, and those interested in health informatics, including the provision of information through the web to support decision-making and policy development. Although there are some examples where research focuses on both clinical and policy domains (Bernard et al., 2007), generally scholars tend to focus on one or the other. In both areas, however, e-health solutions introduce a range of emerging challenges such as the digital divide, information security, privacy, regulation and policy design, information technology infrastructure and regulatory compliance, all of which need to be considered from the perspective of each stakeholder. In this regard, Kluge (2007) argued that ethical considerations should play a fundamental role in the structuring and implementation of e-health. Wickramasinghe et al. (2005) provided a framework for increasing the potential of e-health through four critical pre-requisites:

- Information communication technology infrastructure;
- Standardised policies, protocols and procedures;
- User access and accessibility; and
- Government regulation.

Each element is considered critical as they combine to create an overall favourable environment. For example, irrespective of accessible internet, a poor physical infrastructure or a high level of bureaucracy will lead to significant impediments to formation of a successful e-health initiative. However, the Annual Survey of Health Information Exchange (2009) published recently in the United States, concluded that addressing privacy and confidentiality issues were among the main challenges of successful implementation. There is some evidence to suggest that a large number of individuals are not willing to engage in e-health because they do not trust the underlying information and communication technologies (Baur, 2008; Croll & Croll, 2007). This effect was also observed by Katsikas et al. (2008), who argued that the lack of appropriate security measures is one of the most severe restraining factors for the proliferation of e-health initiatives in Australia. The literature has also identified a growing digital divide in terms of access to digital technologies depending on ethnicity and income (Ahern, 2007).

There is a need to apply different approaches to deliver practical, secure and comprehensive solutions that protect privacy but improve accessibility. For instance, Baur (2008) suggested a user-centred approach that places the needs, preferences, capacities, values, and goals of e-health users at the core of e-health initiatives. Baur also identified the following components of a successful e-health implementation:

- Consumers with diverse perspectives, circumstances, capacities, and experiences are included in the design of and have meaningful access to evidence based e-health tools with strong privacy and security protections;
- Diverse consumers have the skills and support to evaluate, choose, and use e-health tools to derive benefits for themselves and those they care for;
- Healthcare organizations and practitioners use the full range of e-health tools to engage and support diverse consumers in their own health management as a routine element of care;
- Local, state, and national policies and programs support the sustainable development and dissemination of evidence-based consumer e-health tools to diverse individuals and communities, including those served by safety-net providers;
- Alliances and partnerships facilitate sustained consumer access to and use of e-health tools, consistent with the value propositions and perspectives of each participating stakeholder; and
- Appropriate funding and incentives exist in public policy and the market to enable sustainable business models for tools with demonstrated effectiveness.

If e-health is to represent a successful solution to the process of developing healthy and competitive cities, these challenges need to be thoroughly considered and addressed. The result would be e-health implementations that have the potential to contribute to a country's economy and future growth, thus creating healthy and competitive cities. E-health may be an efficient tool for increasing the competitiveness of a country, city or a region, particularly if it becomes part of an evidence-based and informed dissemination and a participatory decision-making process.

CONCLUSION AND FUTURE RESEARCH DIRECTIONS

This chapter has identified some challenges and constraints associated with using an e-health GIS-based DSS to contribute to planning for healthy and competitive cities. By proposing the use of spatial information organised in a meaningful framework, we have bridged an existing gap in the knowledge about how to make decisions that can increase a city's competitiveness. We introduced an overarching framework for conceptualising the e-health-based DSS and described how a shared framework could become an integral component of the healthy cities planning process. We also discussed how access to effective information could be enhanced by e-health approaches, particularly if issues of security and privacy are addressed. E-health could be an effective solution to enhancing the competitiveness of a city, particularly given that it acts as a 'disseminator' of local information. However, increasing accessibility to effective health information through e-health initiatives may not be sufficient, unless health planning is also being practiced in a collaborative manner over the planning process. Thus, we suggested a practical model to guide the planning process (i.e., PAR) to ensure participation and collaboration of all stakeholders in designing an e-health-based DSS. It has been noted that to overcome health inequities and create cities that are liveable for all citizens, *"we should direct our attention towards the development of adequate tools and strategies"* (Beyer & Rushton, 2009). The tools and strategies we have outlined in this chapter will contribute to planning competitive cites by promoting public health and effective health systems.

ACKNOWLEDGMENT

This study is part of a broader Australian Research Council (ARC) project: Coalitions for Community Health: A Community-based Response to Chronic

Disease. The authors would like to acknowledge the work of Professor Scott Baum and Dr Heidi Muenchberger as chief investigators in this project.

REFERENCES

Ahern, D. K. (2007). Challenges and opportunities of e-health research. *American Journal of Preventive Medicine*, *32*(5), 75–82. doi:10.1016/j.amepre.2007.01.016

Annual Survey of Health Information Exchange. (2009). *Migrating toward meaningful use: The state of health information exchange. A report based on the results of the eHealth Initiative's 2009 sixth annual survey of health information exchange*. Washington, DC: eHealth Initiative.

Baum, S., Kendall, E., Muenchberger, H., Gudes, O., & Yigitcanlar, T. (2010). Geographical Information Systems: An effective planning and decision-making platform for community health coalitions in Australia. *Health Information Management Journal*, *39*(3), 28–33.

Baur, C. (2008). An analysis of factors underlying e-health disparities. *Cambridge Quarterly of Healthcare Ethics*, *17*(4), 417–428. doi:10.1017/S0963180108080547

Bernard, P., Charafeddine, R., Frohlich, K. L., Daniel, M., Kestens, Y., & Potvin, L. (2007). Health inequalities and place: A theoretical conception of neighbourhood. *Social Science & Medicine*, *65*(9), 1839–1852. doi:10.1016/j.socscimed.2007.05.037

Beyer, K. M. M., & Rushton, G. (2009). Mapping cancer for community engagement. *Preventing Chronic Disease*, *6*(1).

Central Northern Adelaide Health Service. (2008). *Central Northern Adelaide Health Service: Social health atlas 2008*. Retrieved November 24, 2008, from http://www.publichealth.gov.au /interactive_mapping/ sa_sha_cnahs_2005/ atlas.html

Conway, M. D., Rizzuto, C. D., & Weiss, L. M. (2008). *A better way to speed the adoption of vaccines*. Retrieved November 20, 2008, from http://www.mckinseyquarterly.com /home.aspx

Croll, P. R., & Croll, J. (2007). Investigating risk exposure in e-health systems. *International Journal of Medical Informatics*, *76*(5-6), 460–465. doi:10.1016/j.ijmedinf.2006.09.013

Duhl, L. J. (1996). An ecohistory of health: The role of 'healthy cities'. *American Journal of Health Promotion*, *10*(4), 258–261.

Duhl, L. J., & Sanchez, A. K. (1999). *Healthy cities and the city planning process. A background documentation on links between health and urban planning*. Copenhagen, Denmark: WHO, Regional Office for Europe.

Dur, F., Yigitcanlar, T., & Bunker, J. (2009, March 26). *A decision support system for sustainable urban development: the integrated land use and transportation indexing model.* Paper presented at the Postgraduate Infrastructure Conference, Brisbane, Australia, Queensland University of Technology.

Flynn, B. C. (1996). Healthy cities: Toward worldwide health promotion. *Annual Review of Public Health*, *17*(1), 299–309. doi:10.1146/annurev.pu.17.050196.001503

Gudes, O., Kendall, E., Yigitcanlar, T., Pathak, V., & Baum, S. (2010). Rethinking health planning: A framework for organising information to underpin collaborative health planning. *Health Information Management Journal*, *39*(2), 18–29.

Gudes, O., Yigitcanlar, T., & Pathak, V. (2009). *A community health support system for the planning of healthy cities*. Paper presented at the Postgraduate Infrastructure Conference, Brisbane, Australia, Queensland University of Technology.

Healthy Cities Illawarra. (2010). *Working with the community for a healthier, safer, greener and more caring Illawarra.* Wollongong, Australia: Illawarra city council. Retrieved on October 10, 2010, from http://www.healthycitiesill.org.au / contactus.htm

Israel, B. A., Schulz, A. J., Parker, E., & Becker, A. B. (2001). Community-based participatory research: Policy recommendations for promoting a partnership approach in health research. *Education for Health, 14*(2), 182–197. doi:10.1080/13576280110051055

Johnson, J. (2002). A conceptual model for enhancing community competitiveness in the new economy. *Urban Affairs Review, 37*(6), 763–779. doi:10.1177/107874037006001

Katsikas, S., Lopez, J., & Pernul, G. (2008). The challenge for security and privacy services in distributed health settings. *Studies in Health Technology and Informatics, 134*, 113–125.

Kipfer, S., & Keil, R. (2002). Toronto Inc.? Planning the competitive city in the new Toronto. *Antipode: A Radical. The Journal of Geography, 34*(2), 227–264.

Kluge, E. H. (2007). Secure e-health: Managing risks to patient health data. *International Journal of Medical Informatics, 76*(5-6), 402–406. doi:10.1016/j.ijmedinf.2006.09.003

Maeng, D., & Nedovic-Budic, Z. (2010). Relationship between ICT and urban form in knowledge-based development: Empirical analysis of Washington, DC metro region. *International Journal of Knowledge-Based Development, 1*(1-2), 97–117. doi:10.1504/IJKBD.2010.032588

Minkler, M. (2000). Using participatory action research to build healthy communities. *Public Health Reports, 115*(2-3), 191–197. doi:10.1093/phr/115.2.191

Minkler, M., & Wallerstein, N. (2003). Introduction to community based participatory research. In M. Minkler, & N. Wallerstein, N (Eds.), *Community based participatory research for health* (pp. 3-26). San Francisco, CA: Jossey-Bass.

Mooney, G., & Fohtung, N. G. (2008). Issues in the measurement of social determinants of health. *Health Information Management Journal, 37*(3), 26–30.

Moutinho, J. (2006). From technological determinism to regional development: Lessons learned from the Portuguese digital cities program (1998-2006). In Cunningham, P., & Cunningham, M. (Eds.), *Exploiting the knowledge economy: Issues, applications, case studies.* Amsterdam, The Netherlands: IOS Press.

Northridge, M. E., Sclar, E. D., & Biswas, P. (2003). Sorting out the connections between the built environment and health: A conceptual framework for navigating pathways and planning healthy cities. *Journal of Urban Health, 80*(4), 556–568. doi:10.1093/jurban/jtg064

Perlmutt, D. (2001). Bank of America executive to retire after 23 years. *Charlotte Observer*. Retrieved January 31, 2001, from http://www.charlotte.com/mld/charlotte/archives/

Porter, M. E. (1995). The competitive advantage of the inner city. *Harvard Business Review, 74*(6), 61–78.

Ridley, J., & Jones, L. (2002). User and public involvement in health services: A literature Review. Edinburgh, Scotland: Partners in Change, SHS Trust.

Rowe, G., & Frewer, L. J. (2009). Public participation methods: A framework for evaluation. *Science, Technology & Human Values, 25*(1), 3–29. doi:10.1177/016224390002500101

Schulz, A., & Northridge, M. E. (2004). Social determinants of health: Implications for environmental health promotion. *Health Education & Behavior*, *31*(4), 455–471. doi:10.1177/1090198104265598

Shim, J. P., Warkentin, M., Courtney, J. F., Power, D. J., Sharda, R., & Carlsson, C. (2002). Past, present, and future of decision support technology. *Decision Support Systems*, *33*(2), 111–126. doi:10.1016/S0167-9236(01)00139-7

Spatial Vision. (2010). *Spatially enabling government: 'Healthtracks' wins 2010 WA Spatial Excellence Award.* Retrieved November 30, 2010, from http://www.spatialvision.com.au /index.php/latest-news -spatially-enabling- government.html

Stern, E., Gudes, O., & Svoray, T. (2009). Web-based traditional public participation comprehensive planning: A comparative study. *Environment and Planning*, *36*(6), 1067–1085. doi:10.1068/b34113

Turban, E. (1993). *Decision support and expert systems: Management support systems.* Upper Saddle River, NJ: Prentice Hall PTR.

Webster, D., & Muller, L. (2000). *Urban competitiveness assessment in developing country urban regions: The road forward.* Paper presented at the Urban Group, INFUD, World Bank, Washington, DC.

Wickramasinghe, N. S., Fadlalla, A. M. A., Geisler, E., & Schaffer, J. L. (2005). A framework for assessing e-health preparedness. *International Journal of Electronic Healthcare*, *1*(3), 316–334. doi:10.1504/IJEH.2005.006478

World Health Organisation (WHO). (1999). Health: Creating healthy cities in the 21st century. In Satterthwaite, D. (Ed.), *The earthscan reader in sustainable cities* (pp. 137–172). London, UK: Earthscan.

World Health Organisation (WHO). (2008). Framework and standards for country health Information Systems (2nd ed.). *Health metrics network.* Geneva, Switzerland: Health Metrics Network.

ADDITIONAL READING

Ashton, J. R. (2009). From healthy towns 1843. *Public Health*, *123*, 11–13. doi:10.1016/j.puhe.2008.10.024

Baker, G. (2004). The effects of synchronous collaborative technologies on decision making: A study of virtual teams. In Khosrow-Pour, M. (Ed.), *Advanced topics in information resources management* (p. 333). Hershey, PA: IGI Global. doi:10.4018/978-1-59140-253-4.ch016

Birkin, M., Clarke, G. P., Clarke, M., & Wilson, A. (1996). *Intelligent GIS: Location decisions and strategic planning.* Cambridge, UK: Pearson.

Cromley, E. K., & Mclafferty, S. L. (2003). GIS and public health. *Health & Place*, *9*, 279. doi:10.1016/S1353-8292(02)00059-X

Dooris, M. (1999). Healthy cities and local agenda 21: The UK experience—Challenges for the new millennium. *Health Promotion International*, *14*(4), 365–375. doi:10.1093/heapro/14.4.365

Duffy, H. (1995). *Competitive cities: Succeeding in the global economy.* London: Routledge.

Forester, J. (1989). *Planning in the face of power.* Berkeley, CA: University of California Press.

Hancock, T., & Duhl, L. (1988). *Promoting health in the urban context.* Copenhagen, Denmark: WHO Regional Office for Europeo.

Healey, P. (2003). Collaborative planning in perspective. *Planning Theory*, *2*(2), 101. doi:10.1177/14730952030022002

Higgs, G., & Gould, M. (2001). Is there a role GIS in the 'New NHS'? *Health & Place*, *7*, 247–259. doi:10.1016/S1353-8292(01)00014-4

Igbaria, M., & Guimaraes, T. (1994). Empirically testing the outcomes of user involvement in DSS development. *Omega*, *22*(2), 157–172. doi:10.1016/0305-0483(94)90076-0

Ishida, T. (2000). *Understanding digital cities. Digital cities: Technologies experiences and future perspectives* (pp. 7–17). Berlin: Springer.

Kushner, J. A. (2007). *Healthy cities: The intersection of urban planning, law and health*. Durham, NC: Carolina Academic Press.

Liu, Y., Wong, S. Y., & Jin, T. (2009). Equality of spatial access to primary health services for Singapore's baby boomers. *Asian Population Studies*, *5*(3), 171–188. doi:10.1080/17441730902992091

Murray, D. J. (2006). *A critical analysis of communicative planning theory as a theoretical underpinning for integrated resource and environmental management*. Unpublished PhD thesis, Griffith University, Brisbane, Australia.

Philips, R. L., Kinman, E. L., Schnitzer, P. G., Lindbloom, E. J., & Ewigman, B. (2000). Using geographic Information Systems to understand health care access. *Archives of Family Medicine*, *9*(10), 971–978. doi:10.1001/archfami.9.10.971

Rushton, G. (2003). Public health, GIS, and spatial analytic tools. *Public Health*, *24*, 43–56. doi:10.1146/annurev.publhealth.24.012902.140843

Scott, A. J. (2006). Creative cities: Conceptual issues and policy questions. *Journal of Urban Affairs*, *28*(1), 1–17. doi:10.1111/j.0735-2166.2006.00256.x

Smith, L. T., Johnson, D. B., Lamson, E., & Sitaker, M. (2006). A framework for developing evaluation tools used in Washington state's healthy communities projects. *Preventing Chronic Disease Public Health Research, Practice, And Policy, 3*(2).

Theseira, M. (2002). Using Internet GIS technology for sharing health and health related data for the west midlands region. *Health & Place*, *8*, 37–46. doi:10.1016/S1353-8292(01)00034-X

Zwarenstein, M., & Reeves, S. (2000). What's so great about collaboration? We need more evidence and less rhetoric. *British Medical Journal*, *320*, 1022–1023. doi:10.1136/bmj.320.7241.1022

KEY TERMS AND DEFINITIONS

Competitive Cities: Competitive cities sustain and offer its residents high standards of quality of life and place. As a result, it attracts more financial capital and business opportunities, which in turn, can be invested back into improving the city infrastructure and society well-being.

Decision Support Systems: Decision support systems are information communication technology tools, including spatial-based information systems (e.g. GIS) that provide the mechanisms to help decision-makers, policy makers and related stakeholders to assess complex problems and solve these in a meaningful and effective way.

E-Health: E-health encompasses a range of medical informatics applications based on Information and Communication Technology. Utilisation of e-health, is considered to be one of the trendiest ways to increase accessibility to health information.

Geographic Information Systems: Geographic Information Systems is a decision support tool for making informed decisions by stakeholders, policymakers and decisions-makers. The tool uses a mix of spatial and attributes data, queries, analytical procedures and analysis methods to leverage end-users' spatial knowledge.

Healthy Cities: Healthy cities are associated with the following characteristics: Creating and improving those physical and social environments conditions; strengthening community resources; and achieving its maximum potential. The healthy cities approach is based on the process, not just the outcome. Healthy cities are committed to health and have the structure and capacity to execute the necessary health planning processes.

Participatory Action Research: Participatory action research is an application of a research method where the researcher is also a co-learner in the research process in conjunction with the community being researched. One of the most important characteristics of participatory action research is the fact that participants whose lives are affected by the research initiative take an active role of its design.

Spatial-Based Collaborative Technologies: Spatial-based collaborative technologies are based on geographical information. Their primary purpose is to enhance the collaborative practice, debate and decision-making processes. These technologies are being used predominantly by urban planners, policy makers, city leaders and stakeholders.

Chapter 7
City Branding and the Role of the City Websites in Building City Brands

Erkan Özdemir
Uludag University, Turkey

ABSTRACT

In today's globalized world, the competition experienced intensively in many fields is being experienced amongst cities as well. Competition is taking place not only in the traditional environment, but in the online environment as well. In order to differentiate their cities from the rest and build an image and identity for themselves in this atmosphere of intensive competition, city administrators are taking advantage of the branding techniques used by private sector companies. At this point, the city websites have a key role for building and communicating the city brand. The aim of this chapter is to evaluate the role of the city websites in building city brands and promoting their brands. In order to fulfill the aim, the city branding, related concepts, and the role of the city websites in building city brands are introduced in the background section. Then, randomly selected city websites from the City Brands Index are evaluated using content analysis, and the results of this analysis are presented and discussed respectively.

DOI: 10.4018/978-1-61350-174-0.ch007

INTRODUCTION

In today's globalizing world, competition is an intensively hot issue not only between businesses but also amongst the countries or even between the cities that are the centers creating the wealth for the economies of those countries. The competition between the cities varies around such issues as the city being an important centre of tourism attraction, attracting investments into the city, the headquarters of important institutions and businesses being located in the city, organizing important porting events in the city, the city being the center of economical and cultural activities, the city being an important center of education for students (Gordon, 1999; Lever & Turok, 1999; Muñiz-Martínez & Cervantes-Blanco, 2009; Mishra & Rolland, 2009; Zhang & Zhao, 2009). One of the most important consequences of economical change is the fierce competition between the cities for both public and private sector investments (Berg & Braun, 1999; Short et al., 1993). The cities especially compete with one another for the market share likely to be obtained from those investments. The fact that investment in today's world is active and moving makes this competition all the more severe (Waitt, 1999).

The emergence a number of relations termed as a network society as a result of globalization that has intensified the competition for investments between the countries and regions (Huerta-Carvajal & Luna-Reyes, 2009), and the internationalization of markets cause confusion about the identities of the cities as well. The cities as the places or centers where these countless trends of change are joined make attempts to establish their own special identities in order to become different from the others. Some cities have their own specific identities due to their well-known or unique characteristics. The identities of some other cities, on the other hand, may also be impacted by the works and efforts of the public sector. Therefore, many city administrations make efforts to establish new structures that will form new city icons and symbols, and promote their cities (Muñiz-Martínez & Cervantes-Blanco, 2009).

City branding is attracting the attention of public administrators and the other stakeholders more and more everyday. The reputation of a city is a very important factor in attracting financial and other resources into the city and retaining them there. Therefore, the brand of a city also needs to be managed just like that of a company brand. If one does not pay enough care to it, the brand of a city may be damaged due to the information circulated around through the modern channels of communication (Kerr & Johnson, 2005). Using branding techniques for the cities is increasing more and more because of the global competition faced by the countries and cities in today's local and international markets. Countries and cities need to adopt informed branding practices in order to be able to compete more effectively on the global scale (Dinnie, 2004). In addition, the great changes in the social and political texture of the modern society require the city branding as a prerequisite in the 21st century to be implemented over the cities through a public-oriented approach (Adult, 2005). The commonly accepted duty of city administrations is to constitute (to reconstitute, if necessary) conditions of the city in the most effective way in order to be able to attract the potential investments into the city (Rogerson, 1999).

In this day and age when the whole world is becoming globalized in terms of social, economical and cultural aspects, the city administrations try to become branded cities in order to get ahead in competition. The concept of brand and branding, which is an important concept in marketing are now being applied to cities as well. Many benefits of the internet such as communication and interaction which it provides thanks to its globalization feature has helped city administrations, just like the private sector companies, to get involved in the efforts of online city branding. Online environment provide important opportunities to the cities using different communication strategies to contribute to the building city brand. City websites have an important function on e-branding of the city or online city branding. Therefore, business websites' online branding efforts can be also applied to the branding of cities.

The aim of this chapter is to evaluate the role of the city websites in building city brands and promoting their brands, and to offer suggestions on e-branding to the city administrators who wish to build branded cities. Thus, this chapter is intended to contribute to the city administrators, and professionals, students and academicians working in the field of city e-branding. In order to fulfill the aim mentioned above, the rest of the chapter is organized as follows: The second section of this chapter is to introduce the city branding, related concepts, and the role of the city websites in building city brands. Additionally, the encountered difficulties in building the city brand are examined in this section. In the third section of this chapter, randomly selected city websites from the City Brands Index which is constituted by Simon ANHOLT and GfK ROPER Public Affair & Media are evaluated using content analysis from e-branding point of view, and the results of the content analysis are presented and discussed respectively. In the conclusion section, recommendations about city e-branding are suggested to the city administrators and professionals in the field of city e-branding.

BACKGROUND

City Branding

Branding is one of the major issues taken up within product strategies in the marketing literature (Kotler, 2000). According to the definition of American Marketing Association, a brand is a name, a term, a sign, symbol or a combination of them that aim to identify the products and services of a seller or a seller group, and to differentiate it from the competitors (Fernandez-Cavia & Huertas-Roig, 2009, p. 28; Kerr, 2006, p. 276; Kotler, 2000, p. 404). A brand identifies the seller or maker, and principally constitutes the promise of the seller to the buyers that it will provide the buyers with the benefits and the services consistently (Kotler, 2000).

All public and private organizations can engage in branding. Products, programs, and services, as well as organizations and organizational units can be branded (Tschirhart, 2008). The use of branding techniques is not only limited to the branding of products and services; places such as countries, regions and cities can be branded as well (Freire, 2005). *City branding*, in line with the economical, social, political and cultural development, can be defined as the application of branding strategies and other marketing techniques in the private sector to the cities (Kerr, 2006). Cities as places are also called as mega products consisting of the combination of tangible and intangible features (Florek et al., 2006).

City marketing applications are largely based on the structuring, communicating and managing the image of a city. Therefore, the issue of city marketing is related to the image of a city, and this constitutes the starting point of city branding (Kavaratzis, 2004). City branding is based on the perceptions and images of the people, and these factors constitute the center of the appropriate activities designed to give a shape to the city and its future (Kavaratzis & Ashworth 2005). City branding in which the image of a city is vital is becoming more and more important today as a factor that affect the economies of the cities (Butler, 1999). Cities obtain an image and identity as a result of their being branded. The image and identity of cities are built in general on the basis of such sources as history, demography, culture, custom, economy, industry, landscape, environment, politics, policies, etc. (Inn, 2004; Zhang & Zhao, 2009).

The aim of branding is to affect the brand preferences of the customers by adding value on the product and services, and get the customers to be loyal to a particular brand. What this particular point of view which is considered highly in the marketing literature implies is that the brand is more than the name given to the product. Brand in reality is a total whole composed of physical, socio-psychological features and beliefs. The key factor in successful branding is to establish a relationship between the brand and the customers inclusive of a close harmony between customers' own physical and psychological needs and the functional features and symbolic values of the brand. In this context, cities satisfy the functional, symbolic and emotional needs as well (Kavaratzis, 2004). Since the brands lessen the perceived risks during the product purchasing, they fulfill a function of risk reducer, and cut down the researching costs of the customers (Hankinson, 2004). Especially the positive experiences of customers about the online brands enhance their familiarity with those brands and significantly affect customer satisfaction and brand trust (Ha & Perks, 2005). Brands also provide benefits such as 1- affecting the customer decisions; 2-presenting intangible values with such characteristics as added value, reliability, warranty; 3- indicating the reliability of the product's origins and making as promise of a good performance; 4-facilitating the process of making a choice for the customer. All these benefits can be associated with city brands as well. For instance, cities such as New York, Paris or Rome clearly define their city brands and each of them offers some obvious and different promises helping the process of making decisions (Parkerson & Saunders, 2005).

The complicated structure of any city product due to the combination of very different factors makes city branding and building branded cities all the more difficult and complicated (Blichfeldt, 2005). Factors such as the fact that the public sector has more dominant responsibilities, that the geographical domain has very different aspects, the difficulties in defining the city as products, the necessity to succeed in fulfilling the political and social objectives in addition to economical ones are the important factors that differ the branding of cities from the branding of business products and services (Hankinson, 2001). In addition to these factors, the inability to ensure unity of objectives in city management, insufficiency in the support of the central authority and the lack of marketing experience are the other difficulties encountered in the branding of cities (Pryor & Grossbart, 2007). Furthermore, while businesses may make easily changes on the products that they offer, develop the products, re-launch them or repositioning them, many of those options are not available for the cities (Vitiello & Willcocks, 2006). For instance, some of the intrinsic characteristics of cities such as their nature, climate or geographical location can not be altered. However, factors such as the infrastructure, leisure, culture and education, healthcare, economy, businesses and their employees, public security, administration and policies, social atmosphere and the development of the civil society can be altered (Muñiz-Martínez & Cervantes-Blanco, 2009).

A city brand has an impact on whether individuals will continue to live there or intend to live there, whether investor will make investments in that city and whether people will visit that city (Gold, 2006). Cities should associate their own brands with industries, life styles and the other unique features of the city as well. A successful city brand should be the total whole of that city's infrastructure, inhabitants, industries and quality of life, and city brand should demonstrate the city's vision and get the support of the broad stakeholders (Kerr, 2006).

The Role of the City Websites in Building City Brands

The internet has become an important strategic weapon within today's competitive business framework. This approach aimed at marketing, communication and service delivery has changed the structure of the local and international markets (Simeon, 1999, 2001). The internet that has had a transformational effect on businesses has caused the balance of power to shift from companies in favor of costumers. Due to the unique characteristics of the internet, businesses have had to redefine their marketing and branding strategies (Simmons, 2007). It is because the internet offers for brands the opportunities of both growth and customer loyalty (Müller et al., 2008).

With the globalization of economies and broad scale acceptance of e-businesses, a growing number of companies worldwide have established websites in order to promote themselves to the public, to provide better services to their customers, to assist information exchanges with their customers and suppliers and increase their sales with online channels (Xi *et al.*, 2007). Therefore, the company websites have become an important channel to reach their customers/the public (Maynard & Tian, 2004). Similar conditions are true for cities as well. It is because the internet offers many advantages to the city administrations. The following are the advantages offer by the internet: it is cheaper to design a website when compared to advertising or buying media space; it is also possible to reach different target audiences through the internet with constantly updated information. Business can update their websites with fraction of maintenance costs thanks to user-friendly software. These update changes also include the messages placed on the website and what's more, the number of visitors to the website can also be monitored. Due to its interactive nature, visitor to a website have the chance to leave a feedback by, for instance, filling in an online survey form or a direct "contact us" email. This interactive nature of the internet also enables the place decision makers to get in tough with the target audience and ask for their opinions, for instance, about a new logo, symbol or slogan (Avraham & Ketter, 2008). City administrators have also obtained places in the online environment in order to differentiate themselves from the other cities and to sustain the interest of its customers (residents, tourists, investors etc.) and stakeholders.

City websites can be used by the city administrators as an advertising tool as well in addition to being an important source of information. Besides, these interactive websites also have an important role in the delivery of services offered by the city and city administrators, and attracting the attention of users (Fernandez-Cavia & Huertas-Roig, 2009). In fact, the internet together with the other technologies may also assist organizations in re-organizing their services by taking advantage of existing information and communication technologies. These kinds of services based on information and communication technologies are termed as e-services (Mishra & Rolland, 2009). The city websites offer great conveniences to many different kinds of people such as city residents, tourists and business administrators. For instance, given the sector of tourism which is an information-intensive sector, more and more customers today use the relevant websites as their main tool to plan, book and pay for their holiday and trips (Diaz-Luque, 2009). As is clearly seen, the fact that a city has a website provides opportunities both to offer information about the city and to interact with different city-based institutions (Urban, 2002). The city websites that serve as a tool of information collection, at the same time, have an important impact on the decision making process of the consumers (Prats-Planaguma & Camprubi, 2009).

The number of studies analyzing the city websites in terms of city branding is limited. In one of those limited number of studies, Florek and the colleagues (2006) examined the place branding characteristics of city council websites in New Zealand and evaluated 22 different characteristics in their study. In this study, the place brand identity contents, offered in the city council websites, were classified into three groups of brand design, brand behavior and brand communication. The first group of visual presentation and identification, *brand design*, is composed of slogan, logo, arms, video, web cam, picture gallery and maps. The second group of *brand behavior*, which is linked to the decisions and actions expressing the formulation of place product, is composed of such activities as news, public notices, projects, plans, policies, reports, bylaws, event calendar, customized information, contact, community board members and sister cities. The third group of *brand communication* has been examined in terms of its communication with the external environment (e.g. foreign tourists, potential residents and foreign investors) and with internal environment (e.g. inhabitants, local entrepreneurs, and local students). Within this framework of communication structure, characteristics in the websites such as general information/overview of city, customized information for target audiences (e.g. brochures), contact information, feedback tools, and links to related websites have been examined. In terms of brand type, repetition, and consistency, in her study where he evaluated the use of brand in the city government websites of the ninety-seven cities in the United States, Tschirhart (2008) evaluated the use of brand in the city websites, the symbols, logos and the slogans delivered through these websites. In his study in which he examined urban branding of 67 cities in the USA in terms of city homepage imagery, Grodach (2009) examined the images found in the 30 categories on the official city homepages and visitors bureau website.

Some of the evaluation criterion used in the marketing (Okazaki & Škapa, 2009; Simeon, 1999, 2001) and tourism studies (Baloglu & Pekcan, 2006) aimed at evaluating the websites of businesses, can be used for the evaluation of city websites as well. When the criterion used in these studies is considered, it can easily be observed that brand website functions such as language option, FAQs, free download, customer feedback, links, news release, sitemap, search engine (Okazaki & Škapa, 2009); key website components inclusive of informing, positioning and delivery such as company news, company history, executives listed, downloadable reports, search engines, links to other firms, and discussion groups (Simeon, 1999, 2001); and the evaluation criterion used by Baloglu and Pekcan (2006) for the content analysis of the websites of 4 and 5 star hotels in Turkey for interactivity (online survey, online information request form, online comment form), for navigation (links to other sites, index page, search capabilities), and for functionality (download facilities, date last updated, multilingual capabilities) can be used for the evaluation of city websites. In conclusion, while considering the role of the city websites in building city brands, it is clearly observed that two important elements come to the fore. One of those is the use of fundamental elements of the brand; and the second one is the use of the fundamental elements of the website. Therefore, bringing together the fundamental elements of the brand with the required elements of the website, city administrators are able to bring into the open a successful city e-brand. In this way, some cities that have built a brand in the offline environment continue to be a successful brand in the online environment as well thanks to the successful city websites.

EVALUATION OF THE CITY WEBSITES USING CONTENT ANALYSIS

The main aims of this chapter are to introduce the city branding in competition between cities and to evaluate the role of city websites on the branding cities. In order to fulfill this main aim, firstly, city branding and the role of the city websites building city brands to be included in the content analysis and establish the boundaries for this research are presented in the background section. In this section of the chapter, randomly selected city websites from the City Brands Index are evaluated using content analysis from e-branding point of view.

Methodology

This research aims at understanding how city authorities in cities involved in city branding activities use city websites to communicate and thus support the development of city brands. Do the city authorities use their websites to communicate all of the brand characteristics? Do they use features that are unique to their places in order to differentiate them from competing cities? To understand the branding activities of cities, this study is framed with the following research objective: to examine and evaluate city brands through their websites. This will be investigated by answering two research questions: (1) what characteristics of city brand in the city websites are most widely used? and (2) how do the city authorities use online sources related to the city?

This study utilizes content analysis to evaluate the websites of a select group of cities in terms of city website characteristics and branding practices on the Internet. The content analysis is an "observational research method that is used to systematically evaluate the symbolic content of all forms of recorded communications" (Kolbe & Burnett, 1991, p. 243). Content analysis is also a common method to investigate written documents such as web pages and has been extensively used in marketing (Cheung & Huang, 2002; Liao et al., 2006; Maynard & Tian, 2004; Okazaki & Škapa, 2009; Perry & Bodkin, 2000; Simeon, 1999, 2001; Virtsonis & Haridge-March, 2008) and tourism (Baloglu & Pekcan 2006; Cai et al., 2004; Wan, 2002). Methodological recommendations from prior research were adopted, to establish a high reliability (Baloglu & Pekcan, 2006; Florek et al., 2006; Okazaki & Škapa, 2009; Simeon, 1999, 2001).

To create a dataset, the city brands index which is constituted by Simon Anholt and GfK Roper Public Affair & Media (2009) was used. This index is based on a global survey in which respondents from across 20 major developed and developing countries are asked to rate their agreement with statements about each of 50 cities. The ranking is developed by averaging city scores across the index's six categories: presence (knowledge of city and perception of its global contribution), place (cleanliness, aesthetic qualities and climate), pre-requisites (affordable accommodations and quality of public amenities), people (friendliness, personal encounters and cultural diversity), pulse (interesting events, activities and lifestyles) and potential (perception as good place to do business, to find a job and to go to school). The cities measured include:

- **North America:** Boston, Chicago, Dallas, Los Angeles, New York, San Francisco, Seattle, Toronto, Vancouver
- **Western Europe:** Amsterdam, Barcelona, Berlin, Brussels, Copenhagen, Dublin, Edinburgh, Geneva, Helsinki, London, Madrid, Milan, Paris, Rome, Stockholm, The Hague, Vienna
- **Central/Eastern Europe:** Budapest, Istanbul, Moscow, Prague, Warsaw
- **Asia Pacific:** Auckland, Bangkok, Beijing, Hong Kong, Melbourne, Mumbai, Seoul, Shanghai, Shenzhen, Singapore, Sydney, Tokyo
- **Latin America:** Buenos Aires, Mexico City, Rio de Janeiro
- **Middle East/Africa:** Cairo, Dubai, Jeddah, Johannesburg

In order to select the cities to constitute the sample of the study out of the groundmass forming the list, a number has been assigned to all the listed cities. Fifteen cities were randomly selected from this city brands index list and evaluated using content analysis. Two factors were paid special attention regarding the examination of cities mentioned in this study. Firstly, in order to ensure cross-national data equivalency, only the pages in English on the city website were examined. Secondly, only the city municipalities or city official websites were evaluated by content analysis. Therefore, other city-related websites (tourism organizations, commercial operators, business websites, etc) were not considered in this research. While deciding whether the relevant websites or web pages are within the boundaries of municipality websites, only the home page of the municipality and the extensions of the home page (the domain name of the city municipality and its extensions) were examined; and in cases where the links of the other relevant websites had different domain names from those on the municipality's home page, they were excluded from the study.

As the content analysis proceeded, a detailed coding sheet was first developed. The coding categories were adopted from Baloglu and Pekcan (2006), Florek et al. (2006), Okazaki and Škapa (2009), and Simeon (1999, 2001). Each variable was measured on a categorical dichotomy as to the existence of each feature on a city website. Values of "1" and "0" were assigned for answers of "Yes" and "No", respectively. For example, city websites that had appropriate information associated with "city logo" were assigned "1" for this feature. Those that did not were assigned "0". Finally, a structured form consisting of 32 checkpoints was developed in order to evaluate the websites of brand cities (See Appendix A). All data were coded manually and entered onto a spreadsheet. As a single coder was involved in the analysis, inter-coder reliability was not a concern. The coding and checking was performed during the last week of May 2010. The data were analyzed using SPSS 13.0 for Windows. While analyzing the data in this study, some descriptive analyses techniques in general were used and the frequencies and percentage distribution of the data on the tables were taken into consideration. As websites are a dynamic means of communication with audiences, it is acknowledged that the content of the websites examined would change during and after the research. Their content and structure can change daily or even hourly and this must be taken into account when interpreting the results.

Results

As was mentioned in the methodology section, the official websites of 15 cities randomly selected from the City Brand Index list was evaluated through content analysis. The city websites evaluated are the following; Amsterdam, Auckland, Barcelona, Chicago, Edinburgh, Helsinki, Istanbul, Johannesburg, Los Angeles, Prague, Seoul, Sydney, Tokyo, Toronto and Vienna. The official websites of these cities were examined in terms of the characteristics of brand design, brand behavior and brand communication using content analysis, and it was revealed as a result of content analysis which characteristics were used and to what extent. Besides, the websites of those cities examined using content analysis were listed according to percentage of meeting the criterion.

The brand design characteristics examined in this study and the results of the evaluation are illustrated in Table 1. When Table 1 is analyzed, it is clearly seen that all the cities evaluated used the city logo in their websites; on the other hand, it is also clear that the percentage of the use of a slogan related to the city was not as high (40%). The study also revealed that especially the administrators of the cities such as Amsterdam, Seoul, Sydney and Johannesburg paid great deal of significance to these factors of brand designing, had some special endeavors in this issue, and the results of these endeavors were reflected in their websites. For instance, the following cities chose and used the following slogans; the city of Seoul "Hi Seoul, Soul of Asia", the city of Amsterdam "Iamsterdam", the city of Sydney "city of villages" and Johannesburg, the host city of 2010 World Cup used the slogan "A world Class African Host City". Istanbul, on the other hand, uses the slogan of "European Center of Culture" in its website. It is a great shortcoming not to use a slogan, one of the factors of brand building, extensively in the city's website. Furthermore, starting with the home page, this slogan should appear on every page. Especially the logo and the slogan help the

Table 1. Brand Design Characteristics

Brand Design Characteristics	Number of city websites evaluated	1=Yes		2=No	
		n	%	n	%
Slogan	15	6	40	9	60
Logo	15	15	100	0	0
Video	15	10	67	5	33
Web cam	15	8	53	7	47
Picture gallery/City Pictures	15	11	73	4	27
City Maps	15	11	73	4	27
City history	15	13	87	2	13
Number of the total brand design items evaluated	105	74	70	31	30

cities to have a unique character when compared to other cities in the offline and online environments; and since these two factors help the cities to gain an identity in their competition with the other cities, those cities become more distinctive.

In today's world where and when the information and communication technologies are developing fast, it is surprising to see that some cities do not still use some technological features such as videos, web cams and city maps. In the city websites where there is a web cam, this particular feature is frequently used for traffic control. However, in some other cities, (e.g. Istanbul), it is observed that there exit some web cams showing the different parts of the city used tourism (promotional) purposes. Most of the city websites where the city map link is used are linked up to the search engine of Google Earth City. Some other city websites where there is not a detailed map of the city or the tourist attractions are not indicated, the city map is only shown as a picture. In the majority of the city websites (87%), there is information on the city history mostly on the home page and under a different menu.

The results of the content analysis related to brand behavior characteristics are illustrated in Table 2. These characteristics made up of 10 items are composed of applications reflecting the actions and decisions related to the city. When Table 2 is analyzed, it is seen that the city administrators were successful in putting information about the decision and actions related to the city on the city website. 93% of the city websites chooses to announce the decision and actions under the news and public notices/announcements menu or heading. By a similar percentage, the city administrators announce to the whole world especially their annual plans, the policies determined and the reports of the activities carried out in their websites. Similarly, many activities such as festivals and symposiums organized in the city are advertised under the heading of Events or Event Calendar primarily to the residents, tourists and to all those concerned. All the city websites examined have a similar home page layout in line with customized information. City websites in general have some sub menus under the headings of city government, residents, tourists/visitors, business/investors and e-services. This facilitates for the target audience easy access to the information and services as well. However, these menus are sometimes linked up to the websites of other institutions such as the Ministry of Tourism and The Chamber of Trade and Commerce.

There is detailed information about the city administration in the city websites. Usually under the heading of city hall or city government, there is the Mayor and the deputies as well as the community board members. Besides, under the same heading, there is the Mayor's introductory speech,

Table 2. Brand Behavior Characteristics

Brand Behavior Characteristics	Number of city websites evaluated	1=Yes		2=No	
		n	%	n	%
News	15	14	93	1	7
Public notices	15	14	93	1	7
Projects	15	13	87	2	13
Plans/Policies	15	14	93	1	7
Reports	15	14	93	1	7
Event calendar/Events	15	14	93	1	7
Customized information	15	15	100	0	0
Community board members/List	15	14	93	1	7
Sister cities	15	11	73	4	27
E-services	15	13	87	2	13
Number of the total brand behavior items evaluated	150	136	91	14	9

the distribution of duties and responsibilities of those in the management, and contact information of the members of the community board. 73% of the city websites analyzed also indicate their sister cities. The sister cities are sometimes indicated under the headings of twin cities or partner cities. Under this menu, there is some information about the sister cities. According to the information based on the content analysis, it is observed that the cities with a sister city indicated in their websites have more than one sister city and these sisterhood goes back to as far as 40 or 50 years. For instance, the city of Tokyo has 11 sister cities and the first one is the New York City (USA) that started in 1960. The services offered by most of the cities in the offline environment are provided in the online environment as well and those services are by and large collected under the heading of e-services.

One of the most important functions of websites is the two-way communication. With these characteristics in mind, the city websites were examined in terms of brand communication characteristics such as general communications, interactivity, navigation, and functionality, and the results obtained are illustrated in Table 3. When Table 3 is examined in terms of general communication features, it is observed that contact information and customized information for target audiences is available in all websites. However, it is surprising to find that the FAQs menu used in most of the private sector websites was not as commonly available in the city websites (only 67% have it). It is because when the users fail to find what they are searching for wish to use this menu and get the information to resolve their problems from this menu again. In addition to simplicity and speed, this particular feature also gives satisfaction to the website users.

The city websites examined had the following parentages of feedback tools; About the two third of the websites (67%) had online survey, 73% online information request form, 73% online comment form. These characteristics enhance the interactivity feature of city websites and help the users to convey their wishes and complaint more easily and efficiently through the websites. However, these generally come in the shape of a form and are mostly not placed on the home (main) page. Therefore, it can sometimes be very difficult to find these pages in some websites. Besides, as these items are designed for the residents of that city, they were composed according to the language of those residents in generally.

Table 3. Brand Communication Characteristics

Brand Communication Characteristics	Number of city websites evaluated	1=Yes		2=No	
		n	%	n	%
General Communication					
General information/overview of city	15	14	93	1	7
Customized information for target audiences	15	15	100	0	0
Contact information	15	15	100	0	0
FAQs	15	10	67	5	33
Interactivity					
Online Survey	15	10	67	5	33
Online Information Request Form	15	11	73	4	27
Online Comment Form	15	11	73	4	27
Navigation					
Links to related websites	15	15	100	0	0
Search Engines	15	15	100	0	0
Index Page/Site maps	15	15	100	0	0
Help menu	15	5	33	10	67
Functionality					
Download Facilities	15	15	100	0	0
Date Last Updated	15	7	47	8	53
Language Option	15	12	80	3	20
Security and/or privacy info	15	11	73	4	27
The number of total brand communication items evaluated	225	181	80	44	20

The navigation feature of brand communication is used very efficiently in the city websites. In the entire city websites examined, there are features such as links to related websites, search engines, and index page/site maps. These features help the web users to have easier access to the information and spare them from the trouble of having to open up and examine all the menus of the city website. These features are available on the top right corner of the city website. Links to related websites are usually related to such topics as tourism, business/investment etc. Although these three features related to navigation are available in all websites, it is surprising that the Help menu is nor available in the city websites. However, for the users who fail to find what they are searching through the search engine or on the index page/site maps, the Help menu will be of great help to the users. One of the biggest problems encountered during the content analysis is that there is no Home Menu related to navigation. In most of the city websites examined, when the submenus are clicked, there is no Home menu or back button that will take the user back to the home page. This, in turn, turns out to be time-wasting for the users. As a result of the fact that this menu does not exist, the satisfaction of the websites users may drop and they may abandon the website more easily and quickly.

As far as the functionality of the city websites are considered, the entire city websites examined proved to have the download facilities. The issues which are downloadable are especially comprised of the documents such as reports, projects carried

Table 4. Total brand characteristics of the city websites

City Website Brand Characteristics	Total number of items evaluated	1=Yes		2=No	
		n	%	n	%
Brand design	105	74	70	31	30
Brand behavior	150	136	91	14	9
Brand communication	225	181	80	44	20
- General communication	60	54	90	6	10
- Interactivity	45	32	71	13	29
- Navigation	60	50	83	10	17
- Functionality	60	45	75	15	25
TOTAL POINTS	480	391	81	89	19

out by the city administration, budgets, plans and other activities. For instance, "Auckland city business and economy report, 2009" in the website of the city of Auckland, and the "Council Budged, 2009-2010" in the website of the city of Edinburg can be downloaded. The information of "Date Last Updated", one of the functional features of city websites is one of the least used (47%) features in the city websites. This information which reveals the up-to-dateness of the city websites is not available in the home page of the websites of many cities (e.g. Auckland, Seoul, Chicago, Helsinki, Johannesburg and Vienna). On the other hand, in the websites of such cities as Sydney, Istanbul, Prague, Toronto and Barcelona, the information is available and frequently updated in the home page. The cities that open themselves up to the whole world through their websites, also make available the choice of language menu in their home page in order for the voices heard the world over. The websites of very few cities (Auckland, Los Angeles and Johannesburg) do not have the language choice in home pages. The menu of security and/or privacy info in the city websites is one of the other important functional features used in the websites of cities. This particular feature is especially significant in terms of the use of e-services. Security and/or privacy info link is generally available in the home page of city websites. However, there are still some city websites (Seoul, Istanbul, Johannesburg and Helsinki) in whose home pages this link is still not available.

The percentages of brand characteristics of the websites of 15 cities evaluated through content analysis are illustrated in Table 4. The 7 items comprising brand design characteristics of websites of 15 cities, the 10 items comprising the brand behavior characteristics and the 15 items comprising brand communication characteristics were evaluated through content analysis. The results of content analysis revealed that 70% of design characteristics, 91% of brand behavior characteristics and 80% of communication characteristics were available in the websites of 15 cities. The result of brand behavior characteristics demonstrated that the city administrators offered the decisions made about the city and actions taken for the city to the users effectively (91%). It is also observed that the city websites were used efficiently in ensuring brand communication (80%). However, the use of the city websites dropped in terms of interactivity (71%) and functionality (75%) characteristics. The least common brand characteristics in the city websites examined is those of brand design (70%). When analyzed in general, it is also observed that the brand characteristics of websites of 15 cities evaluated were found to be 81%. This is really a successful result. In other words, city administrators use the city websites successfully in order to build a brand in the online environment.

Table 5. Evaluation results of the city websites examined

Rank	Cities	The evaluation dates of city websites	Total number of criterion evaluated	Total number of the criterion met	
			n	n	%
1	Sydney	May 29, 2010	32	31	97
2	Toronto	May 27, 2010	32	31	97
3	Barcelona	May 28, 2010	32	30	94
4	Istanbul	May 30, 2010	32	30	94
5	Chicago	May 28, 2010	32	29	91
6	Seoul	May 29, 2010	32	28	88
7	Los Angeles	May 28, 2010	32	28	88
8	Edinburg	May 31, 2010	32	26	81
9	Auckland	May 30, 2010	32	26	81
10	Johannesburg	May 30, 2010	32	24	75
11	Amsterdam	May 29, 2010	32	22	69
12	Prague	May 27, 2010	32	22	69
13	Helsinki	May 29, 2010	32	22	69
14	Vienna	May 31, 2010	32	21	66
15	Tokyo	May 31, 2010	32	21	66

However, it should also be remembered that those cities were selected from the list of City Brand Index. Therefore, when the websites of other cities not in this list are examined, the results are more likely to be different.

Table 5 illustrates the presence of 32 criterions in the city websites comprising the evaluation criterion and the sequencing of city websites based on the percentages. In other words, this sequencing is the sequencing of cities using their brand characteristics in their websites in the best way.

On the top of this sequencing in Table 5 are the websites of the cities of Sydney and Toronto. The website of the city of Sydney contains the entire brand characteristics in the most efficient manner. The only shortcoming of the website of Sydney is the fact that the Help menu is not available on the home page. This, on the other hand, is a feature that can easily be completed. An important difference differentiating the website of the city of Sydney is the fact that menus such as Community, Library and Environment are available in the home page. These menus do not exist in the websites of the other cities examined. The website of the city of Toronto was built very successfully in terms of website layout. It is impossible for the user to get lost in this city's website. There is almost no dead link in the city website (only one dead link was detected). The only shortcoming of the website of the city of Toronto is the lack of a slogan of the city. This is a vital shortcoming. Since a slogan helps the city to gain an identity and uniqueness, this is an important brand factor that needs to be dealt with. There is no slogan and help menu in the website of the city of Barcelona and there is no Help menu and the link of Security and/or privacy info in the website of the city of Istanbul. The Help menu and the link of Security and/or privacy info link can easily be placed in the city website. However, the slogan, unless determined previously, is an important brand element that needs to be exerted efforts on.

The websites of those cities such as Tokyo, Vienna, Prague and Amsterdam that are falling behind in sequencing have some shortcomings in terms of interactivity. In the websites of some other cities, on the other hand, the important part of the criterion of brand design was not allowed for. For instance, out of the seven brand design characteristics, Edinburgh had only one, Helsinki 2, Tokyo 3 and Johannesburg have 4 criterion in their city websites. The feature of brand design is an important factor that requires the use of branding techniques in marketing and in the city website, and a factor over which more efforts have to be exerted. In this sense, it is vital that cities and their websites, in order to get ahead of the competitor cities and their websites, pay attention to all the city branding characteristics with the brand design characteristics leading the way, and work more intensively on the city website in terms of branding. Finally, given the dynamic feature of websites, it should be remembered that this is a study aimed at evaluating the city websites at a certain time or point in time, and the evaluation revealed the profile existing during the period in which he study was carried out. Therefore, the ensuing interpretations belong to a particular time period.

CONCLUSION

The competition underway respectively amongst countries, regions, businesses and people has also been going on intensively between the cities today. Competition is taking place not only in the traditional environment, but also in the online environment as well. At this point, the city administrator have, in the face of the competition with other cities, started to use the marketing and branding techniques utilized by the private sector for their own cities as well. However, the fact that cities are mega products consisting of tangible and intangible element, that city administrators fail to bring about some of the changes made accomplished by the private sector for their own cities, and that the social objectives are prioritized and the lack of knowledge and experience of the city administrators about city branding are just some of the fundamental difficulties encountered by city administrators in city branding. In addition to all these factors, the fact that there is the necessity to manage city branding in the online environment as well makes the management of city branding all the more difficult. However, in the face of competition with other cities, it is a must today for the city websites to be used efficiently in order to build a city brand for the communication of city brands.

As a result of this study in which the websites of 15 cities randomly selected from the city brand index were examined using content analysis, it was revealed that based on the evaluation criterion, the city websites were efficiently used (81%). However, despite the fact that the cities evaluated were selected randomly from the city brand index list, it turned out that they had some shortcomings in some of the branding factors. The most important of these shortcomings is that no slogan related to the city is used in the city websites (60%). While many elements such as webcam, video and city maps are put into practice readily related to brand designing, finding a slogan for the city, on the other hand, is a brand element requiring time and some thinking and planning. Finding an appropriate slogan in line with the features of the city and placing that slogan in the city's website is very important in differentiating the city from the other cities. As a result of the study, it also tuned out that the cities examined were equally successful in terms of brand behavior characteristics. In other words, what the city administrators have managed to accomplish are passed on to the public in the form of news and public notices. Activities such as festivals and other social and cultural activities organized in the city are announced to the public under the heading of events. Similarly, the plans, policies and projects illustrating the decisions and applications of the city administration are made available for the scrutiny of public in the form of reports and this way, the transparency of the city administration is emphasized. Similarly, this transparency is also emphasized by the presence of the duties and responsibilities of the city management on the community board members list in the city websites. The results of this study demonstrate that cities take advantage of today's information and communication technologies (ICT) and deliver services through the city websites. It has also become clear that city websites are successfully utilized in relation to brand communication and play an important role in promoting city brands. However, this study also revealed that city websites possessed an important shortcoming in terms of interactivity. City websites need to be used more efficiently in terms of interactivity that makes two-way interaction possible.

City administrators should make up a team in order to be successful in building a city brand and ensuring brand communication. This team should be composed of city administrators, professional with experience in brand building and website designers, and should operate in a coordinative manner. The ideas and opinions of the residents can also be resorted to in finding a slogan and logo for the city. Some contests can be organized to this end. In order for this team formed to be successful, the support of the administration of the city is crucial as well. Besides, while accomplishing the activities of this team, branded cities that have successfully developed their brands in their city websites and that have used their city websites efficiently both in terms of building a city brand and ensuring brand communication should be emulated. In this sense, the results of the study carried out in this chapter will be of great use to the city administrators and to the brand building and developing teams to be formed.

This study, just like many other studies, has some limitations. The first one is the fact that limited numbers of city websites have been examined. The second one is that the groundmass of this study is composed of cities found in the world-wide study comprising the city brand index. This list selected as a base for the sample group is composed of branded cities. Therefore, only the successful cities constitute the groundmass and eventually the sample group. Finally, the study reflects the existing profile only during a specific point in time. These limitations, in fact, also bring along in a way some suggestions for future research. In this context, future studies may carry out their researches on a sample group of more city websites and likewise, compare the websites of branded and unbranded cities. Besides, development ensuing at the end of a particular process can be examined through repeated evaluations.

REFERENCES

Adult, S. (2005). Some important distinctions in place branding. *Place Branding*, *1*(2), 116–121. doi:10.1057/palgrave.pb.5990011

Avraham, E., & Ketter, E. (2008). *Media strategies for marketing places in crisis - Improving the image of cities, countries and tourist destinations*. Burlington, MA: Elsevier.

Baloglu, S., & Pekcan, Y. A. (2006). The website design and internet site marketing practices of upscale and luxury hotels in Turkey. *Tourism Management*, *27*(1), 171–176. doi:10.1016/j.tourman.2004.07.003

Berg, L. V. D., & Braun, E. (1999). Urban competitiveness, marketing and the need for organising capacity. *Urban Studies (Edinburgh, Scotland)*, *36*(5-6), 987–999. doi:10.1080/0042098993312

Blichfeldt, B. S. (2005). Unmanageable place brands? *Place Branding*, *1*(4), 388–401. doi:10.1057/palgrave.pb.5990036

Butler, C. J. (1999). Cities in competition: Equity issues. *Urban Studies (Edinburgh, Scotland)*, *36*(5-6), 865–891. doi:10.1080/0042098993259

Cai, L., Card, J. A., & Cole, S. T. (2004). Content delivery performance of World Wide Web sites of US tour operators focusing on destinations in China. *Tourism Management*, *25*(2), 219–227. doi:10.1016/S0261-5177(03)00095-5

Cheung, W. M., & Huang, W. (2002). An investigation of commercial usage of the World Wide Web: A picture from Singapore. *International Journal of Information Management*, *22*(5), 337–388. doi:10.1016/S0268-4012(02)00026-9

Diaz-Luque, P. (2009). Official tourism Web sites and city marketing. In Gascó-Hernández, M., & Torres-Coronas, T. (Eds.), *Information communication technologies and city marketing: Digital opportunities for cities around the world* (pp. 152–183). Hershey, PA: Information Science Publishing. doi:10.4018/978-1-60566-134-6.ch008

Dinnie, K. (2004). Place branding: Overview of an emerging literature. *Place Branding*, *1*(1), 106–110. doi:10.1057/palgrave.pb.5990010

Fernandez-Cavia, J., & Huertas-Roig, A. (2009). City brands and their communication through Web sites: Identification of problems and proposals for improvement. In Gascó-Hernández, M., & Torres-Coronas, T. (Eds.), *Information communication technologies and city marketing: Digital opportunities for cities around the world* (pp. 26–49). Hershey, PA: Information Science Publishing. doi:10.4018/978-1-60566-134-6.ch002

Florek, M., Insch, A., & Gnoth, J. (2006). City council websites as a means of place brand identity communication. *Place Branding*, *2*(4), 276–296. doi:10.1057/palgrave.pb.6000036

Freire, J. R. (2005). Geo-branding-Are we talking nonsense? A theoretical reflection on brands applied to places. *Place Branding*, *1*(4), 347–362. doi:10.1057/palgrave.pb.5990033

GfK Sverige AB. (2009). *GfK Roper and Anholt-City brands index*. Retrieved May 3, 2010, from http://www.mynewsdesk.com/ se/view/ pressrelease/ gfk-roper-and-anholt-city- brands-index-287066

Gold, E. R. (2006). Intellectual architecture as place brand. *Place Branding*, *2*(3), 220–228. doi:10.1057/palgrave.pb.5990059

Gordon, I. (1999). Internationalisation and urban competition. *Urban Studies (Edinburgh, Scotland)*, *36*(5-6), 1001–1016. doi:10.1080/0042098993321

Grodach, C. (2009). Urban branding: An analysis of city homepage imagery. *Journal of Architectural and Planning Research*, *26*(3), 181–197.

Ha, H. Y., & Perks, H. (2005). Effects of consumer perceptions of brand experience on the Web: Brand familiarity, satisfaction and brand trust. *Journal of Consumer Behaviour*, *4*(6), 438–452. doi:10.1002/cb.29

Hankinson, G. (2001). Location branding: A study of the branding practices of 12 English cities. *Brand Management*, *9*(2), 127–142. doi:10.1057/palgrave.bm.2540060

Hankinson, G. (2004). Relational network brands: Towards a conceptual model of place brands. *Journal of Vacation Marketing*, *10*(2), 109–121. doi:10.1177/135676670401000202

Huerta-Carvajal, M. I., & Luna-Reyes, L. F. (2009). City boosterism through Internet marketing: An institutional perspective. In Gascó Hernández, M., & Torres-Coronas, T. (Eds.), *Information communication technologies and city marketing: Digital opportunities for cities around the world* (pp. 130–151). Hershey, PA: Information Science Publishing. doi:10.4018/978-1-60566-134-6.ch007

Inn, K. (2004). Plan for city identity establishment and city marketing the case of Kimpo city. *Dela*, *21*, 233–240.

Kavaratzis, M. (2004). From city marketing to city branding: Towards a theoretical framework for developing city brands. *Place Branding*, *1*(1), 58–73. doi:10.1057/palgrave.pb.5990005

Kavaratzis, M., & Ashworth, G. J. (2005). City branding: An effective assertion of identity or a transitory marketing trick? *Tijdschrift voor Economische en Sociale Geografie*, *96*(5), 506–514. doi:10.1111/j.1467-9663.2005.00482.x

Kerr, G. (2006). From destination brand to location brand. *Brand Management*, *13*(4-5), 276–283. doi:10.1057/palgrave.bm.2540271

Kerr, G., & Johnson, S. (2005). A review of a brand management strategy for a small town: Lessons learnt! *Place Branding*, *1*(4), 373–387. doi:10.1057/palgrave.pb.5990035

Kolbe, R. H., & Burnett, M. S. (1991). Content-analysis research: An examination of applications with directives for improving research reliability and objectivity. *The Journal of Consumer Research*, *18*(2), 243–250. doi:10.1086/209256

Kotler, P. (2000). *Marketing management: The millennium edition*. Upper Saddle River, NJ: Prentice-Hall.

Lever, W. F., & Turok, I. (1999). Competitive cities: Introduction to the review. *Urban Studies (Edinburgh, Scotland)*, *36*(5-6), 791–793. doi:10.1080/0042098993213

Liao, C., To, P. L., & Shih, M. L. (2006). Website practices: A comparison between the top 1000 companies in the US and Taiwan. *International Journal of Information Management*, *26*(3), 196–211. doi:10.1016/j.ijinfomgt.2006.02.007

Maynard, M., & Tian, Y. (2004). Between global and glocal content analysis of the Chinese Web sites of the 100 top global brands. *Public Relations Review*, *30*(3), 285–291. doi:10.1016/j.pubrev.2004.04.003

Mishra, B., & Rolland, E. (2009). A strategic framework for city marketing: The SSRM approach. In Gascó-Hernández, M., & Torres-Coronas, T. (Eds.), *Information communication technologies and city marketing: Digital opportunities for cities around the world* (pp. 50–63). Hershey, PA: Information Science Publishing. doi:10.4018/978-1-60566-134-6.ch003

Müller, B., Flores, L., Agrebi, M., & Chandon, J. L. (2008). The branding impact of brand websites: Do newsletters and consumer magazines have a moderating role? *Journal of Advertising Research*, *48*(3), 465–472. doi:10.2501/S0021849908080471

Muñiz-Martínez, N., & Cervantes-Blanco, M. (2009). Identity and marketing of cities. In Gascó-Hernández, M., & Torres-Coronas, T. (Eds.), *Information communication technologies and city marketing: Digital opportunities for cities around the world* (pp. 1–25). Hershey, PA: Information Science Publishing. doi:10.4018/978-1-60566-134-6.ch001

Okazaki, S., & Škapa, R. (2009). Understanding brand website positioning in the new EU member States: The case of the Czech Republic. In Bandyopadhyay, S. (Ed.), *Contemporary Research in e-Branding* (pp. 193–215). Hershey, PA: Information Science Publishing.

Parkerson, B., & Saunders, J. (2005). City branding: Can goods and services branding models be used to brand cities? *Place Branding*, *1*(3), 242–264. doi:10.1057/palgrave.pb.5990026

Perry, M., & Bodkin, C. (2000). Content analysis of Fortune 100 company Web sites. *Corporate Communications: An International Journal*, *5*(2), 87–96. doi:10.1108/13563280010338331

Prats-Planagumà, L., & Camprubí, R. (2009). E-tourism image: The relevance of networking for Web sites destination marketing. In Gascó-Hernández, M., & Torres-Coronas, T. (Eds.), *Information communication technologies and city marketing: Digital opportunities for cities around the world* (pp. 184–202). Hershey, PA: Information Science Publishing. doi:10.4018/978-1-60566-134-6.ch009

Pryor, S., & Grossbart, S. (2007). Creating meaning on main street: Towards a model of place branding. *Place Branding and Public Diplomacy*, *3*(4), 291–304. doi:10.1057/palgrave.pb.6000080

Rogerson, R. J. (1999). Quality of life and city competitiveness. *Urban Studies (Edinburgh, Scotland)*, *36*(5-6), 969–985. doi:10.1080/0042098993303

Short, J. R., Benton, L. M., Luce, W. B., & Walton, J. (1993). Reconstructing the image of an industrial city. *Annals of the Association of American Geographers. Association of American Geographers*, *83*(2), 207–224. doi:10.1111/j.1467-8306.1993.tb01932.x

Simeon, R. (1999). Evaluating domestic and international Web site strategies. *Internet Research: Electronic Networking Applications and Policy*, *9*(4), 297–308. doi:10.1108/10662249910286842

Simeon, R. (2001). Evaluating the branding potential of Web sites across borders. *Marketing Intelligence & Planning*, *19*(6), 418–424. doi:10.1108/02634500110405432

Simmons, G. J. (2007). I-Branding: Developing the Internet as a branding tool. *Marketing Intelligence & Planning*, *25*(6), 544–562. doi:10.1108/02634500710819932

Tschirhart, M. (2008). Evaluation of brand use on city government websites: A search for brand type, repetition, and consistency. *Journal of Nonprofit & Public Sector Marketing*, *19*(1), 35–53. doi:10.1300/J054v19n01_02

Urban, F. (2002). Small town, big Web site? Cities and their representation on the Internet. *Cities (London, England)*, *19*(1), 49–59. doi:10.1016/S0264-2751(01)00045-2

Virtsonis, N., & Haridge-March, S. (2008). Website elements in B2B online communications: A case from the UK print industry. *Marketing Intelligence & Planning*, *26*(7), 699–718. doi:10.1108/02634500810916672

Vitiello, R., & Willcocks, M. (2006). The difference is in the detail: Its potential as a place branding tool and impact upon perceptions and responses. *Place Branding*, *2*(3), 248–262. doi:10.1057/palgrave.pb.5990061

Waitt, G. (1999). Playing games with Sydney: Marketing Sydney for the 2000 Olympics. *Urban Studies (Edinburgh, Scotland)*, *36*(7), 1055–1077. doi:10.1080/0042098993097

Wan, C. S. (2002). The Web sites of international tourist hotels and tour wholesalers in Taiwan. *Tourism Management*, *23*(2), 155–160. doi:10.1016/S0261-5177(01)00048-6

Xi, Y., Zhuang, Y., Huang, W., She, C., & Zhang, Z. (2007). The quality assessment and content analysis of corporate Web sites in China: An empirical study. *International Journal of Information Technology & Decision Making*, *6*(2), 389–405. doi:10.1142/S0219622007002472

Zhang, L., & Zhao, S. X. (2009). City branding and the Olympic effect: A case study of Beijing. *Cities (London, England)*, *26*(5), 245–254. doi:10.1016/j.cities.2009.05.002

KEY TERMS AND DEFINITIONS

Brand: Brand is a name, a term, a sign, symbol or a combination of them that aim to identify the products and services of a seller or a seller group, and to differentiate it from the competitors. A brand identifies the seller or maker, and principally constitutes the promise of the seller to the buyers that it will provide the buyers with the benefits and the services consistently. Brand is a total whole composed of physical, socio-psychological features and beliefs. Products, programs, and services, as well as organizations, organizational units and cities can be branded.

City Branding: City branding can be defined as the application of branding strategies and other marketing techniques in the private sector to the cities. Cities as places are also called as mega products consisting of the combination of tangible and intangible features. City branding is based on the perceptions and images of the people, and these factors constitute the center of the appropriate activities designed to give a shape to the city and its future. Cities obtain an image and identity as a result of their being branded.

City Brands Index: The city brands index which is constituted by Simon Anholt and GfK Roper Public Affair & Media (2009) is based on a global survey in which respondents from across 20 major developed and developing countries are asked to rate their agreement with statements about each of 50 cities. The ranking is developed by averaging city scores across the index's six categories: presence (knowledge of city and perception of its global contribution), place (cleanliness, aesthetic qualities and climate), pre-requisites (affordable accommodations and quality of public amenities), people (friendliness, personal encounters and cultural diversity), pulse (interesting events, activities and lifestyles) and potential (perception as good place to do business, to find a job and to go to school).

City Marketing: City marketing applications are largely based on the structuring, communicating and managing the image of a city. City marketing is related to the image of a city, and this constitutes the starting point of city branding.

City Websites: City websites are aimed to provide information about the city to residents of the city, tourists and investors. City websites, e-services and information relevant to the city offers a visually appealing manner. Therefore, the city websites have become an important channel to reach their customers and the public. These interactive websites also have an important role in the delivery of services offered by the city and city administrators, and attracting the attention of users.

Content Analysis: The content analysis is an observational research method that is used to systematically evaluate the symbolic content of all forms of recorded communications. Content analysis is also a common method to investigate written documents such as web pages.

Website: A group of similar web pages linked by hyperlinks and managed by a business, organization, or person. A website may include text, graphics, images, audio and video files, and hyperlinks to other web pages.

APPENDIX A: CITY WEBSITE EVALUATION FORM

Site URL:			
Country name:			
City name:			
CITY WEBSITE BRANDING CHARACTERISTICS	**1=Yes**	**0=No**	**Explanations**
BRAND DESIGN			
Slogan			
Logo			
Video			
Web cam			
City pictures/picture or photo gallery			
City maps			
City history			
BRAND BEHAVIOR			
News			
Public notices			
Projects			
Plans and/or policies			
Reports			
Event calendar / Events			
Customized information			
Community board members			
Sister cities			
E-services			
BRAND COMMUNICATION			
General Communication			
General information/overview of city			
Customized information for target audiences			
Contact information			
FAQ			
Interactivity			
Online survey			
Online information request form			
Online comment form			
Navigation			
Links to related websites			
Search engines			
Index page/site maps			
Help menu			

continued on following page

Functionality			
Download facilities			
Date last updated			
Language option			
Security and/or privacy info			

Date:

Comments:

Chapter 8
Competitiveness and Creative Cities:
Technologies of Neoliberal Urbanization in Perspective

Evren Tok
Carleton University, Canada

ABSTRACT

This chapter examines the role of technologies of neoliberal urbanization in reproducing new urban spaces that are oriented towards securing a competitive edge. As such, this chapter explores questions about how urban technologies influence the parameters of the reproduction of city spaces and which social representations, actors and networks are involved the process. This chapter contends that competitiveness is not limited to the realm of economics. It also contends that in neoliberal times, "new technologies of power" enable the creation of more polycentric urban areas.

These emerging technologies and applications necessitate an analysis of their place in mobilizing low urban politics. This analysis should be attuned to nuanced articulations of public interests and private initiatives that are enriched by representations of the spatial realities of cities. We are more likely to realize the importance of urban technologies and "street level" everyday urban practices if, rather than treating them as divorced from each other, we ontologically recognize their marriage. Such an approach will help to better diagnose the role that technologies play in moving beyond dualities like public and private.

DOI: 10.4018/978-1-61350-174-0.ch008

Timelines -- Trinity Square Park, Toronto.

...this project chronicles the lives of the homeless through wall- mounted timeline vines in bright green, orange and purple. The timelines, embedded with audio speakers talking about people's connections to the park and their lives, also display photographs of their subjects (opencityprojects.com).

—Aaron Szymanski, OCAD University

Junction Box -- TTC Shelter, Toronto.

...this project proposed turning a bus shelter into an interactive touch screen where Bluetooth-bearing residents and riders can access information about local restaurants, events and music, or add their own thoughts, pictures or videos to the screen's database (opencityprojects.com).

—Calvin Lee, OCAD University

INTRODUCTION

Perhaps, the above-mentioned projects by OpenCity Projects consultancy in Toronto tell us more than anything about the limits of how cities search for creativity. These projects not only stretch the limits of creativity, but also inform us about the neoliberal trajectories pursued by "globalizing" cities and their excessive focus on sustaining competitiveness. This chapter looks at these trajectories by specifically focusing on the interstices of creative cities and competitiveness and the role of played by communication technologies. Speaking of creativity, Richard Florida's "Creative City" (2008, 2006, 2005, and 2002) thesis is a fertile analytical approach for understanding how cities, especially in periods of neoliberal globalization, have become central to the attainment of competitive advantage. Resulting from the internalization of neoliberal values by national governments, local authorities, municipal leaders and other actors, it is now possible to detect a form of "hollowed out" nation state (Jessop, 2002), or the downloading of the capacities, responsibilities and authority of nation states onto cities. Consequently, as Florida's "Creative City" thesis[1] illustrates, the real economic competition today plays out among cities, not nation states. Florida's thesis positions the three T's - Technology, Talent, and Tolerance – as responsible for attracting the "Creative Class", contributing to the national economy and establishing the requisite framework for sustaining competitive advantage. While communication technologies should not be perceived as the only determinant of "creativity", they are indicative of the "new technologies of power" which have arguably become the momentum behind the intensification of a neoliberal ethos in urban spaces. As such, Florida's creative city thesis and its accompanying buzzwords, "creative class" and "creative economy", celebrate the increasing hegemony of the neoliberal market mentality.

Nonetheless, this hegemony is a contested one, albeit in different forms, intensities and technologies. The mechanisms portrayed as making cities more competitive also inform us that, while competitiveness is a key neoliberal concern in the production of urban spaces and creative cities, innovative technologies work in two directions: (1) deepening the hegemony of markets by marshalling creativity and communication technologies towards the deeper entrenchment of market logic; and (2) impacting the way urban space is imagined, instituted and produced by actors outside the realm of the state. This chapter will focus to a greater extent on the latter dimension, which acknowledges that communication technologies make use of a new conception of space that is not under the hegemony of state actors, but influenced by non-state, civil actors as well. Here, it should be noted that the second conception of communication technologies is not at loggerheads with the first, and does not negate the role of markets in capitalist societies. In fact, it indirectly supports Florida's creativity thesis by showing that urban communication technologies, when utilized by private non-state actors, civil organizations and non-governmental initiatives, also generate creativity.

After explaining these theoretical accounts in more detail, this chapter will proceed to examine the role of communication technologies and new technologies of power in reproducing new urban spaces that are oriented towards gaining competitive advantage. Three key questions will motivate our discussion: "how are technologies instrumental to producing new city spaces; how do technologies influence the parameters of the reproduction of city spaces; and which social representations, actors and networks are involved in this process". These questions can be approached both theoretically and empirically. As such, instead of offering only an in-depth analysis of case studies, this chapter will both delineate the central theoretical issues and subsequently assess examples from Toronto to better demonstrate the use of new technologies of power by non-state agents in efforts to increase city competitiveness.

The examples from Toronto, Ontario, Canada will comprise the empirical section of this chapter and both will refer to the critical role of technologies in reproducing the city space in accordance with the changing rules of the increasingly neoliberal market economy. Both will also show that efforts to foster more creative cities can benefit from innovative technologies. While public and/or private actors can be the drivers such efforts, Toronto is a city of non-state actors equipped with efficient forms of information communication technologies who, by using strategic design, grapple more effectively with common urban challenges, such as safety, diversity, sustainability, aesthetics, and design spaces that people use and care about. Even though, in many cases, the neoliberal urbanization experience is deprived of effective civil actors equipped with strong organizational capacities who are influential in making use of information communication technologies to address urban problems. OpenCity Projects as an exemplary analysis will pave the way for future studies of a comparative nature and will establish the ground for a platform of learning and policy-transfer among cities, and inter-urban networking.

NEOLIBERAL URBANIZATION IN PERSPECTIVE

First, let us start with what neoliberalization means for the city and why the city is itself important to processes of neoliberalization? Neoliberalization is a range of policies intended to extend market discipline, competition and commodification throughout society. It is a search for securing the "vital cycle of economic growth". Neoliberal mindset needs quality economic foundations, such as quality of life, financial capital, business climate, so and so forth, reassuring Florida's main thesis so that new people and ideas in the form of creativity fosters this cycle. In the current context, neoliberal doctrines have brought about the deregulation of state control over industry and markets, assaults on organized labor, the downsizing and/or privatization of public services and assets, the dismantling of welfare programs, the enhancement of international capital mobility and the intensification of interlocality competition. Its implementation has relied on national restructuring projects (Peck et al, 2009), and the dominance of competitive logic over redistributive objectives has opened up new spaces of contestation among economic actors, institutions, municipalities, nation states and civil actors. This has had significant implications for localities given the risks and responsibilities of downloading.

When we stand back and try to decipher the dynamics behind the increasing hegemony of neoliberalism in redesigning, restructuring and reproducing urban spaces, it becomes possible to conceptualize this phenomenon as a state strategy for creating new conditions of capital accumulation. However, as Brenner and Theodore (2002, pp.

120-130) note, although neoliberal projects take place at different and tangled scales, it is cities and regions in which the contradictions and tensions of actually or already existing neoliberalism are manifested. Cities become the venues on which these tensions and contradictions are concretized and managed (or attempts at management of these tensions and contradictions are made). There is now, for example, increasing city level competition across national borders for foreign direct investment and for trade in city based services and urban assets. Increasingly, urban economic performance matters more than national economic outcomes.

These dynamics correspond to 'a gradual shift away from distributive policies, welfare considerations, and direct service provision' and mark the competitive state's movement 'towards more market-oriented and market dependent approaches aimed at pursuing economic promotion and competitive restructuring' (Swyngedouw *et al.* 2002, p.200). Through this process, supply-side economic logic privileges the production of urban 'spaces of spectacle' in which the consumption of, and access to, goods by the urban population exists as a source of political motivation for neoliberal ideas. In this sense, rather than attempting to shift the social inequities brought about by the neoliberal political economy, policy concerns are directed towards ensuring urban centers appear attractive to mobile-capital and elite consumers with the explicit goal of improving "the tax basis of the city via a sociospatial and economic reorganization of metropolitan space" (Swyngedouw *et al.* 2002, p. 204).

This situation, however, should not lead us conclude that this systemic dynamic unfolds without agency and that the attainment of economic competitiveness is disembedded from other spheres of life. Competitive embeddedness within a society and cultural context does not necessarily mean that competitiveness is undervalued. In contrast, embeddedness provides institutional thickness and engenders sociospatial dynamics that mobilize urban actors for the purpose of attaining competitiveness. Competitiveness itself is a socially produced phenomenon and in neoliberal times, as Mahon and Keil highlight, it is the urban scale as well that is "socially produced and reproduced through myriad, sometimes purposeful, sometimes erratic social, economic, political and cultural actions" (Mahon and Keil, 2006, p.5).

This increasing emphasis on the rise of cities and urban centers as the new spaces of capitalist accumulation (Brenner, 2004) also brings into the fold a variety of new or more diverse actors that collectively produce and reproduce urban spaces. Thus, the rise of cities and the way in which cities are perceived as the new engines of neoliberal capitalism should be seen as a product of agency. Perhaps, the centrality of agency can be better perceived if we take into account a sense of neoliberalism not only as prescriptions and strategies imposed from "outside" or "above", but also as a push from "below." This latter dimension enables the possibility of thinking about neoliberalism as involving processes that produce spaces, states, and subjects in multiple forms and accordingly generates the "freedom", "empowerment", and "choice" that are rarely acknowledged and even more rarely theorized (Larner, 2003).

This leads us to approach the attainment of competitiveness as part of the urban socio-spatial dialectic that is not limited to the realm of economics, but that exists in the realm of culture as well. For this reason, as the two cases below will demonstrate, the use of new technologies of power is an effective tool for analyzing this intermingling and articulations between economy and culture so that they become instigators of the processes reproducing the urban space jointly through high and low urban politics. Viewed in this way, communication technologies build a crucial linkage between "high urban politics" and "low urban politics." If we concentrate on this linkage, questions such as "how are decisions about cities' social and economic futures made, which actors or institutions are empowered to make these decisions, and whose interests are central to these decisions" gain importance. Here it is critical that we perceive not only urban growth coalitions, but also other social forces beyond the urban scale, as relevant.

In considering the strategies of urban politics in the city, when the city is seen primarily as an economic community and a spatial actor, much attention is then given to local governments and notions of hierarchy – that is urban "high politics." There are, however, important unnoticed fields of urban low politics unfolding simultaneously. The fields of urban low politics are mostly constituted by the activities of civil actors and non-state organizations established and institutionalized around certain traditions of participation or social mobilization at a variety of scales. These involvements bring people together in a variety of informally or more formally organized associations. This means that there is an enlargement of the field of urban politics, including the construction of spaces of identification that go beyond the politics primarily concerned with economic strategies of rescaling. This involves imagining the construction of a city as a multiplicity of lived places. Central to such constructions are the different city visions or narratives that appear as points of discursive reference in the creation of different strategies of urban development.

This understanding advocates a "multiplex" city that should be understood in terms of a density, coexistence and difference that characterize the city and its multi-layered activities. As this involves issues related to low politics, we need to be more sensitive to the everyday participatory practices situated within civil society and actively involved in the creation of urban spaces. In other words, spaces of engagement are actively produced through everyday participation and dynamics of the spaces of dependency. This is a bottom-up effort in that rather than a top-down hierarchy enforcing or mobilizing interests, the participation of a variety of different actors and social groups contributes to creating the capacity and collective effort needed to construct and reconstruct the city space. Here, low politics becomes the cornerstone of urban politics. In a way, looking at the dynamics of low politics at the urban scale helps us to better address questions such as "when do spaces of dependence turn into spaces of engagement" and "through which processes does this transformation unfold". These questions are relevant to the internal dynamics, which lead local agents to mobilize and search for strategies and spaces of engagement at extra-local scales, and pertain to the role of new urban technologies as a vital element in these processes.

The importance of low politics also manifests itself in the idea of constructing the city from the perspective of everyday practices or "scaling from below". Florida's creativity theory finds a strong connection between places that successfully harness talent and technology and places that are open to immigrants, artists, gays and lesbians and racial integration. It rests on the capacity of everyday experiences to influence the structural dynamics of the urban space. In other words, low urban politics play a "decisive role in constructing the rules of interlocal competition' and are thus distinct insofar as they shape 'the very metrics by which regional competitiveness, public policy, corporate performance or social productivity are measured" (Peck, 2001, p.40).

There is no doubt that new urban technologies are important in harnessing creative capacities for the sake of producing more vibrant, more competitive and more livable cities. The role of communication technologies is indicative of structural dynamics in relation to the unfolding of the neoliberal vocabulary, which asserts that cities matter for accumulation of capital. Florida's creative city thesis, however, goes one step further and emphasizes that it is not just that cities matter, but that particular strengths and qualities of cities matter. There are greater incentives for cities and city regions to actively promote positive locational characteristics, including patterns of linkage and innovativeness, which can then assist cities in attaining competitive advantage and responding flexibly to changing market conditions. (Florida, 2006, p.11)

Urban Space and Communication Technologies (CTs)

Having indicated that competitiveness is not limited to the realm of economics and that new technologies of power in neoliberal times designate more polycentric urban areas, it is now important to bring the role of communication technologies in mobilizing low urban politics into the discussion. In other words, to examine the everyday life or the dynamics of the city as a lived space in order to demonstrate that the search for competitiveness is being articulated to these rhythms. While political power is perhaps more dispersed than ever, the questions remain, "how is it dispersed and how do non-state actors become involved in instituting this process". Broadly, we are interested in how communication technologies penetrate the interstices of economy and culture and how this involvement produces competitiveness. The following quote offers an intriguing starting point for approaching these questions:

The city is a result of many players and forces that operate through global, regional, and local spheres of influence, utilizing tools and techniques that are permanent, ephemeral, invisible, and strategic. They are knitted together through both real and virtual networks of physical connection, telecommunications, social relationships, and political positioning (Page and Phillips in Firmino et al, 2008, p.80).

As Castells contends, "the unfolding promise of CT opens up unlimited horizons of creativity and communication, inviting us to the exploration of new domains of experience..." (1991, p.4). Castells then encourages us to explore these domains, both as the emergence of new modes of socio-technical organization (which he calls the informational mode of development), and through the restructuring of capitalism, which involves significant economic and institutional dynamics and constitutes a new stage in capitalist development (Grondlund, 1999).

Castells' title, "Informational City," captures a dual understanding. On one hand it recognizes that the city, via increasing interactivity on a global scale, resembles the image of society. On the other hand, this image is composed of the "cyberspace of the city" and "non-cyberspace city – the city of streets, facades, bodies in the public space and real life activities" and it is this latter dimension which can be seen as information interfaces and information processes. To put it differently, information communication technologies and "street level" everyday urban practices are not divorced from each other. At this point it is pertinent to examine how the image of the city in neoliberal times is projected as a venue of diverse processes, asymmetries, struggles, contradictions and contestations internalizes communication technologies, and more importantly, to examine its sociospatial and socioeconomic representations.

As characterized by Ghere and Rismiller (2010, pp. 39-41), we can approach the influence of new forms of CT on the character of an urban fabric in multiple ways. Broadly, these forms make up the types of CT strategies (instruments), the desired functions that require augmentation, such as local government or community, and who will have access to these technologies – corporate entities, individuals, neighborhood associations, workers? Ghere and Rismiller identify four areas of convergence between the sociospatialities of cities and CTs. Here, the sociospatial dialectic of the city is primarily related to forms of societal interaction, participation, and imagining the city as a form of social collaboration and deliberation.

First, this societal collaboration and grassroots input is desired as a strategy of the devoluted state to increase its effectiveness in local governments. Thus, maximum participation is desired from a "top-down" perspective. Secondly, CTs influence the traditionally established patterns of "bureaucratic authority relationships." Thirdly, and not only within the realm of bureaucracy, CTs designate new mechanisms of interaction between governments and people by infusing new forms of "cyber power" into established patterns of state-society relations. Perhaps, concepts such as cyber-democracy, e-government, and wired government have successfully captured this trend. Lastly, as Ghere and Rismiller (2010, p.40) emphasize, CT opportunities involve democratic ideals, aspirations and opportunities for localities, local communities and networks, all of whom can benefit from these technologies by finding a "voice" and "capability" in being part of the city. In other words, CTs can more effectively generate voices "from below" and engender more participatory, more democratic and more civic urban spaces.

The key dilemma in this context appears to be the ways in which the above-mentioned "from below" input is accommodated and establishes a wider framework for increasing participation. At this point, it is important to note that there emerges a series of issues pertaining to the questions "whose technology for whom" and "whose technology for what" (Ghere and Rismiller, 2010, pp. 51-52). These questions are beyond the scope of this chapter, but it would suffice to note that from a political economy perspective, the interests of capital and the interests of communities retain their complexity, although the former is still privileged over the latter. In terms of the capacity of CTs to generate progressive change and offer a solution to urban social problems, it may be unrealistic to expect local governments, who are in the midst of fiscal retrenchment, budget reductions and an increasing reliance on neoliberal modes of attaining competitiveness, to privilege the diffusion of CTs for wide-spread community involvement, participation and engagement.

The key question then is whether it is government, local apparatuses or regional establishments that bring forward network systems, CTs and enhanced social innovations. Or is it the capability and willingness "from below" that matter here, or perhaps both? In fact, non-governmental, privately owned, and civil-society-oriented efforts do exist and suggest that it is possible to utilize CTs that are tailored by these efforts. These initiatives are critically important to understanding the multifaceted relations between CTs and urban communities.

Such relations revert our attention back to the ways through which cities, as complex spatial entities, involve integrations of physical space and cyber space. The restructuring and redesigning of cities under neoliberalism are intensively informed about these integrations. As cities experience significant social, political and economic transformations, these articulated and integrated spaces offer crucial insights into the reproduction of urban space, the design of urban policies, planning the prosperity of cities and, most importantly, the rethinking of city public spheres. Our discussion so far has hinted at the fact that these augmented public spaces are not under the sole supervision of the state. While local authorities are still agents in the provision of infrastructure and institutional platforms for the unfolding of CTs for the sake of building sustainable communities, reaching more participatory and deliberative forms of democracy, and planning urban spaces, this reproduction of urban spaces demands other vocalities.

These attempts at producing and reproducing urban spaces in the augmented city, as Firmino indicates, require a delicate balance between commercial and civic interests and the interfaces between them. Although it is true that commercial interests are privileged by city administrations, as the interests of capital are entrenched in the minds of municipal decision makers, our discussion implies that "the way municipalities incorporate information and communication technologies in their planning processes and their urban agendas could indicate how aware urban planners are of the augmented city and how they face the challenges of debating and planning urban environments where physical and electronic elements are intermingled" (Firmino *et al.*, 2008, p.80). Recognizing the "from below" dimension in this context urges robust interpretations of augmented spaces in urban spaces that are produced-reproduced in relation to both this dimension and the sociospatial realities of cities. CTs, doubtlessly, enact new meanings and representations of these realities in neoliberal times. That which we have earlier described as the "new technologies of power" refers to these new meanings and representations that have become visible in cities.

In fact, the social construction of urban spaces now needs the "from below" dimension and increasingly, the pervasiveness of CTs is being considered a significant pillar in socially constructed urban prosperity. It could also be argued that attaining competitiveness in a neoliberal era demands a more intensive promotion of public interests by private actors, since the urban space is a social construction and this construction should not be without the input of "everyday practices" and local-cultural characteristics.

In summary, urban planners and metropolitan governance mechanisms need civil inputs, which are now (more than ever) equipped or transmitted through well functioning CTs with enriched representations of the spatial realities of cities. These technologies are brought into the picture by actors in a way that reflects social and cultural embeddedness, which in turn has important implications for the design and implementation of urban policies. While the impacts and importance of CTs for the production and reproduction of urban spaces needs further analysis and empirically enriched accounts, the remaining thrust of this paper is to look more closely at case studies demonstrating the role of non-state actors in filling the gap between urban planners, municipal administrations, metropolitan governance mechanisms and other urban decision makers.

Before delving into these cases, we should also note that Florida's understandings of creativity, and his broadly endorsed indexes, operationalize this concept as a vehicle for sustaining economic growth in conjunction with the strategies of urban growth coalitions. Nevertheless, what has been portrayed as the role of social and cultural embeddedness, and the penetration of CTs in the interstices of culture and economy, indicate that creativity might take different trajectories, and expose different forms and articulations in different localities. Perhaps, it would not be unfair to argue then that Florida's "elite centered" creativity making-creativity sustaining story could be enriched with what is happening "on the ground" through the everyday practices of urban spaces. Consequently, it is clear that we need to be sympathetic to these "bottom up" or "from below" dynamics bolstered through non-elites.

TORONTO'S OPENCITY PROJECTS

An alternate title for this subsection could be "what can be learned from Toronto's urban politics?" Even though the political economy of Toronto's neoliberal experiences exhibits complex patterns, for the sake of illustrating the theoretical underpinnings outlined earlier, this subsection will be limited to a narrow aspect of these dynamics. In light of our previous discussions, this subsection will concentrate on the sociospatialities of cities, especially with respect to the above-mentioned claim that urban planners and metropolitan governance mechanisms need civil inputs, which exhibit nuanced articulations between public interests and private initiatives and enriched representations of the spatial realities of cities.

In this case the focus will be on the use of technologies by OpenCity Projects, a Toronto-based consultancy that uses a unique approach, informed by research and public engagement, to view place through the eyes of people who use it and create more inclusive, vibrant public spaces that help promote a sense of community. From this perspective, OpenCity Projects fills a gap between urban planners, architects and developers. What kind of an urban space does OpenCity seek to produce? How does it situate itself within the sociospatialities of Toronto and how does it relate to the new technologies of power? How does such an initiative become part of urban realities and take part in the reproduction of urban space? How does it situate itself within the interstices of economy and culture? Perhaps, the overarching question that encompasses this series of sub-questions is how do OpenCity Projects enter into the interstices of economy and culture in imagining new urban spaces? Before engaging with this question, it is equally important to explain how OpenCity Projects works, what are the technical issues involved and how does the process works?

OpenCity Projects defines its mission as working with prospective clients that involve the challenges, but with the idea in mind that these challenges could also be considered as opportunities in the later stages. Therefore, OpenCity Projects identify partnerships that combine the right expertise and viewpoints. The process involves understanding the current context, place, assets, and stakeholders through desk research, asset audits and precedent research. The next stage for OpenCity Projects is to verify the research and understand the situation from all relevant stakeholder's point of view. In other words, "people who will use and influence the experience" and become progenitors of urban transformation, initiated and implemented both virtually and physically. Using a range of research tools tailored to the challenge, OpenCity Projects engages stakeholders to understand their needs and also expectations through interviews, journaling and ethnographic research. Based on research and engagement insights, OpenCity Projects creates an experience by innovating, improving, and combining information gathered. This is broadly defined as OpenCity Projects' process.

What is striking about this process is that, it positions creativity as a public good, and this basically means that creativity is sought in a collaborative consultation process. Similar to Florida's arguments presented earlier, it is conveyed that cities need creativity, and it is the task of OpenCity Projects to help foray the creativity that is inherent in the urban space, with its socio-spatial dynamics. Therefore, the projects seek for the potential for creativity, and turn it to real urban policies, utilize from the spill-over impacts of creativity and in turn, using creativity for the public. As mentioned in the OpenCity Projects web site, "creativity belongs to everyone". An important implication of this phrase is that, as previously mentioned, creativity is a public good and the subtly of this public good is that, it is a product of both public and private interests. It has a "fluid" character, which is made possible through CTs and their catalyzing impacts. This in turn makes people centered designs possible and through the use of CTs, OpenCity Projects helps Torontonians care about their cities. Also, helps them be more vocal and influential about the fortune of urban spaces. Whether initiated by a public authority or a private agent, the virtual space created by-OpenCity Projects designate a crucial role for "*city*.zens" to enhance the identity of their cities in a bottom-up fashion.

I prefer to call them "*city*-zens" with emphasis on the "city" due to the fact that cities such as Toronto are considered as global cities. Their role in global trade and investment flows, finance, culture and politics is significant. The urban identity is a developed one and often city identity transcends the national one. To stretch further, one can argue that city identity could be seen as a new form of citizenship. In the neoliberal era, the increasing role and function of cities in producing economic growth, local governments being more and more responsible from redistribution and recognition of diversity issues, cities have been downloaded with significant responsibilities. Especially in a city like Toronto, as the most densely populated city of the Ontario province, the city could be considered as Canada's global face, jumping from the urban scale to international, supra-regional, transnational and global scales. Doubtlessly, Toronto's CT infrastructure and the ways through which it is reflected on urban spaces is an important stimulant in this process. Online mobilization and as in the case of OpenCity Projects, establishing online platforms, such as blogs and Facebook groups are interpreted as the rising virtual spaces which are used for furthering "*city*.zen" participation and inputs. In other words, this dimension explains the "dot" between "city" and "zens". As it is used in the world wide web, the "city" being the domain and the "zen" being the extension, as if "city" becomes the central substance of national identities.

Overall, the city benefit from its virtual richness, which is created by CTs and private initiatives such as OpenCity Projects.[2] Initially, let us consider the formation of OpenCity Projects' "virtual" urban space. The virtual space created by the OpenCity Projects has a special role in increasing awareness and consciousness among Torontonians. The web site of OpenCity Projects is organized in such a way that it allows for receiving inputs of the community members, youth and various other social actors. For instance, the Facebook group and the web site host many urban related discussions, new municipal arrangements and policies that affect urban spaces. The OpenCity Project web site, with slightly more than hundred thousand hits since it is founded in 2006, provides an avenue, especially via its Facebook group (almost 800 members), which reveal that Torontonians can have more meaningful places and spaces, as well as experiences through strategic design. As it is argued by the founders of OpenCity Projects, communication technologies are crucial tools to address urban challenges that arise and it is the task of OpenCity Projects to combine viewpoints and partners and create people-centered experiences. One of the striking features of this approach is that, it can easily flow across virtual and physical spaces, affect policies, impact public opinion and transform places. More importantly, it is a tool for empowerment for the public, a bottom-up and democratic mechanism for sustaining livelihoods indicating that urban space possesses power, which could be initiated by different social forces. It can commence from anywhere, and it can encompass any issue within the boundaries of the urban realm that has impacts on the "*city*.zens". For instance, most recently, the debate on the OpenCity Projects web site and the Facebook group on the "politics of bike sharing program", entitled as BIXI, indicated that the online tools provided by OpenCity Projects and Facebook group work efficiently, and it is possible to build linkages and articulations between public and private realms such that urban space is better equipped with more effective designs and utilize more from technologies.

The role of OpenCity Projects in articulating private and public forms of power involves partnerships and collaborations with organizations ranging from the City of Toronto's Urban Development Services to OCAD University (OCADU) and Luminto. The discourse of OpenCity Projects rests on its focus on people-centered experiences. In an augmented city, these experiences show that physical and virtual spaces interact and intermingle, which in turn, structure new urban realities, everyday practices and urban forms. Broadly, the sociospatiality of the city is being transformed. By penetrating the augmented spaces, that is the intermingling of physical and cyber dimensions, these projects engender new spatial imaginaries that are not necessarily implemented in a top-down fashion, nor do they resemble contestations to the restructuring of cities according to neoliberal capitalist ideals. These projects have a mediating role between the public and the private, between physical urban forms and the informational city, and between pure economic motives and social/cultural concerns. Therefore, rather than thinking in terms of dualities, we grasp a sense of the city in terms of articulations, interminglings, and co-constitutions, which find existence in new forms of technology, and the changing technologies of power.

One recent project, Toronto Unbound, illustrates the ways in which everyday practices, such as daily needs, neighborhood experiences and local issues can be addressed. In partnership with OCADU and The City of Toronto, OpenCity Projects turned the classroom into a creative lab working with design students to identify opportunities for neighbourhoods enhancement. OpenCity Projects equipped students with a range of research tools to immerse themselves in neighborhoods and understand residents' needs. Research tools included desk research to learn about neighbourhood demographics and history as well as site visits and observation in urban public spaces. Students conducted one-on-one interviews with key stakeholders, including residents, small business owners and students, who will use the design solution. Based on research and engagement insights, they developed personas representing fictional portraits of people interviewed and mapped stakeholders' journeys in neighborhoods to help pinpoint opportunity areas (opencityprojects.com). What makes these efforts significant is twofold. First, a collaboration with OCADU students paved the way for the design of experience solutions and the secondly, a special jury composed of partners from The City of Toronto's Urban Development Services team and the Office of the Mayor selected the top six projects, 3 of which were implemented in 2009-2010. These projects demonstrated that improving the quality of life in urban neighborhoods is feasible and the "from below" dimension, which is in this case composed of students and neighborhood residents, could be articulated to both the interests of the public and private, with the city administration acting as the catalyst in the process.

The case of Toronto and the way OpenCity Projects unfold in the urban landscape should not limit us thinking beyond Toronto. In other words, OpenCity Projects provide insights for us to travel across multiple urban spaces and have varying resonance in different contexts. For instance, one recent ongoing OpenCity project pertains to the innovation of public space design, as illustrated in Figure 1. This specific project informs us regarding the use of CTs in building more sustainable communities. The use of CTs, as the case of OpenCity Projects highlights, could foster public engagement and deliberation. In the case of the park design in progress, the strategy as mentioned in the web-site, relates to an 8-acre park as part of a new neighborhood development in the Town of Markham. This location is one of Canada's most culturally diverse communities. Working with Markham's Planning and Urban Design Department and a landscape architect partner, OpenCity Projects challenged traditional planning standards to create an experience that is relevant across cultures for a population that is 57% foreign born. (www.opencityprojects.com). Strategic design and the use of CTs would support the planning process, make it more sensitive to the needs and expectations of the local inhabitants, but more importantly, envisages a new understanding of planning that would reflect the local realities.

This project emphasizes an important characteristic of OpenCity Projects, which is the articulation of public and private interests through urban spaces. In fact, this could be considered as a key quality of OpenCity Projects in general. For instance, in the case of Markham, the good neighbor program of the Town of Markham as elaborated in its official web site is an indicator that urban spaces need social cohesion. It is further signaled that providing social cohesion in urban spaces need "good" urban planning. In this context, Opencity Projects and the Town of Markham aspire to innovate public space design so that diversity of the community, along with "good neighbor policy", jointly strive for designing urban spaces that are sensitive to the needs of the public.

It is also important to note that, due to Canada's multicultural constitution and being a country of immigrants, as well as the socio economic level reached, local initiatives, especially stemming from the public realm have better opportunities to wind up with CTs. Especially many cities from the developing country context lack sufficient infrastructure to bridge local needs and the use of CTs. Furthermore, the national or federal structures

Figure 1. Source: www.opencityprojects.com

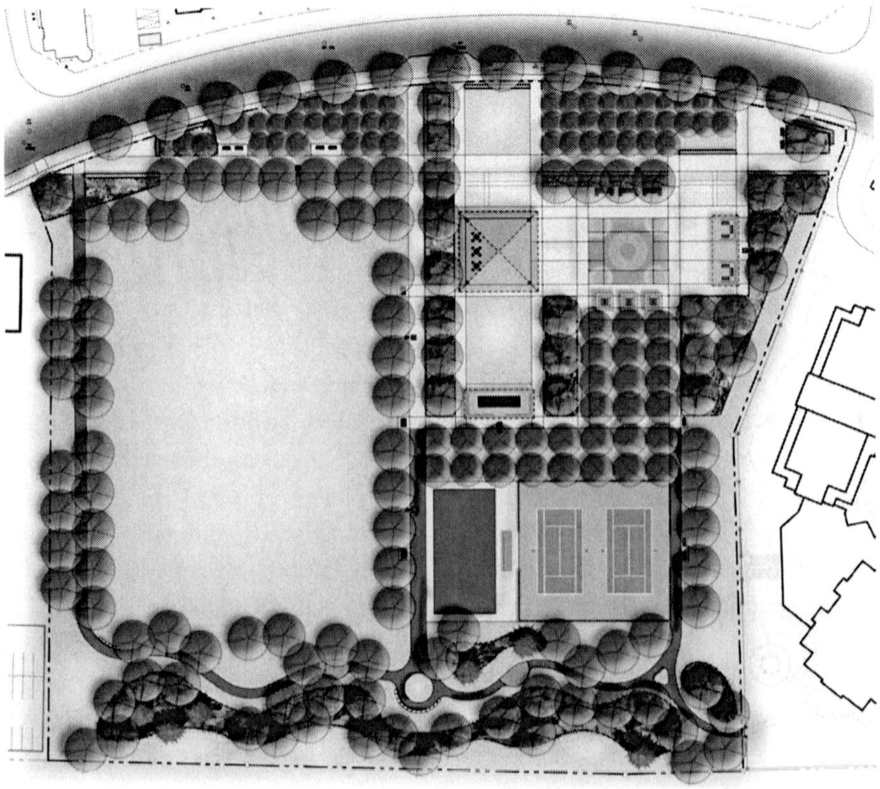

should support local attempts. Overall, it should be recognized that the CTs are not always convenient for local civil actors to implement and integrate into their claims pertaining to their civil rights, identities and the prospects of their cities. One of the influential mottos in OpenCity Projects posits that Toronto is a city, "unfinished and full of possibility" – similar to many other globalizing cities, whether they are located in North America, Europe, Asia, South America, Africa or Australia. Urban spaces and CTs render possibilities for different social segments, operate as a meeting place for public and private interests and help unleash potentials for creativity, co-existence and better governance.

CONCLUSION

Both the "Timelines" and "Junction Box" projects highlighted at the beginning reveal that CTs can be vehicles for recognizing the different sociospatial dialectics within cities. Of course, OpenCity Projects' work is not limited to that discussed here. As a collaboration of students, urban neighborhoods, municipalities and private capital, OpenCity Projects reveals that the way technology is used by the information society varies. While CTs are vital parts of daily experiences, they might play other roles as capitalism restructures and reproduces itself in urban spaces as they are indicative of broader changes and transformations in terms of the escalating prominence of new technologies of power. We must be more attuned to the sociospatial characteristics through which CTs are exposed in cities, not only for the sake of the "more and more economic growth" motto, but also because of their embodiment of social and cultural meanings. Creative cities in this sense are those that better identify and concretize these potentials and OpenCity Projects are illuminating examples in this regard.

By identifying what we have called nuanced articulations of public interests and private initiatives that are enriched by representations of the spatial realities of cities, we are more likely to realize that CTs and "street level" everyday urban practices are not isolated phenomenon. Rather, their marriage helps us to better diagnose the role that information and communication technologies play in going beyond dualities such as public and private, and physical and cyber. It should also be noted, especially for the purposes of urban planning, that how CT dissolves these dualities and fertilizes diverse articulations is indicative of the ways in which the actual use of CT is not isolated from daily lives of different social segments. As OpenCity Projects and similar cases illustrate, there is a strong "bottom up" or "from below" component. Social and communitarian use of CTs then not only fill the gap between urban planners, architects and developers and address common urban challenges, but play a key role in fostering democratic ideals through neighborhoods, communities and in general everyday practices. This may hold the most significant potential for the transformative power of urban realities, as Henri Lefebvre eloquently portrayed, "urban politics is a dynamic and thoroughly contradictory social space, caught up in the contradictions between the macrostructures of capital and state and the microworlds or everyday life, urban politics is no mere local affair, but rather multiscalar, potentially universalist and most importantly transformation" (Lefebvre in Keil, 2002, p. 586).

Perhaps, understanding what everyday life means for the city and internalizing the concerns of different voices in the city now becomes as critical as seeing the city as an economic growth machine through a neoliberal lens. Urban planners, metropolitan decision making mechanisms and city administrations should take into consideration the notion that information and communication technologies are not only important for increasing economic returns and maximizing urban rent, but also for their vital role in the everyday lives and livelihoods of ordinary citizens and building more inclusive and cohesive neighborhoods and communities, which nowadays constitute key nodes in livability indexes and creative city indicators.

REFERENCES

Castells, M. (1991). *Informational city: Information Technology, urban restructuring and the urban regional process*. Oxford, UK: Blackwell Publishers.

Cox, K. R. (1998). Spaces of dependence, spaces of engagement and the politics of scale, or: looking for local politics. *Political Geography*, *17*(1). doi:10.1016/S0962-6298(97)00048-6

Evans, G. (2009). Creative cities, creative spaces and urban policy. *Urban Studies (Edinburgh, Scotland)*, *46*(5-6). doi:10.1177/0042098009103853

Firmino, R. J., Duarte, F., & Moreira, T. (2008). Pervasive technologies and urban planning in the augmented city. *Journal of Urban Technology*, *15*(2). doi:10.1080/10630730802401983

Florida, R. (2002). *The rise of the creative class*. New York, NY: Basic Books.

Florida, R. (2005). *Cities and the creative class*. New York, NY: Routledge.

Florida, R. (2006). *The flight of the creative class*. New York, NY: HarperCollins Publishers.

Florida, R. (2008). *Who's your city?*New York, NY: Basic Books.

Ghere, R. K., & Rismiller, L. S. (2001). Information Technology's potential to improve urban neighborhoods: Some citizen planning dilemmas'. *Journal of Urban Technology*, *8*(2). doi:10.1080/106307301316904781

Glaeser, E. L. (2000). The new economics of urban and regional growth. In Clark, G. L., Feldman, M. P., & Gertler, M. S. (Eds.), *The Oxford handbook of economic geography*. Oxford, UK: Oxford University Press.

Grundy, J., & Boudreau, J. A. (2008). Living with culture: Creative citizenship practices in Toronto. *Citizenship Studies*, *12*(4). doi:10.1080/13621020802184226

Hamel, P., Boudreau, J. A., Keil, R., & Jouve, B. (2006). Arrested metropolitanism: Limits and contradictions of municipal governance reform in Los Angeles, Montreal and Toronto. In Heinelt, H., & Kübler, D. (Eds.), *Metropolitan governance in the 21st century: Governing capacity, democracy and the dynamics of place*. Milton Park, UK: Routledge.

Jessop, B. (2002). Liberalism, neoliberalism and urban governance: A state-theoretical perspective. *Antipode*, *34*(3). doi:10.1111/1467-8330.00250

Johansson, B., Karlson, C., & Stough, R. (Eds.). (2006). *The emerging digital economy: Entrepreneurship, clusters, and policy*. Berlin, Germany: Springer-Verlag.

Keil, R. (2002). Commonsense neoliberalism: Progressive conservative urbanism in Toronto, Canada. *Antipode*, *34*(3). doi:10.1111/1467-8330.00255

Kim, T. J. (2008). Planning for knowledge cities in ubiquitous technology spaces: Opportunities and challenges. In Yigitcanlar, T., Velibeyoglu, K., & Baum, S. (Eds.), *Creative urban regions: Harnessing urban technologies to support knowledge city initiatives*. Hershey, PA: IGI Global. doi:10.4018/978-1-59904-838-3.ch013

Laguerre, M. S. (2005). *The digital city. The American metropolis and Information Technology*. New York, NY: Palgrave MacMillan.

Larner, W. (2000). Neo-liberalism: Policy, ideology, governmentality. *Studies in Political Economy*, *63*, 5–26.

Lefebvre, H. (1991). *The production of space*. Cambridge, MA: Blackwell.

Lewis, N. M., & Donald, B. (2010). A new rubric for Canada's creative cities. *Urban Studies (Edinburgh, Scotland)*, *47*(1). doi:10.1177/0042098009346867

Mahon, R., & Keil, R. (2006). *The political economy of scale: An introduction* (manuscript).

McCann, E. (2007). Inequality and politics in the creative city-region: Questions of livability and strategy. *International Journal of Urban and Regional Research*, *31*(1). doi:10.1111/j.1468-2427.2007.00713.x

OpenCity Projects Inc. (2007). *Town of Markham* (case by W. Gold, & J. Rutgers). Retrieved on November 29, 2010, from http://www.opencity-projects.com

Peck, J. (1998). Geographies of governance: TECs and the neoliberalisation of local interests. *Space and Polity*, *2*(1). doi:10.1080/13562579808721768

Peck, J. (2007). Banal urbanism: Cities and the creativity fix. *Monu*, *7*, 35–47.

Peck, J., Theodore, N., & Brenner, N. (2009). Neoliberal urbanism: Models, moments, mutations. *SAIS Review*, *29*(1).

Peck, J., & Tickell, A. (2002). Neoliberalizing space. *Antipode*, *34*(3). doi:10.1111/1467-8330.00247

Polanyi, K. (2001). *1957). Great transformation: The political and economic origins of our time*. Boston, MA: Beacon Press.

Sands, G., & Reese, L. A. (2008). Cultivating the creative class: And what about Nanaimo? *Economic Development Quarterly*, *22*(1). doi:10.1177/0891242407309822

Sohn, J., Kim, T. J., & Hewings, G. (2004). Intra-metropolitan agglomeration, Information Technology, and polycentric urban development. In Cappelo, R., & Nijkamp, P. (Eds.), *Contributions to economic analysis*. Amsterdam, The Netherlands: Elsevier.

Swyngedouw, E., Moulaert, F., & Rodriguez, A. (2002). Neoliberal urbanization in Europe: Large-scale urban development projects and the new urban policy. *Antipode*, *34*(3). doi:10.1111/1467-8330.00254

ENDNOTES

* Special thanks to Wendy Gold, President of OpenCity Projects, for her patience, support and guidance for the case study section of this chapter. I also would like to thank two anonymous reviewers for their constructive comments and Erika Shea editing and commenting on the earlier drafts. Finally I would like to extend my gratitude to Yelda Gulderen Tok for her support through out the process.

1 The creativity of cities is attributed to various factors, including openness to diversity, level of tolerance and the peaceful coexistence of differences in a society. Levels of technological innovation and a population comprised of talented and innovative people are also believed to have a direct impact on the creativity and hence, development of a city. In his latest book, Florida (2008) asserts that people can now choose the cities in which they want to live in, and this choice is shaped by the characteristics of the cities. In an earlier text (Florida, 2002), two population indexes are developed. The creativity index measures the number of artists, musicians, and painters living in a city, (also called the bohemian index). The diversity index is itself composed of two indices, the first one is the melting pot index, the number and ratio of foreign born people and immigrants, and the second is the gay and lesbian index, measuring the tolerance of the community, as well as the region.

2 It should be noted that even though CTs are integral elements of creating virtual spaces, the role of OpenCity Projects should be contextualized within a broader understanding of technologies. More specifically, even though OpenCity Projects benefit from the CTs, their role has important implications in relation to the changing and transforming forms of urban governance.

Section 2
Applications for Improving Urban Subsystems

Chapter 9
High Speed Rail and Regional Competitiveness

Lara Brunello
Queensland University of Technology, Australia

Jonathan Bunker
Queensland University of Technology, Australia

Sandro Fabbro
Università degli Studi di Udine, Italy

Franco Migliorini
Istituto Nazionale di Urbanistica, Italy

Renzo Ferrara
Rete Ferroviaria Italiana, Italy

ABSTRACT

The presence of High Speed Rail (HSR) systems influences market shares of road and air transport, and the development of cities and regions they serve. Changes in accessibility have led researchers to investigate the effects on the economic and spatial derived variables. Contention exists when managing the trade off between efficiency and access points, which are usually in the range of hundreds of kilometres apart. In short, it is argued that intermediate cities, bypassed by HSR services, suffer a decline in their accessibility and developmental opportunities. This chapter will analyse possible impacts of HSR infrastructure by considering small and medium agglomerations in the vicinity of HSR corridors which are not always served by HSR stations. The benefits of accessibility and distribution will be quantified and investigated in a model that accounts for locations where an HSR station cannot be positioned and different rail transit services offered, including (i) cadenced, (ii) express, (iii) frequent or (iv) non-stopping services. This theoretical approach linking infrastructure, accessibility, distributive patterns, services, and competitiveness is applied to a case study in the North-Eastern Italian regions. Results indicate that benefits derive from well informed decisions on HSR station positioning and the appropriate blend of complementary services in the whole region to interface an HSR infrastructure. The results have sig-

DOI: 10.4018/978-1-61350-174-0.ch009

nificance for all countries in Europe and worldwide, not only for investing in an HSR infrastructure, but mostly in terms of building territorial cohesion, while seeking international recognition for developing successful new technology and systems.

1. INTRODUCTION

High Speed Rail (HSR) systems have the ability to significantly impact the market shares of road and air transport. However to obtain competitive speeds and travel times, HSR systems need to trade off station spacing and therefore accessibility. As a consequence, decisions need to be made regarding whether to serve intermediate cities, and which of those would be capable of generating substantial patronage. For this reason, bypass strategies have often impacted on the accessibility of many small or medium agglomerations in the vicinity of HSR corridors. Limiting access to these destinations thereby reduces their ability to participate in development opportunities facilitated by HSR infrastructure.

Effects on spatial distribution and the economic development of cities and regions after the deployment of HSR infrastructure have been investigated from the viewpoint of accessibility. In fact, it seems appropriate to consider the quality of accessibility to the HSR service as one of the main factors to win competition from other modes. Methods to measure accessibility have been expanded in the attempt to explain the land use and transport interaction by either classifying stations or ranking cities served by HSR.

A review of the literature shows that direct integration between HSR and complementary transit systems is long overdue and is vital to understanding which level of service quality would best serve this outcome. Evidence from the literature shows that the mere presence of a HSR station is not sufficient to entice development (Troin, 1995), while the spatially derived benefits of HSR tend to be concentrated around access nodes, and are co-dependent on ancillary investments by way of uplifting of station areas (Brons, Givoni, & Rietveld, 2009; Bertolini, 2008). The need for greater integration with local networks is also highlighted as a means to distribute benefits. However, there is a lack of tools available for cities and regions to analyse the best means of integrating HSR infrastructure into their transport networks with the aim of spatially optimising benefit distribution and increasing competitive advantages. Thus, there is need for careful planning of local land use and transport links to develop strategies to work as an interface with HSR infrastructure, not just as a feeder, but as a complementary high quality peer of HSR to improve accessibility of cities and regions, and to allow them to thrive.

In this chapter, the gaps above mentioned will be tackled and a new tool will be offered to understand which accessibility strategy might improve HSR benefit distribution at a regional level. The methodology adopted is useful to evaluate alternative rail transit services in relation to HSR infrastructure. Four railway systems with a regional scope have been selected for this study. Through comparative analysis of their efficiencies and accessibility, the methodology aims to demonstrate which enhancements would best suit the area under study, and which would produce network effects capable of influencing benefit distribution across hierarchical networks. Thus, technology improvements could be evaluated in light of their ease in exchanging benefits. Furthermore, measuring the spillover between networks would provide an understanding of the gains derived from the HSR investment, through the interface strategy chosen for the city and its region.

Often many small and medium sized cities and their regions lose out or are prevented from participating in the benefits produced despite their proximity to HSR infrastructure. Assessing the competitiveness of a region, based on the efficiencies and accessibility provided by local transit networks to complement hierarchically higher links, such as HSR corridors, is a valid instrument for both planners and policy makers. Such assessments are applicable to international contexts where HSR interests are shared, while catering for their own regional good. The relevance of building a stronger regional transport fabric lies in the ability to accommodate HSR corridors that do not drain resources toward bigger cities or shift resources away from the region itself.

To ground the research, a case study has been selected presenting concurrently HSR links, planned or under construction, and a regional transit network tested to enter service. Generalised travel costs are taken into consideration in terms of total journey times, accounting for the impedances due to service quality, such as waiting times and transfers, with constraints due to infrastructure capacity. Then, the advantages of transit network development are weighed with population densities to account for distributive effects of expanding HSR catchment areas.

Although depicted for a single region in North-Eastern Italy, the case study results are significant for all countries considering HSR in their transport plans, and which need greater information regarding the regional distribution of potential spatial benefits derived by integration of HSR with local transit networks. Therefore, the international relevance of the research project will be affirmed and explained. The provision of additional tools to value HSR infrastructure against its wider regional benefits and their distribution, beyond the actual HSR stations, would have great implications, both politically and technically, in moving towards new dimensions of HSR evaluation and development. On the one hand, results could provide benchmarks upon which to build coalitions and balance shares among different stakeholders, from political cooperation to joint industry efforts in the management of an integrated regional system. On the other hand, technological development of transit systems could be tested upon their effectiveness in improving accessibility, thus allowing implementation of the strategy best suited for the context under analysis.

Better managed regions would be able to face greater challenges, when competing for resources at interregional or international levels. Efficient local and regional transit systems working as an interface for HSR would bring benefits to intermediate and regional cities that do not directly enjoy the presence of a HSR station. These benefits would help in overcoming negative impacts, such as concentration or drainage effects. The process of improving regional accessibility could enter the implementation agenda, even without any HSR link planned, and ultimately establish the foundation for its eventual development.

This chapter intends to explain the relation between HSR systems and the competitiveness of cities and regions, specifically drawing a nexus between local and regional rail strategies to complement such a supporting role. To start with, the following section will present an overview on territorial competition theories and their relation with infrastructure provision (Section 2). Afterwards, the effects of HSR deployment will be considered in relation to competition (Section 3), followed by a review on some regionalisation trends originated to overcome HSR critical issues on regional accessibility (Section 4). Section 5 constitutes the core of the chapter, linking accessibility to competitiveness and presenting the research methodology. Section 6 presents results, grounding the theory to the case study, followed by their discussion, offering solutions and recommendations (Section 7). Future research directions (Section 8) and conclusions (Section 9) close the chapter.

2. TERRITORIAL COMPETITION AND THE ROLE OF INFRASTRUCTURE

It is postulated that two characteristics determine the pace of change within cities and regions. Firstly, cities and regions are complex and large systems, thus they are inertial. For this reason, spatially orientated policies take time to produce effects. Secondly, globalisation is putting pressure on cities and regions to deal with external agents. This has a consequence of higher levels of competition. Under a scenario of greater movement of capital and labour, cities and regions have more opportunities to market their capabilities and develop their territories, attracting activities. However, challenges lay on national governments' restraints, fragmentation of governmental departments or shifting of political actors, which slow the public decision-making processes. Evidence from Cheshire (2009) shows that institutional arrangements for local government play a major role in the success of a city, while policies themselves have at most a minor role.

According to this point of view, competitiveness could be defined as a governance product that enables cities and regions to offer different levels of service quality and regulation, condition operating costs, innovation and economic environment. Other factors, such as education levels and geographical scale, could reflect a region's capacity to confront the actual changing economic and productive environment and its capacity to manage such changes (Cuadrado, 2008). Thus, whether a city or region engages in territorially competitive policies seems to depend substantially on the capacity to form and maintain a vision, while understanding the strength of cooperation.

Within this scenario, accessibility and transport have an influence on growth, even though they are not enough to explain the whole of a city's competitiveness. Accessibility determines competitiveness in terms of infrastructure availability and usefulness. Accessibility, as a principal outcome of a transport system, determines the locational advantage of a city or region in relation to neighbours. Trivially, travelling distances requires transport infrastructure, which implies costs, and economies of scale further increase the advantage of infrastructure provision. Thus, the role of transport in spatial development implies that regions with better access to locations of input-materials and markets will, *ceteris paribus*, be more productive, more competitive and hence more successful than more remote and isolated areas (Linneker & Spence, 1996). Moreover, this implies that within a globalised market, as there are greater opportunities of growth, there might also be greater risks of marginalisation.

Therefore, the relationships forming to gain competitive advantages have lead to trends towards agglomeration to ease the process of acquisition of resources and skills. Agglomeration is a strategy whereby producers ease the tasks of transactional interaction because proximity translates into lower costs and wider opportunities for matching needs and capabilities (Scott & Storper, 1992). To explain geographical clustering of urban and regional agglomeration, it is necessary to take into account external economies, highlighting the advantages of networks, where relationships assume configurations of transport and communication interactions, further resulting in infrastructure networks. Regional agglomerations make a difference to the competitive advantage of constituent cities through economies of scale. Thus, territorial competitiveness could be better understood as the performance to provide services, thanks to integration and cooperation to uphold a territory.

This view is shared by Krugman (1994) who sees positively all forms of international economic integration, including trade, capital and labour mobility. In his analysis, externalities and imperfect competition highlight the potential gains in the *new trade theory* (Krugman, 1979), where effects are the uneven distribution of benefits and the adjustment of costs. Accordingly, the existence of excess returns in imperfectly competitive networks may advance some in comparison to others with different proportions. Thus, advantaged regions might benefit more than others from network integration. However, costs in one network could work as investment into another. Thus, an improvement into a network would bring benefits to neighbouring ones, as a sort of compensation, or a spillover from an over concentrated and specialised market into other sectors. Distributive patterns could be used to measure performance, and thus provide a tool to weigh gains or to direct investments where they are most needed. Similarly, Maggi (1992) believes that infrastructure holds a key role in the economic development of cities and regions. However, she stresses the fact that networks have to provide a high quality service and have to be evaluated on their performance.

The relationship between infrastructure and economic development is well acknowledged. Andersson and Andersson (2008) link socio-economic development to networks. They link waterways to commerce along rivers and coastal areas; sailing and banking to trade with the East and West Indies; steam and then internal combustion engines to the industrial revolution, then to railways, and later to cars; computerisation to logistics and non-material flows. In line with this, Weisbrod (2008) reviews methods for assessing economic impacts deriving from transport projects and the list is not short, nor is it exhaustive. Critiques and a variability of consensuses pertain to the field, especially on the direction of causation. For example, Munnell (1992) demonstrated positive and statistically significant effects of public capital investment on private sector output and productivity. However, analyses have often concentrated on production functions and only more recently they have looked into the locational advantages of businesses, especially with the increasing opportunities brought about by the internationalisation of economies. Regarding this, it is debated whether competition should be *perfect*, with constant returns and a small range of external economies, as suggested by Marshall (1920), or *imperfect*, driven by increasing returns and external economies. Questions pertain to the limits of returns, crowding-out effects between public and private investments, and levels of competition, often more concentrated at local and regional scale, rather than nationally or internationally (Thompson, 2003).

Concerning our theme, competition on production costs has added to the discussion the whole production chain, affecting balances in the economy, society and the environment. New competitors could be found anywhere in the world and knowledge economies are intensifying around major urban areas. In this context, territories are not geographically determined, but become processes where relations between regions develop, impacting on production, growth patterns and, above all, on mobility. The concept of competition itself has overtaken traditional limits of industrial competition and has entered more intangible and less defined dimensions as cities and regions have stepped into competition for services, capital and employment. In this way, cities and regions have become susceptible to global changes and will have to assume responsibility for their actions in response to new opportunities and risks. In this scenario, transport infrastructure plays a key role as connective fabric to the supply of services, in terms of incentives, networks and technologies, in order to be adequate for such competitive challenges.

Thus, this chapter focuses on territorial competitiveness from the transport viewpoint and investigate problems deriving from the relation between land use and infrastructure. The following section will deepen the analysis on a more specific type of infrastructure, explaining reasons and issues.

3. HIGH SPEED RAIL SYSTEMS: NICHE OR BROAD COMPETITION?

Historically, railways have been linked to economic development both as a cause and a consequence. However, after losing massive market share, particularly to road and air modes, from the early 20th century until now, few studies have investigated the relationship between railways and economic development. Highways have substituted railways' former role as a strong direct link to land uses. Railways have undergone a period of recession, being unable to keep pace with innovation and development. Policies have not always been supportive and many railway branch lines have been closed, trimming networks and services (Haywood, 2009). This has happened diffusely in Europe and to varying extents worldwide depending on the levels of rail development. Often, shorter haul tramways have been the first to be dismantled.

Nowadays, the relationship between highways and development has reached critical levels in many regions and external costs imposed by road transport itself are impacting heavily on the economy. In this context, an increasing number of studies are concentrating efforts to investigate sustainable transport modes. Among these, railways are regaining lost attention, in particular light rail, metro rail and HSR, with their respective increased haul distances and catchments. HSR systems are the focus here, being substantially different from conventional rail in terms of speed and coverage, and competing with both long haul road and air transport.

3.1 High Speed Rail and Other Modes

To better understand the effects on city-regions, it is necessary to examine HSR competition with other transport modes. This could be extended to both European and non-European countries who are interested in HSR development. French and Japanese experiences suggest that HSR traffic generation could overcome expectations, with increases in the order of 150 to 200 per cent, after only a few years of operation (Plassard, 1992). In addition, HSR could maintain competition over a long period. For example, French TGV (Train à Grande Vitesse) carried over 80 million passengers in 2006, four times more than Air France (2010) for the same year on domestic flights. Thus, HSR is more likely to have a non-marginal effect on relative speeds and generalised costs within the transport system (Vickerman, 1997).

However, there is also evidence of thresholds relating to time and distances, within which HSR has a clear advantage. Journey times up to four hours represent a limit to perform activities within a day, thus distances in the order of 200 to 800km are commonly considered as competitive for HSR travel. Below the lower bound, road and in a few cases metro rail flexibility dominates the market, while above the upper bound air travel offers faster overall speeds. In any case, market size and shares are mostly determined by the purpose of trip, and thus by the value of time assigned for that activity. This yields a difference between business traffic and leisure travel. Therefore, new demand forecasting models have been developed and specifically designed for HSR (Couto & Graham, 2008; Yang & Sung, 2010).

However, some risk exists in confining HSR to a niche market of expensive and classist travel. This depends mainly upon the image marketed and fare levels proposed. Notwithstanding the slowing of the world's economy at the time of authorship, it is significant that Asian countries are investing significantly in HSR development. The first Chinese HSR line opened in 2003 linking Qinhuangdao to Shenyang, 400km apart; soon followed by the South Korean KTX (Korea Train eXpress) in 2004 and the Taiwanese THSR (Taiwan High Speed Rail) in 2007. Certainly, there are favouring factors as determinants of HSR ridership, such as densely populated areas, fast growing cities, high transport demands, that impact positively on the deployment decision.

Moreover, other successful examples could be found where high-speed rail services have been provided in transforming competition between air transport and HSR into cooperation through supportive relations to provide high quality connections between city centres and airports (Arduin & Ni, 2005). Some airports (e.g. Paris' Charles de Gaulle International Airport and Frankfurt Airport) have an integrated HSR station and airlines sell integrated ticketed flights with connecting rail segments.

The use of HSR for airport connection could provide a sleek and glamorous image that sits comfortably with air travel, however it might be considered less likely for an individual to take a long distance trip by rail before or after a flight. Since distances between "out of town" airports and cities average in the range of 20 to 30km, the need for frequencies is higher than the need for speed to reach downtowns. Thus, airports often choose to be served by more frequent services provided by extensions to conventional metro rail networks (e.g. San Francisco's Bay Area Rapid Transit, New York and Brisbane's AirTrains, Sydney's AirportLink and many others), or more recently by innovative systems, such as personal rapid transit (PRT) vehicles on dedicated routes (Daily Mail Reporter, 2009) and automated systems (e.g. Paris' OrlyVAL), in a market otherwise dominated by taxis.

Nevertheless, trends are pushing airlines to operate HSR services in partnership with a rail service provider (Greenair, 2008). Reasons for considering HSR in unison with air might be found in rising oil prices, increasing airport congestion, environmental concerns, cases of bad weather and conflicting events on the groundside access road network. For their part, HSR operators are readying themselves to position HSR travel as a faster, more reliable, more punctual and greener alternative to short-haul flights (Buyck, 2008).

Thus, competition dynamics are evolving toward more collaborative models with the establishment of alliances and partnerships.

3.2 High Speed Rail Impacts

HSR projects are progressively becoming more accepted and deployed worldwide, with an expected growth reaching 35,000 track kilometres by 2025 (Aliadière, 2008). China alone is planning more than 12,000km of HSR track to link its provincial capital cities in the next ten years (PRC Ministry of Railways, 2010).

This massive expansion of the world HSR network is likely to bear large impacts on cities and regions during the process. In this regard, changes in spatial distribution and economic development following HSR implementation have been increasingly investigated from the viewpoint of accessibility. Increases in speeds have shrunk travel times (Spiekermann & Wegener, 1994), impacting on accessibility maps and allowing trips not previously possible. Thus, economic and spatial effects have been investigated as derived variables of changes in accessibility. This approach leaves room to consider symbiosis between development and infrastructure.

Studies have examined issues deriving from HSR on economic integration (Blum, Haynes, & Karlsson, 1997; Haynes, 1997), location choice of activities (Willigers & Floor, 2007), territorial cohesion (López, Gutiérrez, & Gómez, 2008), and city image and residential development (Garmendia, Urena, Ribalaygua, Leal, & Coronado, 2008). Digressing discussions have been found on the appropriateness of accessibility indicators (Gutiérrez, 2001) or their synthetic indices (Martín, Gutiérrez, & Román, 2004; Martín & Reggiani, 2007). Attempting to draw a correlation between HSR and territorial competition, some studies have ranked cities based on accessibility changes (Bruinsma & Rietveld, 1998), which harks back to the issue of competing cities on an infrastructural basis.

Most research has identified accessibility inequalities due to polarisation effects around HSR access nodes. On the one hand this result indicates the importance of access time and convenience factors when accessing the HSR network; on the other hand, it shows a problem regarding accessibility analyses at different geographical scales, where effects appear different whether analysed at the local, regional or national levels. Studies have addressed this problem in the attempt to harmonise results (Kwan & Weber, 2008).

However, regardless of the methodology adopted for the analysis, HSR projects highlight a real discrepancy on accessibility patterns. This may be understood by viewing HSR infrastructure as a sort of tunnel crossing a territory, without engaging in any exchange aside from rare points where stations are located (Plassard, 1992). This configuration is sure to impact on any benefit deriving from HSR, by concentrating effects on access points, generally located in major urban areas, leaving regional areas uncovered.

Undoubtedly, to optimise benefit distribution, as the gateway to the HSR network and its services, station location plays a major role. Several studies have looked into improving accessibility around stations and have classified stations based on land use and transport relationships (Reusser, Loukopoulos, Stauffacher, & Scholz, 2008), on customer satisfaction (Brons, Givoni, & Rietveld, 2009; Givoni & Rietveld, 2007), on surrounding functions' balance (Bertolini, 2008), on intermodal access time (Burckhart & Blair, 2009; Tapiador, Burckhart, & Marti-Henneberg, 2009) or on the quality of service and performance (Debrezion, Pels, & Rietveld, 2009).

However, positioning of HSR stations and their integration with local land use and transport networks is still a promising but fairly sparsely researched field of study, in contrast to such integration of its urban cousin station types such as light rail transit (LRT), bus rapid transit (BRT) and metro rail. It is therefore necessary to understand the relationships between HSR links, land use and more local transport networks to address possible benefit distribution that may influence success and competitiveness of city-regions.

The next section presents some developing trends to tackle critical issues on HSR accessibility at the regional level.

4. TRANSIT REGIONALISATION TRENDS

As presented in the previous section, most effects deriving from HSR deployment can be traced back to station location. Access to the HSR network is definitely a critical factor for considering competition. However, there are some limitations. Coverage expansion conflicts with efficiency in terms of travel times, load factors, stop frequency, etc. Therefore, these constraints impact on station location choices when attempting to maintain high quality standards of service.

Figure 1 shows how station spacing influences average speeds of HSR service. The impact on average speed increases sharply as the distance between stops falls below the mean value of 103km (Schneider, 1993).

Consequently, HSR operators face new challenges when searching to increase markets. This to avoid lost market share in the major markets through increased journey times resulting from stopping HSR trains too often (Vickerman, 1997). Also, Vuchic and Casello (2002) demonstrate that speed increases in maximum speed have decreasing marginal gains in travel time savings depending on the length of run between stations. Thus, HSR corridors have seen the necessity to by-pass intermediate cities that are considered not capable of generating sufficient patronage to justify the capital investment for stations, nor to compensate the tradeoffs on journey times. However, by-pass strategies negatively impact on the accessibility of small agglomerations and surrounding areas along HSR corridors, reducing their ability to participate in the development opportunities facilitated by HSR infrastructure. Moreover, potential benefits have tended to concentrate into major agglomerations (Martin, 1997), which are

Figure 1. This time-distance graph presents average speeds of HSR service affected by station spacing (acceleration: 0.1g; station dwell time: 3 minutes). Adapted from Schneider (1993) with permission.

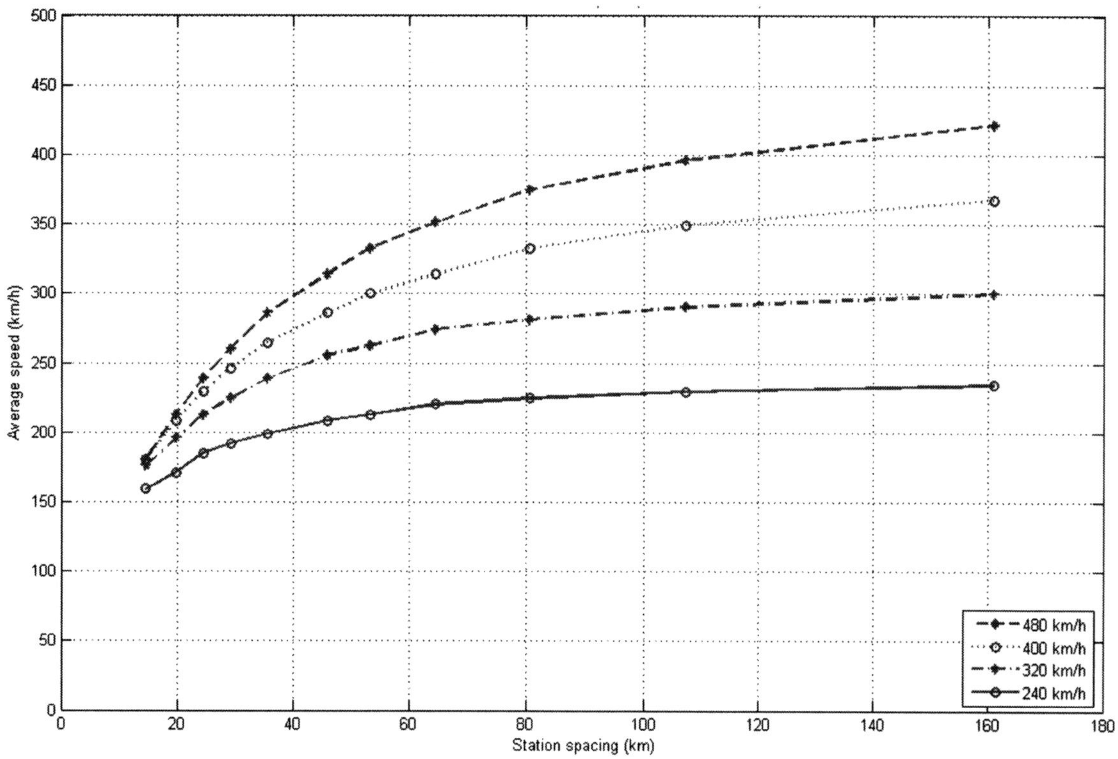

often better prepared to distribute them through urban systems. Only a few studies have analysed the benefits derived from a direct HSR connection to smaller intermediate cities that happen to be interspersed on a new HSR corridor (Garmendia, Urena, Ribalaygua, Leal, & Coronado, 2008; Urena, Menerault, & Garmendia, 2009).

A knowledge gap therefore exists in the study of regional accessibility of smaller centres in the vicinity of a HSR corridor, but bypassed by the HSR service. Notwithstanding the recognition of a tunnel effect impacting on the distribution of benefits from non-users to users (Plassard, 1992) or the necessity of good access infrastructure to reach the HSR stations (Vickerman, 2008), no precise correlation has ever been made between HSR infrastructure and the competitiveness of such cities not directly served by a HSR network.

Clearly, the regional level is most controversial, where some cities might be served by HSR while others may not. Thus, conflicting relationships amongst parts of the same region give rise to core-periphery patterns (Sasaki, Ohashi, & Ando, 1997; Vickerman, Spiekermann, & Wegener, 1999). Imbalances are strengthened by time-distance reductions on interregional relations, while they are exacerbated at the local level by accessibility issues of small cities and towns that are bypassed by HSR, risking hinterland segregation (Geurs, van Wee, & Rietveld, 2006; Menerault, 1998).

These considerations beg the following question:

Could HSR be implemented or interfaced in such a way that would improve accessibility at the intra-regional level, with the aims of spreading benefits through cities, around conurbations, and bringing regions together without interfering with the HSR specific competitive advantages of high quality services and time savings on medium-long distance relations?

The hypothesis is then whether a blend of different but integrated transport services could cover larger portions of territory and thus increase patronage, while working in synergy with the HSR service. Thus, technical interoperability between high-speed and conventional networks provided the ground on which to select four strategies according to a number of criteria (e.g. speed, suburbanity, capacity, frequency and automation). These strategies are rail-based transit systems, which assume a new role to access the whole of the HSR network as interfaces, instead of simply feeding the HSR station. Selected systems are presented below.

4.1 Regional Metro Rail

Metro systems are rapid transit systems deriving their names from the word *metro(polis)* with the meaning of *urban* service running at grade or grade separated, either in underground or elevated routes. However, metro systems often extend outside urban centres serving both the urban and the regional context. As multifunctional systems, regional metro rail (RMR) could work within major urban areas and around their suburbs for several hundreds of kilometres, and it is common in many cities worldwide (e.g. Brussels, Copenhagen, Milan, Munich, Oslo, Paris, Philadelphia, Washington D.C., Sydney, Vienna, etc).

The RMR strategy was selected for its cadenced service, which usually comprises a fixed interval service frequency in the range of 5-10 minutes downtown and up to 20 minutes in peripheries (30 minutes at night and on weekends). This type of service allows for coordinated interchanges between trains and other transport modes, such as buses.

4.2 Regional High Speed Rail

The regionalisation of HSR services and networks as a new suburban transport mode started in France in the 1990's. Awareness grew that HSR, conceived to connect city-pairs in distant regions in competition with air transport, was failing to provide much benefit *along* its corridors (Hall, 1999). Thus, during the construction of the high speed line TGV-Nord, political actors decided to link the whole of the Nord-Pas-de-Calais region within an hour travel time to Lille. *Train Express Régionaux Grande Vitesse* were introduced as French regional high speed rail (RHSR) service to exploit a technical opportunity to improve train occupation of high speed trains (Torchin, Grilly, Combes, Hasiak, & Menerault, 2008). A key success factor was the combination of short travelling times with low fares, especially for the commuting market sector, facilitating competition with the car mode over relatively short distances.

The RHSR concept entails greater integration with the land use patterns, according to regional needs and objectives (Garmendia, Urena, Rivas, Coronado, Menendez, Gallego, & Romero, 2009) and its implementation could take place by (*i*) increasing the number of HSR stops, (*ii*) introducing regional trains on HSR lines or (*iii*) extending HSR services onto conventional tracks once off the high speed mainline. Examples of this sort of HSR evolution can be found increasingly around the world, where service is modified from direct connections to intermediate stops (e.g. Belgium, France, the Netherlands, Spain, UK, US). Most RHSR services are introduced thanks to upgraded infrastructure or the use of tilt trains (Fröidh & Nelldal, 2008). Technical characteristics relate principally to speeds in the range of about 110km/h up to 320km/h depending on style of train and route travelled. Thus, the RHSR strategy was selected mainly for its express service implemented on a mixed traffic operating model.

4.3 Light Rail Transit

Light rail transit (LRT) is a form of urban transport which sits in between metro rail systems and traditional on-street running trams. In particular, LRT systems are capable of combining the advantages of different service types thanks to interoperable vehicles. Thus, they could share infrastructure with commuter rail and cover suburban and regional destinations (Priemus & Konings, 2001). Archetypes of this twofold nature are the urban trams running over a regional railway infrastructure in Karlsruhe, named *tram-trains*, and the regional trains running over a urban tram network in Zwickau, named *train-trams*. This integration of urban and regional networks offers the added value of a reduction in the number of transfers required of a passenger.

Running small train sets, the LRT strategy was selected for its frequent service, with headways usually in the range of 1-5 minutes.

4.4 Continuous Railway System

Continuous railway systems (CRS) were included in the comparative analysis as an atypical alternative to extend the bounds of the research. CRS do not currently exist in practice and can be traced mostly through patents (see Brunello, Bunker, & Ferreira, 2008 for a review), a limited scientific literature (e.g. Gudehus, 1974; Schulte, 1971; Pottgiesser, 1968), and few past real world applications which have been almost forgotten. A partial CRS operation was conducted for a period of about 150 years from the mid 19th century, when nearly all railway companies throughout the United Kingdom ran some form of non-stop trains to optimise and expedite rail service. The practice served intermediate stops, without stopping the train, but detaching one or more railway cars from the tail of that train and mechanically braking them to rest at the designated station. The total time taken to complete a journey could then be reduced, while a higher average speed could be maintained. This practice was also considered to be highly energy efficient, with an estimated 12,000 fewer engine kilometres travelled in 1914 by the London, Brighton & South Coast Railway (Fryer, 1997). However, the practice of detaching railway cars could only serve alighting passengers, while boarding of passengers could only be performed by stopping the whole train.

Some other experiments have been conducted in France, such as the *Aramis* (acronym of *Agencement en Rames Automatisées de Modules Indépendants dans les Stations*) project sponsored by the *Régie Autonome des Transports Parisiennes* (RATP) between the 1960s and 1980s to develop a prototype of automatic controlled cars running bumper to bumper (Latour, 1993); and in the Netherlands with the ubiquitous *koplopper* (literally *front walker*) trains that couple with one another to share route and adjust capacity upon necessity. Until 2005, these trains were noteworthy in that they had doors at their extremities to allow passenger to walk through and change convoy according to the portion stopping at their destination.

The possibilities offered by end-to-end headways between railway cars and trains maximise network capacity without the need for an increase in speed, and the absence of transfers enhances the benefits of seamless travel. It is envisioned that modern technology could facilitate a safe working, reliable and modern CRS application. While technical issues are left for future consideration, the CRS strategy was here selected for its non-stopping service and challenging possibilities.

5. ACCESSIBILITY AND CITY-REGIONS' COMPETITIVENESS

This section will present a research method developed to correlate the strategies presented above with potential HSR benefits, drawing their territorial distributive patterns to understand possible impacts on the competitiveness of cities and regions. It is important to note that the analysis will be supported by quantitative content to ground the methodological approach of measuring these relationships.

Reforms in the railway industry have tended to separate responsibilities with a view to improving performance and efficiency of operations. Vertical separation between access infrastructure owners and railway operators, with varying levels of privatisation, versus the once commonplace but now legacy model of government ownership and operation, has become pervasive since the early 1980s. The rationale for this shift has been that the (mainly) government owned and vertically integrated model has been deemed by economists in particular to be weak in response to market demands, over-dependent on subsidies and without opportunities for competition (Guihéry, 2002). This railway reorganisation has lead to a variety of policies, from the closure of inefficient lines, generally rural and regional branch lines with a concomitant shift of service to the road based modes, and in particular to the differentiation and spatial segmentation of networks. Differentiation of services and networks generates a sort of *spill-over effect* in interlinked networks that was not perceived within an integrated structure. These effects are more evident as networks specialise within a hierarchical railway market, where each type of transport would have a corresponding network (e.g. urban, suburban, regional, inter-regional and HSR).

In this perspective, the responsibility of urban and regional networks might be transferred to more local authorities that often operate on franchise, while interregional corridors might experience higher levels of competition for services. Thus, technological externalities take place when there are spillovers from the production function of one network into those of other networks, thanks to an improvement that could be fostered through direct links, cooperation and interoperability. Accordingly, large scale transport infrastructure projects, such as HSR links, could generate proportional effects spreading well outside the limits of their own networks with real welfare and efficiency impacts.

To properly attribute benefits among networks, the role of interconnections is extremely important (Bonnafous, 2002; Burmeister & Colletis-Wahl, 1998), as it is that of any compatible strategy, particularly in the framework of sustainable integration. Risks to be managed include duplications of services and harsher competition on more profitable routes, while the role of feeders might be at stake in more remote areas. Hence, the question on how to distribute benefits among the networks involved is tackled by measuring the quality of service in terms of accessibility, in order to obtain comparable data for the complementary strategies. The principle to guide the process is based on the holistic view that the whole network will drive effects that the mere summing of its parts would not produce. Thus, a method has been developed to assess network effects between hierarchically different networks. Measurement and comparison of network efficiency accessibility improvements will be performed on different strategies selected to complement HSR performance locally and regionally. As such, complementary strategies could be evaluated in light of their ease in exchanging benefits with HSR to provide a competitive interface for the city and its region.

5.1 Accessibility and Productivity

Conceptually, accessibility is renowned within the regional transport debate and significant research has been conducted since the 1950s. Firstly defined as the potential for opportunities of interaction (Hansen, 1959), accessibility has undergone several interpretations, generally linking it to concepts of nearness or proximity. Often considered as the ease of spatial interaction, accessibility was later seen as the potentiality of using the opportunities offered (Domanski, 1979) related to the concept of utility, and thus is better interpreted as a potential or expected benefit (Ben-Akiva & Lerman, 1979; Leonardi, 1978). In recent years, the concept of accessibility has re-emerged as a fundamental element dealing with transport sustainability (Banister, 2008; Marshall, 2001; Nijkamp & Van Geenhuizen, 1997; Reggiani, 1998). In this context, the concept of accessibility can provide a useful framework for the integration of transport and land use planning (Bertolini, le Clercq, & Kapoen, 2005). The arguments for land use transport integration are well-known and discussed (Freilich, 1998; Page, Kelly, May, Jones, & Forrester, 2009).

Within an approach that regards land use and transport integration as a requirement for sustainable development (Meyer & Miller, 2001; Priemus, Nijkamp, & Banister, 2001; Wegener & Fürst, 1999), accessibility can be tied to both the quantities of the transport system, such as travel time, and the qualities of the land use system, such as different destinations (Handy & Niemeier, 1997). Thus, a shift in focus has occurred toward considering accessibility as a means to better understand the interdependences between transport and land use (Banister & Berechman, 2001; Bertolini & le Clercq, 2003; Straatemeier, 2008). In this way, accessibility could be linked to planning goals and could aid in considering different effects on the economy (e.g. providing a firm with access to workers, customers, suppliers), society (e.g. providing people with access to jobs, services, goods and contacts) and the environment (e.g. using resources efficiently) leading to a broader, more holistic view in the evaluation of alternative transport options and policies (Geurs, van Wee, & Rietveld, 2006).

From this point of view, accessibility becomes correlated with economic development as determinant of macroeconomic changes in terms of productivity. Studies have confirmed the existence of very substantial spillover effects associated with transport infrastructures impacting on the production function (Cantos, Gumbau-Albert, & Maudos, 2005). Investment affecting interregional links could provide accessibility benefits similar to those of agglomeration economies (Brocker & Rietveld, 2009; Rietveld & Bruinsma, 1998; Glen Weisbrod & Treyz, 1998), where improvements in transport infrastructure leads to a reduction of transport costs, which in turn may give rise to substantial redistribution effects among economic groups and also among regions (Bruinsma, Nijkamp, & Rietveld, 1992).

Notwithstanding this articulate methodological framework, methods to measure spatial spillover effects of increased transport accessibility are still in their infancy. Spanish studies have been particularly proficient thanks to a great availability of public capital data at the regional level, released by the Spanish government and covering the last forty years (Pereira & Roca-Sagalés, 2003). One of the most recent and interesting studies on spatial spillover effects of infrastructure has been conducted for the road network in Spain (Gutiérrez, Condeco-Melhorado, & Martín, 2010) highlighting regional distribution of economic potential imbalances in terms of accessibility exported or imported gains. On this pathway, studies that have investigated accessibility of HSR infrastructure could be divided into two main currents of thought: those concerned with HSR impacts to the development of cities (Garmendia, Urena, Ribalaygua, Leal, & Coronado, 2008; Urena, Menerault, & Garmendia, 2009) and those investigating HSR accessibility impacts and their measurement (Gutiérrez, Monzón, & Piñero, 1998; López, Monzón, Ortega, & Mancebo 2009; Martin, 1997). In the analysis presented here, both approaches will be taken into consideration to reach a formulation suited to the hypothesis of finding a direct correlation between transit strategies and territorial competitiveness, through the distributive patterns of network effects exchanged with HSR infrastructure.

According to van Exel, Rienstra, Gommers, Pearman, and Tsamboulas (2002), network effects could be calculated using generalised costs and accessibility measures within an interactive land use transport framework. However, to neutralise the ordinary notion of cost and the effect of geographical location, an indicator of network efficiency accessibility (NEA) could be used to indicate the ease of access (Martín, Gutiérrez, & Román, 2004) in terms of service quality. This indicator might be better understood with all its components if presented as an equation, thus the following subsections offer synthetic formulas for the mathematically inclined.

5.2 Network Efficiency Accessibility Indicator

The main reason to present an equation to represent an indicator is to exactly quantify what was defined theoretically. Among the few indicators available in the literature to measure HSR accessibility (Bruinsma & Rietveld, 1998; Gutierrez, 2001; Martín, Gutiérrez, & Román, 2004), a network efficiency accessibility (NEA) indicator appears less influenced by the geographic location of the network nodes than other indicators and offers a better picture of the spatial variations in accessibility as a result of the provided infrastructure (Gutiérrez, Monzón, & Piñero, 1998).

Substantially, a NEA indicator is composed of two parts: firstly, an impedance ratio accounts for the ease of access through the network between origins and destinations; secondly, a weighting factor puts in relation each connection to the value given at each destination. Equation 1 is formulated to reflect NEA as the sum of accessibility efficiencies throughout a regional network in correlation to a specific HSR station served by a selected transit service. These efficiencies are given by the ratio between the real and ideal impedance (or the distance between real generalised travel costs against an ideal seamless trip) for the connection of each regional municipality to the HSR station, weighted by its population density in relation to all others in the network.

$$E_j = \sum_i \frac{\frac{I_{ij}}{II_{ij}} w_{ji}}{\sum_k w_{jk}}$$

Equation 1: E_j stands for the NEA indicator in the HSR station j; I_{ij} is the real impedance to reach said destination j from origin i, which accounts for the generalised costs of travel in terms of time; II_{ij} is the ideal impedance for the same trip, as travel time over a seamless infrastructure; w is the weighting factor to calibrate results at origins i, k.

NEA indicators have proven useful in emphasising contrasts between major transport infrastructure corridors (e.g. HSR links) and the areas located between them, also addressed as *shadow* areas (Baradaran & Ramjerdi, 2001; Gutiérrez, Monzón, & Piñero, 1998; López, Gutiérrez, & Gómez, 2008). Thus, it seems appropriate to use NEA indicators in analysing the unequal relationship between hierarchical networks and more specifically to investigate the interaction between the strategies presented in the previous section and HSR infrastructure. In fact, the fundamental issue lays in valuing regional networks and their development to bring about a mesh of connectedness capable of integrating with networks at higher levels in the hierarchy.

Thus, the importance of *network integration* (Banister & Berechman, 2001), *network efficiency* (Gutiérrez, Monzón, & Piñero, 1998) and *network synergy* (Capello & Rietveld, 1998) could be summed in the quantification of network effects to recognise the potential benefits exchanged between regional transit strategies and HSR, as the contribution of each network to the performance of its neighbouring networks. The following subsection presents the formulation of network effects.

5.3 Network Effect

Network effects are calculated as the aggregate value of NEA indicators, weighted by population densities at destinations, or the potential HSR stations (equation 2).

$$NE_j = \sum_j E_j \frac{P_j}{\sum_r P_r}$$

Equation 2: NE_j stands for the network effect; E_j is the NEA indicator at the HSR station destination j based on efficiencies calculated at origins (see equation 1); P is the weighting factor to calibrate results based upon all the potential HSR stations destinations j, r in the regional network.

The above equation is reversed from the one formulated by López, Monzón, Ortega, & Mancebo, (2009), who calculated efficiencies at destinations and then balanced them with the population at origins to estimate cross-border effects in Spain. To avoid limiting the number of destinations in each origin's NEA, our formulation calculates the NEA at destinations which are HSR station locations and expresses the efficiency to access it from all possible origins in the regional network.

From here, steps to show distributive patterns of HSR benefits into regional networks are easily understood if results are converted into monetised terms. Monetisation according to the HSR investment is given in equation 3.

$$M_j = \frac{M}{NE_j}$$

Equation 3: M_j is the monetised value of network effects; M stands for the investment in HSR infrastructure, while NE_j is the network effect.

The above formula provides a monetised value of HSR investment that could be redistributed throughout the regional network according to the efficiencies of each origin-destination connection. Alternative transit strategies complementing HSR infrastructure would demonstrate their distributive potential through comparison. Having presented the methods to conduct this comparison, the following section will introduce a case study on which to apply the analysis.

6. CASE STUDY: NORTH-EASTERN ITALY, ENCLAVE OR COMPETITIVE?

Having illustrated the research methodology in the previous section, a case study is presented here to ground the theoretical approach. From the European context to the regional area under analysis, we investigate the relationship between infrastructure and competitiveness. Afterward, results from the case study will be presented and discussed.

6.1 Competitive Cities

In line with worldwide trends, population concentration around major conurbations is a significant aspect in Europe, where more and more areas are becoming urbanised. In particular, the pentagon between London, Hamburg, Munich, Milan and Paris has strengthened its relative weight compared to surrounding regions. Outside this pentagon, polarised development rewards political and economic capital and assumes a crucial role in urban and regional policies. Major cities are generally more favourable when they possess a varied mix of important elements, such as transport, industry, tourism, administration, universities and entrepreneurship (ESPON, 2006). Minor cities, on the other hand, are balancing within the dichotomy of concentration and rarefaction. These cities are often regional chief towns, with lower management and congestion costs than metropolitan agglomerations. As long as minor cities are able to network with neighbouring cities, enlarging their sets of relations and expanding their connections, they might play a more profound role on the whole economy, not just as local provider of services, but opening up to market dynamics and economies of scale. In this way, even smaller cities might work as cohesive tissue between local and international links.

Commonly, competitive-based ranking of cities are computed on their performance and several indicators might be applied, including (i) the price of land or its rate of increase; (ii) the penetration of contested markets; (iii) the relative increase in factor productivity; (iv) the rate of growth of population and (v) the rate of growth of real GDP per capita (Cheshire, 2009). However, not all of these indicators are readily usable and data is not always available. The last one is mostly used for the classifications at the European level implying to measure GDP over self-contained city-regions that are delimited as functional urban regions (FURs). At smaller scales, population is often measured to understand prosperity and welfare, however it needs contextualisation.

Some examples of competitive European city-regions include the Randstad and the Rhein-Ruhr polycentric regions. In the Dutch case, the metropolitan configuration lays around a ring surrounding a rural central area, strongly interconnected with efficient transport systems. Urban and regional links connect Schiphol Airport with all ground transport directions. In practice, the Randstad highlights competitive governance and efficient planning, developed nationwide to link all sectors of excellence in the country, which is also the most densely populated of Europe. With the strength of efficiency and organised critical mass, the Randstad is competitively well positioned in Europe, even in comparison to strong neighbours, such as England and Germany.

The German case is somewhat different, where local authorities have had to endure massive conversion efforts to transform the major European mineral and metallurgic industrial site into a liveable city-region. The outcomes are still underway, but demonstrate a clear expansion of the limits of possibility (Knapp, Kunzmann, & Schmitt, 2004).

In the following subsection, we contextualise our case study.

6.2 Context of Case Study

In the case study presented here, the choice to use population density and its rate of growth, as a statistical parameter in the formulation of benefit distribution, reflects the context in which it is set. North-Eastern Italy (NEI) presents a polycentric regional panorama with phenomena of concentration and rarefaction. Agglomerations do not generally have sufficient mass to be considered major cities at the national level, but densities might reach record levels within just a few square kilometres. Thus, densities might work better to highlight differences at small scales, without having to aggregate data on the population. Furthermore, this choice is supported by the absence of gravitational demand in the regional movement of people or goods. Thus, all agglomerations and any surrounding could be considered equally.

According to level one of the NUTS (Nomenclature of Units for Territorial Statistics) classification, NEI includes four regions: Veneto, Friuli-Venetia-Julia, Trentino-Alto-Adige, and Emilia-Romagna. The first three regions are culturally and historically better integrated among each other, and generally addressed as *Triveneto* (nineteenth century expression to highlight the cultural identity of the three *Venetie*). This area is also geographically well defined. The pre-Alpine chains constitute the natural border to any sprawling expansion to the North. Lake Garda separates North-East from North-West, and on the eastern border, the Adriatic Sea joins the mountains. Southern morphology blends on the Po riverside and rural farmers defend their settlement rarefaction against urban contaminations (Figure 2).

NEI cities, especially those in Veneto, are places that better represent the traditional polycentrism of the whole Italian peninsula, even if some similarities could be recalled in the Lombard piedmont, Florence-Prato plain and Neapolitan mainland. These peculiarities refer to the ways suburban and interurban development has intensified along fractured modules, dispersed within

Figure 2. Map of North-Eastern Italian regions with city dimensions by 2015 population (developed on an OpenStreetMap background, 2011)

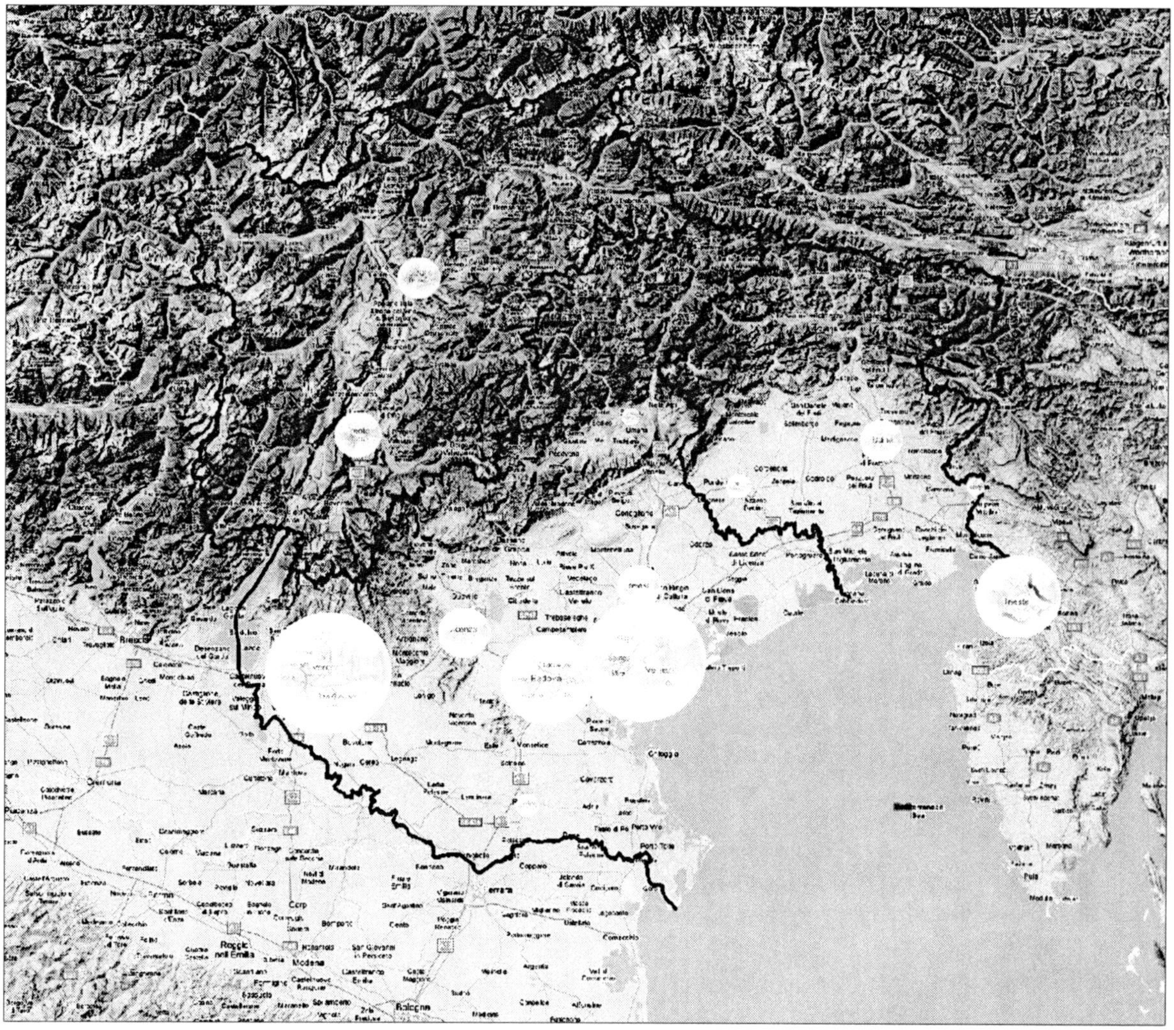

the pre-existing rural grid; while the central triangle, underlined by the cities of Padua, Treviso and Venice and their adjoining peripheries, holds together three million inhabitants. This critical mass makes the central NEI a polycentric regional metropolis (Migliorini, 2007). Even functionally, the spatial segment of Venice-Padua has been classified as a dual-core polycentric metropolitan area (BBSR, 2011 p.94).

It is here, in the heart of NEI, that we find concurrently a peripheral polycentric regional context, the presence of a planned HSR corridor within the trans-European transport network (TEN-T) programming, and the project to develop a suburban/regional rail service, namely SFMR (*Sistema Ferroviario Metropolitano Regionale*). A portion of HSR infrastructure and the core of the SFMR network have been recently completed and constitute our case study (Figure 3).

Within the European HSR design scheme, NEI will be crossed by TEN-T corridors (Figure 4), connecting Lyon to the Ukrainian border on the East-West axis (Corridor V), Berlin to Naples (Corridor I), and the Baltic Sea to the Adriatic

Figure 3. Overview of case study networks. Adapted from Regione del Veneto (2004) with permission.

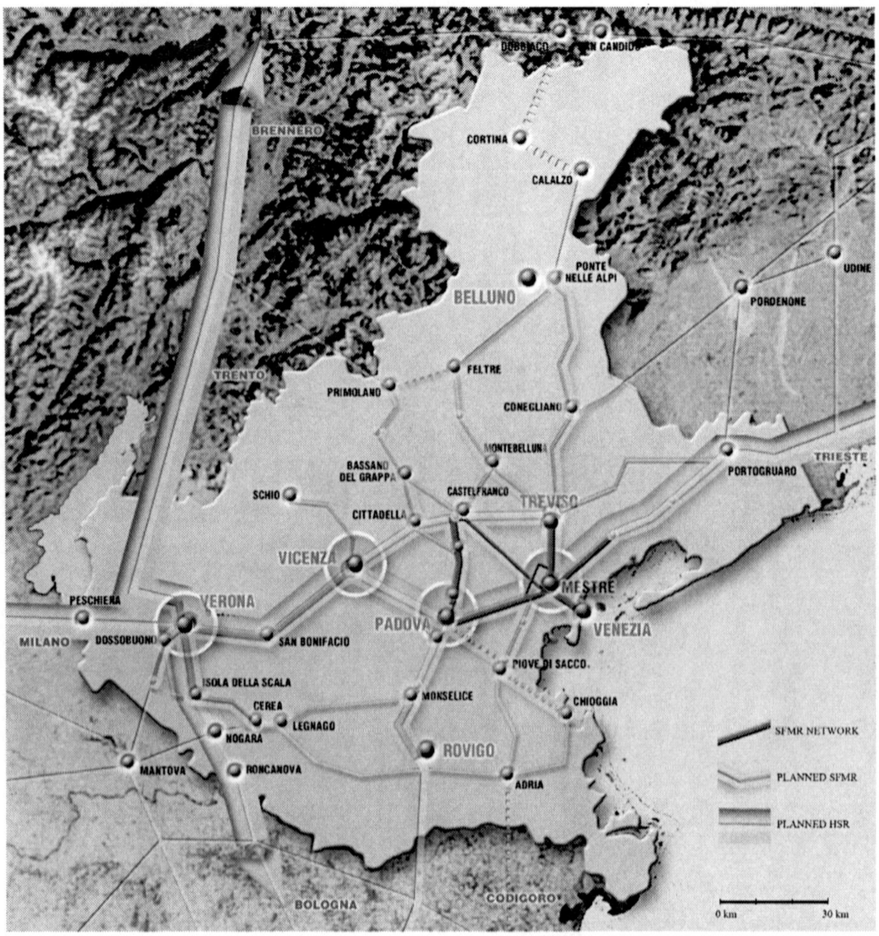

and Ionic Seas on the North-South axes (E.C., 2001, 2005, 2008). Not all of these flow axes bear strong relationships with the NEI regions, but great opportunities seem to lay behind them. In particular, the East-western corridor, from Turin to Trieste, is that which probably best retraces the geography of economic and urban development in Northern Italy. This corridor threads into one of the most densely populated regions in Europe, where residential and industrial sprawl are fast consuming land and where transport links are easily saturated. Its deployment, although born out of a top down decision, would bring greater efficiency, especially for the HSR infrastructure in terms of rail accessibility.

The HSR section between Venice and Padua is a 28km stretch of double track line built alongside the existing double track line. With restructured nodes, the capacity of the network has increased to 700 trains per day. HSR service is foreseen to link Venice to Milan by 2020, while no precise deadline has yet been proposed for connections to Trieste. At the moment, travel times between the furthermost NEI city centres, Verona and Trieste, a distance of 250km, are at least 150 minutes by car and 200 minutes by train. These timings are definitely too long for any daily activity, but HSR plans forecast a reduction to only 90 minutes for a whole journey, enabling activities otherwise infeasible.

Figure 4. Tran-European transport network corridors (© European Union, 2004). Used with permission.

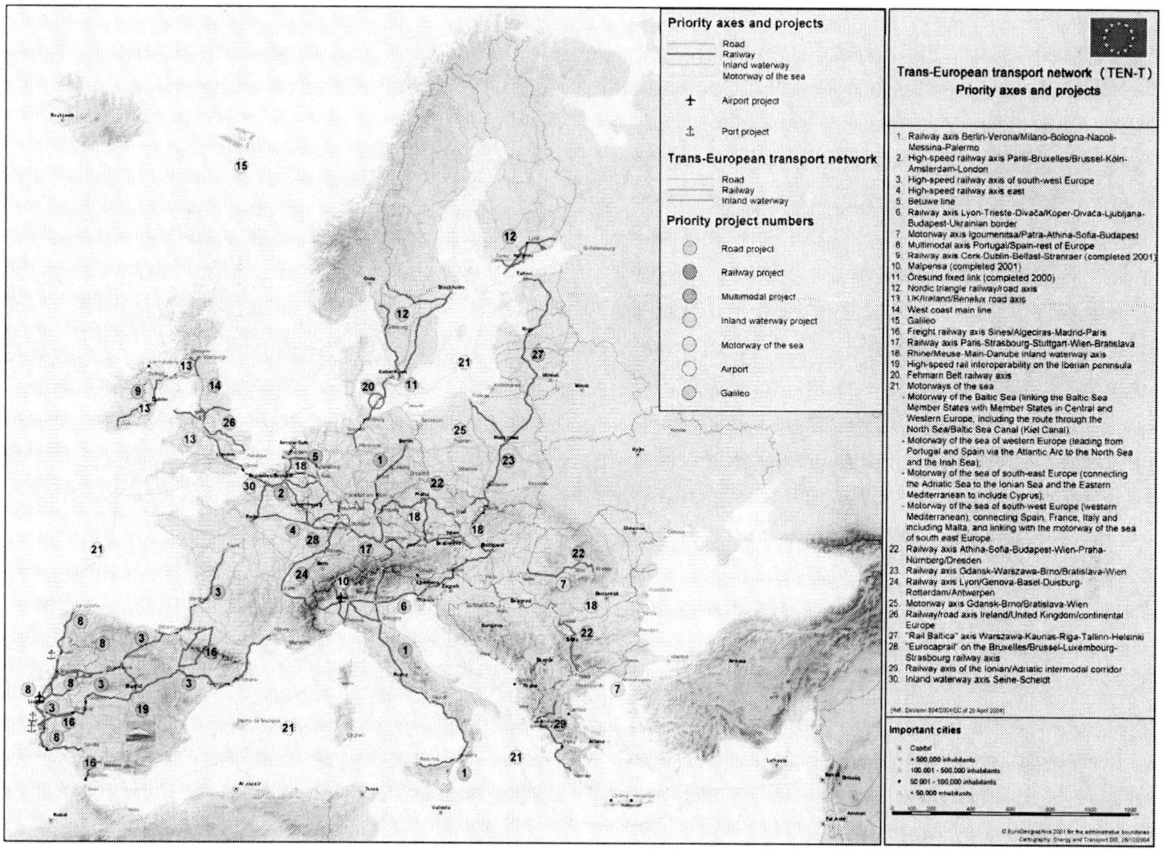

For what concerns the SFMR project, the stated objectives are to provide a sustainable transport alternative to car-mobility in a region that has reached capacity in several if not most of its highways (Regione del Veneto, 2004). With a current network of 150km, SFMR connects Padua, Venice and Treviso. Planned expansion of other 310km will bring Vicenza, Bassano, Montebelluna and Conegliano into the network by 2015. Further extension plans are under investigation to connect Rovigo, Belluno, and eventually other NEI regions (Regione del Veneto, 2001).

The following subsection illustrates some issues regarding the challenge of corridor development through local territories.

6.3 Critical Issues

HSR project implementation in NEI is being slowed down. Among several difficulties, one regards the layout of HSR infrastructure between Verona and Padua, without a definitive project due to Vicenza's case. The city demands HSR connection to expand its recognisability and prestigious city image, however its relatively small size and short distance from Padua (less than 40 km) constitute a problem to HSR efficiency. HSR station location is certainly a major planning issue in NEI, where no more than four HSR interchanges could be implemented without affecting HSR efficiency. These are to be located at the far ends, in Verona and Trieste, and intermediately, in Padua and Venice, while the rest of the network will have to find an alternative solution (Migliorini, 2007).

A second issue deals with limited land availability between Venice and Trieste, where regional authorities have devoted resources to expand highways rather than railways. Let alone environmental issues, this area is subject to increasing traffic flows originating in eastern European countries. Here, economies did not recede during the Global Financial Crisis of the late 2000 decade, thanks to low production costs and continued export of processed goods to the production cycle of more industrialised countries, such as Germany, France, the UK, Italy, and to a lesser extent Spain and Portugal. These goods cross NEI mostly by truck, generating huge external costs. Even though highways are strongly supported, congestion relief is a temporary outcome. Congestion is but a consequence of a missing strategy. In the absence of which, investments are directed as usual to highways to tackle what appears as the immediate problem. In this manner, rail projects and corridors are delayed in time and space.

However, moving from old and tranquil enclave to border land between East and West amongst Southern European countries, NEI should take advantage from its geographical position, but delays in strategic infrastructural provision dump opportunities. Without a strong rail strategy, it would be difficult to offer a competitive and efficient territorial system to bring together capabilities and market them internationally. Difficulties may be encountered for a number of reasons, including fractured decisional centres, whereby mobility is managed on separated levels (e.g. regional and urban), where there is often a lack of communication between different stakeholders and providers, and where solutions are slow to circulate and progress. Local decision-makers seem not to perceive Europe as the scene for territorial competition. Certainly it is a complex issue and, whereas Veneto has initiated the SFMR project, no other complementary transport strategy has yet been agreed upon to integrate NEI regions. The need for a common framework is essential to understand the new local conditions and build shared strategies to enter competition at higher levels.

In the following subsection few indications on the data used in this study are presented.

6.4 Data

The comparative analysis between the alternative strategies was conducted considering the quantities of transport with data acquired from the national railway operator RFI (*Rete Ferroviaria Italiana*), the Veneto Region, the consultancy NET Engineering. This data relates to travel times, waiting times, transfer times, speed, capacity, service quality, and the design of network and stations for the HSR and SFMR projects.

Technical characteristics of transit strategies, such as RMR, RHSR and LRT were readily available, while for the CRS strategy we collected data through Delphi surveys (Nelson & Niles, 1999). Experts were asked to give their input and use their imagination in an educated sense to retrieve possible steps that would lead toward the construction of a desired scenario (also called *back-casting* in (Robinson, 2003)), reflecting the need to obtain data from a preconceived *image of the future* (Banister, Stead, & Hickman, 2008). Given the nature of the technical study, participants were selected from an international and multidisciplinary panel of experts (Linstone & Turoff, 2002). Thus, the survey was published online in five languages (English, French, Italian, German and Spanish) and invitations to participate were sent by email to sector specific recipients selected worldwide from the railway industry, national and regional rail authorities, infrastructure managers and operators, university research organisations, and transit associations. The process respected the following four key characteristics (Rowe & Wright, 1999): (i) anonymity of participants; (ii) multiple rounds of the survey; (iii) controlled feedback; and (iv) statistical aggregation of responses. Running for 12 months, from August 2009 to July 2010, and with 1,000 recipients, the actual number of

Table 1. NEA indicators in relation to specific HSR stations for both years of analysis

	Venice		Padua	
	2010	*2015*	*2010*	*2015*
RMR	4.1451	3.5626	2.1456	2.2856
RHSR	3.9315	3.349	1.932	2.072
LRT	3.5296	3.0636	1.93	2.042
CRS	3.3027	2.818	1.6635	1.5556

participants was 282, with a response rate for completed questionnaires in the first round of 61.8% and 57.2% in the second round.

To account for the quality of land use, such as the distribution and evolution of population densities pertaining to the case study, the demographic data was acquired from the database provided by the Italian Institute of Statistics (ISTAT, 2010). Thus, data was retrieved at the municipal and provincial levels for locations along the SFMR network. To assess each strategy upon network expansion, growth rates account for the change between scenario years 2010 and 2015.

The intent of the remaining subsection is to present results obtained from the analysis of the correlation between HSR and selected regional strategies.

6.5 Results

NEA indicators were calculated for each alternative strategy in base year 2010 and network expansion year 2015. NEA indicators were also distinguished between the specific destination nodes in the network being the HSR station locations, either Padua or Venice. Thus, we obtained results for 16 scenarios (Table 1).

NEA indicators are dimensionless and represent increasing efficiency with diminishing values, being indirectly proportional to travel impedances, measured as the resistance to cover the network distance between origins and destinations, including the amount of travel, waiting and transfer time, fixed and variable costs, effort (reliability and comfort) and constraints due to infrastructure capacity.

Furthermore, NEA values were tested for their sensitivity to the choice of weighting factors. Thus, results (computed on a *density-based* weighting factor, e.g. on population densities of each municipality in the regional network) were compared with others computed on a *gravity based* weighting factor (e.g. on the same population densities discounted by the squared network distances to reach a HSR station). Figure 5 is an example to better understand the differences between the two sets of results (NEA indicators and testing results).

Network effects represent a global effect of the interactions with both HSR stations. They were calculated as the aggregate value of NEA indicators weighed by densities at destinations, thus they also present dimensionless values (Table 2).

Monetising network effects according to the HSR investments in the NEI regions by 2009 (corresponding to € 675 millions) and redistributing them through the regional network provides a more understandable view of the distributive patterns of accessibility benefits facilitated by the transit strategies (Figure 6).

These results will be discussed in the following section.

Figure 5. Comparison of ratios between every strategy's scenario in respect to RMR, the base strategy of the SFMR network, in correlation to the Venice interchange for both years of analysis

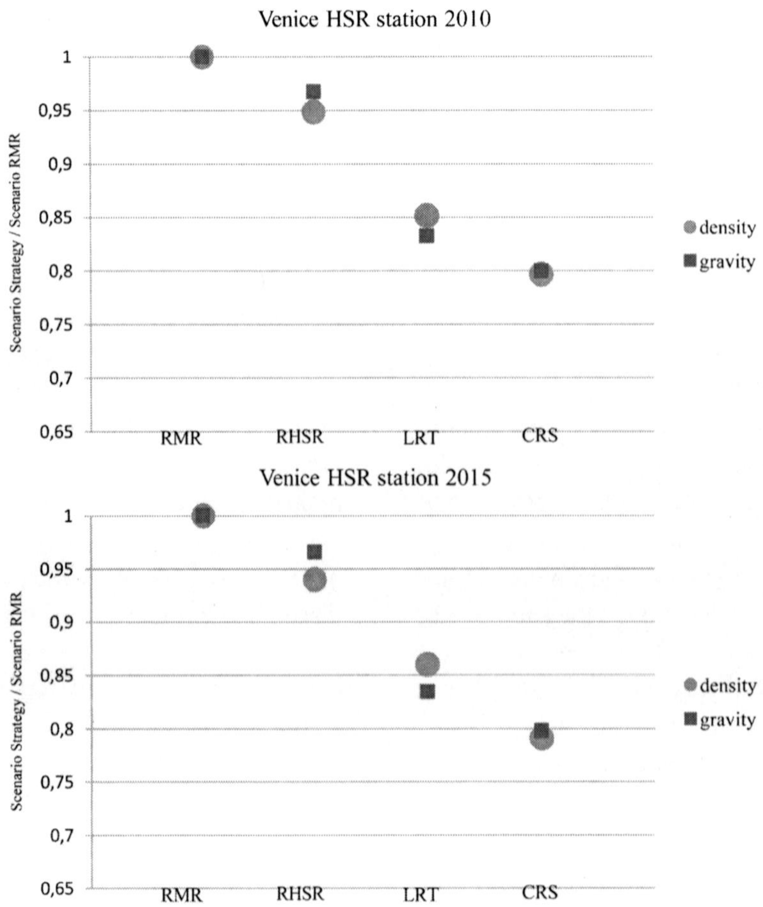

Table 2. Network effects and their monetisation for both years of analysis

	2010		*2015*	
RMR	3.1079	217.2 mil€	2.8978	232.9 mil€
RHSR	2.8944	233.2 mil€	2.6843	251.5 mil€
LRT	2.6999	250.0 mil€	2.5318	266.6 mil€
CRS	2.4525	275.2 mil€	2.1608	312.4 mil€

Figure 6. Monetised and distributed network effects in a portion of network far from any HSR station. Histograms show each strategy's distributive ability in the year 2015.

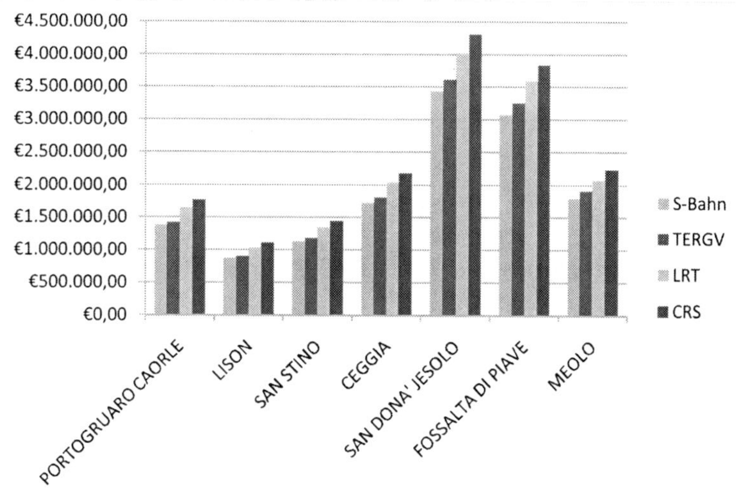

7. SOLUTIONS AND RECOMMENDATIONS

Using a theoretical interpretation of accessibility in relation to city competitiveness and a methodology to capture network effects of infrastructure improvements, such as changes in quality of service and area coverage, this chapter has examined a case study by considering four rail transit strategies as possible regional interfaces to HSR. In this section, analysis results will be discussed to evaluate strategies' ability to provide regional synergies in relation to, and eventually favour economies of scale deriving from, HSR infrastructure.

7.1 Discussion

Table 1 shows a gradual improvement when comparing strategies. NEA indicators improve if RHSR is made a substitute for RMR, further increasing if LRT is made a substitute for RHSR, and even more so if CRS is implemented instead of LRT. Overall, NEA indicators at the Venice interchange improve with network expansion, while they are somewhat contained in Padua. In particular, values in Padua indicate greater accessibility than those in Venice, but network development and densification seem to change the balance. Certainly, network configuration has a weight on the relative position of HSR stations and their influence. Thus, while Padua is more centrally located in respect to the 2010 regional network, Venice gains from eastward extensions in 2015. As a whole, these results confirm that network configurations do have an impact, when accounting for the kind of service provided, as well as its integration with land uses, this not only when investigated from the viewpoint of network design (Derrible & Kennedy, 2010).

Figure 5 compares two different measurement methods adopted to compute NEA indicators. It shows how values weighted on population densities present less variability than values weighted on a gravity-based function. Typically, distance discounts place greater emphasis on relations over short distances. Thus, it is evident how this measurement method lessens the impact of the RHSR strategy with a service quality based on relatively higher speeds and longer inter-stop spacing, while it favours the LRT strategy with a service quality based on frequency. This effect is further accentuated by network densification, bringing more branches and more stops into the service area. Nevertheless these methods do not impact on the relative position of strategies.

Table 2 presents network effects as a weighted aggregation of NEA values deriving from both HSR stations. Thus, it replicates strategy improvements viewed in Table 1 adding a global vision. In particular, the CRS strategy shows the best values in both years, demonstrating great potential to distribute benefits through seamless travel and the least number of transfers, confirming the importance of user benefits deriving from improved ease of travel, accessibility and convenience (Crockett & Hounsell, 2005; Wardman & Tyler, 2000). However, it is to be remembered that this strategy is highly hypothetical and would require a technologically revolutionary system design and operation.

To better evaluate network effects, Figure 6 presents their monetisation and redistribution for each strategy as sum of effects deriving from the interaction with both HSR stations. Histograms in different colour show each strategy's network effects on municipalities along the regional network, providing insight on their performance. It is evident how each strategy provides different shares of benefits following a similar distributive pattern, which appears to be dictated by context conditions. On the contrary, proportions depend on the distributive capability of strategies, thus on their characteristics. According to quality and service attributes, each strategy is able to distribute a portion of the original investment. The more a strategy is efficient, the larger the portion to distribute would be. In theory, it would be possible to forecast a scenario to determine the appropriateness of a specific strategy to a particular context, knowing demographic changing rates and modalities of network expansion/densification.

Overall results indicate that the correlation between HSR and any rail-based strategy in a regional context is supported by time-distance relations, where speed might be offset by frequencies. Therefore, the dilemma facing HSR operators whether to reduce stops in order to offer higher speeds (risking niche market shares) or to improve regional accessibility increasing connections (risking service quality deterioration) does not pose a threat if complemented with transit strategies where the level of service is upheld by short headways, such as in the case of supportive LRT (De Bruijn & Veeneman, 2009; Voskuhl, 1995).

7.2 Limitations

Results cannot be regarded as definitive, nor conclusive of the whole issue of transit's role in city competitiveness, but suggest a pathway that cities and regions might consider when dealing with proximate HSR infrastructure implementation, regional cohesion issues and road transport externalities.

The major limitation of this research analysis regards the assumption that a regional network is arbitrarily connected to only a short stretch of HSR infrastructure and vice versa. Certainly, other network influences could enter as well as exit our case study. However, it was necessary to investigate the problem with manageable data. It would be advisable to extend the analysis to more neighbouring networks, refining the measurement of network effects and possibly extending the knowledge on their interpretation.

These results might be useful elsewhere, although it would be advisable to contextualise them and extend the understanding of micro effects to aid decision-making. A great deal is left to cities and regions to build political and cultural coalitions to improve their competitiveness. These relationships precede any movement flow and are essential to create a vision of the desired territory.

Furthermore, the results analysed here provide a comparison between only four rail strategies, nonetheless they offer evidence as to which implementation strategy might be best suited for the regional area under study, with a view to maximising benefits deriving from proximate HSR infrastructure. In particular, results could help decision-makers as grounded theory, hence as a tool to understand gains and losses derived from different rail strategies to complement HSR service. However, it would be interesting to include in the analysis more strategies, even non-rail based, to account for the 'last mile', and to analyse correlations deriving from a chain of modes. This inclusion would provide insight also on the problem of self-potential (Bruinsma & Rietveld, 1998; Gutiérrez, 2001). Abnormally high values of NEA indicators in localities a short distance from a HSR station, especially if population densities are high, might indicate that a sufficiently efficient strategy is missing to service those areas. It is a problem of efficiency, exacerbated by long waiting times in respect to travel times. Thus, including in the analysis further strategies to improve accessibility at smaller scales would provide tools to complement and integrate the local transport system.

7.3 Recommendations

Intention of movement is difficult without accessibility, which in turn depends on infrastructure. A combination of transit infrastructure provision and territorial cooperation seems appropriate for long-sighted decisions to bring about a revolution toward sustainable mobility. Policies cannot call for drastic behavioural changes when no alternative infrastructure is provided at first. Hence, it would be useful to synthesise the degree of development in terms of infrastructure provided to the population served and the role of accessibility could be extended from a planning indicator to an indicator of promotion of regional development.

The challenge for cities and regions lays in finding the balance between cooperating with each other and the common goals of competitiveness. External agents may enter the field of competition at any time, which might bring increased competition or alternatively opportunities for new alliances. Inner cooperation among local actors is played in a larger framework of competition, both internal and external. It is then up to individual territorial stakeholders to find the best combination of factors to favour success; we consider this to be competitive cooperation.

At this point, in order to culturally or politically interlace relationships among decision-makers and stakeholders towards a shared common vision, decision-aiding tools are needed, both at central and local levels, to link local issues to strategic plans and understand commingling effects. Thus, understanding the relationship between local and strategic transport plans could support the quality of life, sustainability and competitiveness of cities and regions.

The next questions concern essentially how much public policy that targets investment on railways could satisfy an individually oriented mobility demand, and whether regional authorities are willing to adopt a strategy of no return from private to public mobility. This chapter presents these questions for future consideration, and for now recommends that investment in rail-based strategies that effectively distribute HSR benefits and integrate hierarchical networks be considered. To achieve this, it is important to raise regional rail networks to a standard which enables operation of interoperable trains, as for the RHSR and LRT strategies, or upgrade conventional tracks for the operation of metro systems, such as RMR.

Securing new alignments for future operation is also important, as is reopening regional and urban rail corridors for new service. Again, all of this needs strong political will and long-term collective focus by regional authorities to support initiatives directed toward improving accessibility through the development of local and regional transport rail links. This is particularly significant when processes of internationalisation, market enlargements and barrier elimination contribute to greater competition and new opportunities. Thus, gaining access to HSR infrastructure could become an issue of territorial survival, sustainability and competitiveness, where the debate could be brought at higher levels in the political hierarchy. The potential for positive economic impacts seems to be greater when a node becomes served by HSR service rather than being bypassed by it (Givoni, 2006). At this point, the battle is usually won by large metropolitan areas that tend to dominate by expanding their influence both locally and internationally (Rietveld & Bruinsma, 1998). If neighbouring cities do not have the strength alone, networking capabilities at the regional level might form the strong core necessary to face competition. Regions might counteract resource and opportunity losses, thanks to some instruments investigated in this chapter, by offering competitive functions and services, where mobility and accessibility should be considered the main objectives of transport policies as a good of strategic importance. Thus, there are few essential ingredients needed to achieve this goal:

1. A regional metro rail system or similar strategy, along with some shared utilities;
2. Better access to quality long distance transport services such as HSR, air or sea;
3. Training in competitive practice; and
4. Development of strong international interests.

It is important to have a territorial branding to market city-regional recognisability, and thus build an interurban alliance. This needs to be supported by a common vision of territorial strategic functions as a shared tool of regional policies in order to facilitate the concept of a networking metropolis.

With due limitations, the case study results presented above hold international relevance for all countries that are considering HSR in their transport plans and that need greater information regarding the regional distribution of potential accessibility benefits of HSR integration with local and regional rail transit networks. The process to determine network effects could also be useful as a tool to understand how a HSR link could be best incorporated to optimise benefit distribution within the region, rather than just finding a justification for its deployment (Harman, 2006). As such, results might aid in planning the location of a HSR station following the conclusions and recommendation from Brons (2009), which indicate that investments will bear greater benefits if directed toward the most accessible part of the network. Thus, this methodology could be used to evaluate which node in a regional transit network is best suited to accommodate interlinking services with networks of higher hierarchical importance, such as a new HSR link. With this aim, improving regional accessibility through regional rail strategies could be planned for and implemented, even in the absence of a specific HSR plan, but which would ultimately support its development.

Furthermore, in regions where densities or distances are not traditionally considered favourable to HSR implementation, it seems appropriate to proceed gradually, implementing first an alternative transit strategy. In this way, the transit strategy might help to build consensus and greater use of rail infrastructure (Gordon, 2010; Kuby, Barranda, & Upchurch, 2004; Lane, 2008), and with time might foster movement towards HSR.

When considering HSR project evaluation and development, the impacts of HSR for medium and small centres that are distant from a HSR station may be particularly significant should connectedness as discussed be realised, and may therefore bear great implications both socio-politically and technically. Accounting for network effects might also form a basis on which to balance stakeholders' considerations either in network development projects, or in managing a hierarchically integrated transport system.

Finally, assessing the competitiveness of small and medium sized cities and their regions for efficiency and accessibility provided by local transit strategies could prove a valid instrument for both planners and policy makers to build a stronger regional transport fabric, which is capable of accommodating HSR corridors that do not drain resources toward bigger cities or shift resources away from the region itself.

8. RESEARCH DIRECTIONS

This chapter has revealed that the relationship between regional rail transit strategies and HSR infrastructure projects are both a cause and a consequence of city-region competitiveness. Thus, transit networks may be the product of advanced economies, but may also produce accessibility benefits to drive changes in population location and eventually influence economic and social trends. However, to facilitate the exchange of benefits, transit strategies need to be conducive to network effects in association with HSR service. In fact, network effects generally occur as a resonance of accessibility between other modes, sectors or areas and they arise because of economies of scale, densities or consumption externalities (Laird, Nellthorp, & Mackie, 2005). The twofold nature of these spillovers between hierarchical networks is that of giving and receiving in unison, much depending on one's perspective. Thus, spillovers could be identified either as gains and losses, or as imports and exports, from one area to the others. The first viewpoint presupposes a strategy to retain spillovers (Cheshire, 1999), while the second facilitates sharing and collaborating (Pereira & Roca-Sagalés, 2003; Roy, 2003; van Exel, Rienstra, Gommers, Pearman, & Tsamboulas, 2002). In this chapter, this second viewpoint has been pursued with the belief that territorial self-centeredness undermines economies of scale and scope, and thus network effects. The correlation between transport links at the regional and inter-regional levels have been investigated, taking into account synergetic effects of coordinated planning, design and use of advanced infrastructures (Maggi, Masser, & Nijkamp, 1992) such as HSR and innovative complementary transit strategies.

This chapter has limited its analysis to the quantification of network effects exchanged between HSR infrastructure and only four selected transit strategies to complement HSR accessibility at the regional level. A suggestion for further research would be to investigate more options or combinations of strategies to better appreciate benefits deriving from the differentiation and integration of regional and urban services. As an example, it might be interesting to investigate whether the combination of two strategies, such as RHSR (running express trains) and LRT (running frequent trains) might provide even greater accessibility benefits than those highlighted in this chapter.

The sensitivity of network effects could be further tested in relation to network expansion, densification or integration with neighbouring networks to assess whether and where dilution thresholds exist. Thus, future research ought to tackle impedance values (Gutiérrez, Condeco-Melhorado, & Martín, 2010) to cross networks with regard to different strategies, but also with regard to network integration. This would help to understand whether more remote areas would need more efficient transit systems to connect to the city thereby enhancing regional competitiveness.

Moreover, since evaluation methodologies to assess the exchanges of benefits between HSR and regional transit systems are rare or rather limited (Cascetta & Pagliara, 2008), future research and investigation would certainly be worthwhile. A further consideration might be to analyse cost effectiveness of alternative rail strategies against HSR. It is known that high quality, grade-separated transit improvements can be cost effective investments when all economic impacts are considered (Litman, 2007). It would be helpful to planners and decision-makers to better understand when and where a HSR alignment could be more efficiently integrated with regional networks. Such analysis might also account for further benefits that otherwise would have not been considered when looking only at HSR profitability.

Developing a series of worthwhile alternative rail strategies might also generate sufficient interest to direct planning resources and project funding opportunities that, based on historical trends, would otherwise be invested on road infrastructure. In our case study, besides the short stretch of HSR infrastructure that was built between Padua and Venice to solve pre-existing capacity issues, along with the long waited regional metro system in the heart of Veneto, no other rail strategy exists from Verona to Pordenone from which to draw funding. The very same HSR plans are at stake when not supported by long term intermodal and coordinated visions, as is the case between Venice and Trieste, where HSR is considered incompatible with the enlargement of the toll highway along the same corridor. Alternative alignments conflict with local interests and HSR planning has been delayed indefinitely, yielding severe consequences on the implementation of the European railway axis from Lyon to the Ukrainian border.

To overcome this impasse, a suggested research direction would be to deepen the investigation on HSR as a freight carrier (Pazour, Meller, & Pohl, 2010; Plotkin, 1997; Van Ham & Muilerman, 2002) and to analyse the potential relationship between HSR and local transit for goods delivery in view of regional competitiveness. Investigating network effects from a freight perspective would be useful to better appreciate the feasibility of HSR for freight movement, and the most conducive conditions for its territorial integration. The use of tramways for this purpose is still fairly rare (Anderson, 2008) and exploratory (e.g. cargo trams in Dresden and city cargo in Amsterdam). Investigations have already indicated that freight use on rail transit systems may provide a means to lower levels of subsidy, if not altogether deriving profits (Resor, 2004; Sivakumaran, 2010). In Australia, for instance, a number of urban rail systems share slots with freight trains during off peak periods. Even strategies not yet realised, such as CRS, might be more feasible alternatives when freight is considered. Some model applications have been already developed in the literature to increase infrastructure capacity through end-to-end headways, such as the *train-coupling and sharing* system (Böcker, Lind, & Zirkler, 2001). Thus, the field appears promising and welcoming of new studies on transit strategies to provide evidence of benefit distribution and competitiveness, with freight handling a central role.

In summary, research is certainly necessary to provide instruments for cities and regions to manage their territories and bring about sustainable development. This chapter has suggested that, to achieve the desired level of integration between transport provision and regional development, investigation of complementary rail strategies could provide greater benefits when complementing HSR links. However, close cooperation between all stakeholders is most important and the role of the municipal government is indispensable in securing needed synergies (De Bruijn & Veeneman, 2009; Priemus & Konings, 2001). The cities of Padua and Venice, future HSR station locations, seem to have understood this and have both been developing a system of urban trams and collaborating to commence the SFMR project.

9. CONCLUSION

This chapter has presented and analysed the North-Eastern Italian (NEI) regions as a case study for city-regions that are in transition to metropolitan form. Whether interpreted as a precarious balance between rural and urban environments or as a slow move from one economy to the other, it was necessary to investigate the impact of transit networks on regional competitiveness and territorial change. The analysis has examined whether strategic infrastructure, such as high speed rail (HSR), might be better integrated territorially to produce positive effects on local accessibility. Therefore, the research effort was directed to understand the impact of prospective transit systems to interface with HSR and serve regional and urban agglomerations, which are not directly served by HSR.

Although it has been predicted that increases in service quality offered by HSR links may result in increased demand at the HSR stations, results from the applied method to the case study demonstrate that HSR patronage could depend mostly on the ease of connectedness of the HSR stations with the surrounding territory. In particular, network effects, that account for the exchange of spillovers between different networks, relate to the accessibility efficiency of alternative transit strategies to complement HSR service at the regional level.

A further appreciation was possible by monetising transit network effects, brought about by HSR investments in the region, and redistributing them through the regional network, thus quantifying benefits to each municipality served. In so doing, local communities might observe the impact of a distant HSR station on their territory. Also, national and regional authorities might obtain insight on which rail transit strategy best provides ease of connections, as an indispensable condition for territorial cohesion and competitiveness.

With the methodology presented in this chapter, a supporting tool is offered that might help to formulate adequate policies addressing infrastructure, technologies, service networks and land uses. Regional cities are required to collaborate and integrate their capabilities while sharing a vision for integration and equity.

Decision-makers could derive from this chapter a better understanding of some possible complementary solutions to exploit HSR benefits for the competitiveness of cities and regions. Results indicate that building and expanding transit networks could be most effective for small to medium municipalities to improve regional accessibility, and eventually stimulate growth and development through collaboration with neighbouring communities. This chapter has offered notions for further research in relation to whether regional transit systems might induce sufficient demand for rail use, even in small cities, to increase their competitiveness, and eventually allow them to win the location choice of a new HSR alignment and access point in competition with bigger ones, that may be without a rail strategy.

It is the role of political and managerial masters of cities and regions to serve as collective entrepreneurs of territorial and geographic change. Assuming land use as a subjective form produced by their capacity to achieve common visions, it appears most important that regional territories share a strategy to build capability and critical mass for a common metropolitan policy and governance, rather than acting in terms of metropolitan perimeters. These results are useful for examining future directions of territories open to international competition, both in Europe and worldwide.

REFERENCES

Air France. (2010). *Tgv high-speed train*. Retrieved July 26, 2010, from http://corporate.airfrance.com/ en/alliances/competition/ tgv-high-speed-train/

Aliadière, L. (2008). Fast track to sustainable mobility. *Railway Gazette International*, *164*(3), 141–143.

Anderson, N. (2008). Light rail for freight. *Logistics & Transport Focus*, *10*(8), 42–45.

Andersson, D. E., & Andersson, Å. E. (2008). Infrastructural change and secular economic development. *Technological Forecasting and Social Change*, *75*(6), 799–816. doi:10.1016/j.techfore.2007.08.003

Arduin, J. P., & Ni, J. (2005). 40 years of high-speed railways - French TGV network development. *Japan Railway & Transport Review*, *40*, 22–28.

Banister, D. (2008). The sustainable mobility paradigm. *Transport Policy*, *15*, 73–80. doi:10.1016/j.tranpol.2007.10.005

Banister, D., & Berechman, Y. (2001). Transport investment and the promotion of economic growth. *Journal of Transport Geography*, *9*(3), 209–218. doi:10.1016/S0966-6923(01)00013-8

Banister, D., Stead, D., & Hickman, R. (2008). Looking over the horizon: Visioning and backcasting. In Himanen, V., Lee-Gosselin, M., & Perrels, A. (Eds.), *Building blocks for sustainable transport* (pp. 25–54). Helsinki, Sweden: VATT and Emerald.

Baradaran, S., & Ramjerdi, F. (2001). Performance of accessibility measures in Europe. *Journal of Transportation and Statistics*, *4*(2-3), 31–38.

BBSR. (2011). *Metropolitan areas in Europe*. Bonn, Germany: Federal Institute for Research on Building, Urban Affairs and Spatial Development.

Ben-Akiva, M., & Lerman, S. R. (1979). Disaggregate travel and mobility choice models and measures of accessibility. In Hensher, D. A., & Stopher, P. R. (Eds.), *Behavioural travel modelling* (pp. 654–679). London, UK: Croom Helm.

Bertolini, L. (2008). Station areas as nodes and places in urban networks: An analytical tool and alternative development strategies. In Bruinsma, F., Pels, E., Priemus, H., Rietveld, P., & van Wee, B. (Eds.), *Railway development: Impacts on urban dynamics* (pp. 35–57). Heidelberg, Germany: Physica-Verlag.

Bertolini, L., & le Clercq, F. (2003). Urban development without more mobility by car? Lessons from Amsterdam: A multimodal urban region. *Environment & Planning A*, *35*(4), 575–589. doi:10.1068/a3592

Bertolini, L., le Clercq, F., & Kapoen, L. (2005). Sustainable accessibility: A conceptual framework to integrate transport and land use planmaking. Two test-applications in the Netherlands and a reflection on the way forward. *Transport Policy*, *12*(3), 207–220. doi:10.1016/j.tranpol.2005.01.006

Blum, U., Haynes, K. E., & Karlsson, C. (1997). The regional and urban effects of high-speed trains. *The Annals of Regional Science*, *31*(1), 1–20. doi:10.1007/s001680050036

Böcker, J., Lind, J., & Zirkler, B. (2001). Using a multi-agent approach to optimise the train coupling and sharing system. *European Journal of Operational Research*, *131*(2), 242–252. doi:10.1016/S0377-2217(00)00124-7

Bonnafous, A. (2002). The regional impact of the TGV. *Classics in Transport Analysis*, *3*, 480–490.

Brocker, J., & Rietveld, P. (2009). Infrastructure and regional development. In Capello, R., & Nijkamp, P. (Eds.), *Handbook of regional growth and development theories*. Cheltenham, UK: Edward Elgar.

Brons, M., Givoni, M., & Rietveld, P. (2009). Access to railway stations and its potential in increasing rail use. *Transportation Research Part A, Policy and Practice*, *43*(2), 136–149. doi:10.1016/j.tra.2008.08.002

Bruinsma, F., Nijkamp, P., & Rietveld, P. (1992). Regional economic transformation and social overhead investments. *Tijdschrift voor Economische en Sociale Geografie*, *83*(1), 3–12. doi:10.1111/j.1467-9663.1992.tb00604.x

Bruinsma, F., & Rietveld, P. (1998). The accessibility of European cities: Theoretical framework and comparison of approaches. *Environment & Planning A*, *30*(3), 499–521. doi:10.1068/a300499

Brunello, L., Bunker, J., & Ferreira, L. (2008). *Continuous railway systems: An innovative approach to improve high speed rail sustainability*. Paper presented at the 10th International Conference on the Application of Advanced Technologies in Transportation.

Burckhart, K., & Blair, C. (2009). *Urban intermodality: Potentials for connecting the cities' public transport*. Paper presented at the XV Urban Transport Conference.

Burmeister, A., & Colletis-Wahl, K. (1998). Proximity in networks. The role of transport infrastructure. In Reggiani, A. (Ed.), *Accessibility, trade and locational behaviour* (pp. 201–218). Aldershot, UK: Ashgate.

Buyck, C. (2008). *Planes versus trains*. Retrieved July 29, 2010, from http://atwonline.com/ it-distribution/article/ planes-versus-trains-0309

Campos, J., & de Rus, G. (2009). Some stylized facts about high-speed rail: A review of hsr experiences around the world. *Transport Policy*, *16*(1), 19–28. doi:10.1016/j.tranpol.2009.02.008

Cantos, P., Gumbau-Albert, M., & Maudos, J. (2005). Transport infrastructures, spillover effects and regional growth: Evidence of the spanish case. *Transport Reviews*, *25*(1), 25–50. doi:10.1080/0144164100016 76852

Capello, R., & Rietveld, P. (1998). The concept of network synergy in economic theory: Policy implications. In Button, K., Nijkamp, P., & Priemus, H. (Eds.), *Transport networks in Europe* (pp. 57–83). Cheltenham, UK: Edward Elgar.

Cascetta, E., & Pagliara, F. (2008). Integrated railways-based policies: The regional metro system (RMS) project of Naples and Campania. *Transport Policy*, *15*(2), 81–93. doi:10.1016/j.tranpol.2007.11.001

Cheshire, P. (1999). Cities in competition: Articulating the gains from integration. *Urban Studies (Edinburgh, Scotland)*, *36*(5-6), 843–864. doi:10.1080/0042098993240

Cheshire, P. (2009). Spatial policies, planning and urban competitiveness. In Karlsson, C., Johansson, B., & Stongh, R. (Eds.), *The particular case of London*. Cheltenham, UK: Edward Elgar.

Couto, A., & Graham, D. J. (2008). The impact of high-speed technology on railway demand. *Transportation Planning and Analysis*, *35*, 111–128.

Crockett, J., & Hounsell, N. (2005). Role of the travel factor convenience in rail travel and a framework for its assessment. *Transport Reviews*, *25*(5), 535–555. doi:10.1080/01441640500064389

Cuadrado, J. R. (2008). Towards new European peripheries. In Stongh, R. (Ed.), *Innovation, agglomeration and regional competition*. Cheltenham, UK: Edward Elgar.

Daily Mail Reporter. (2009). *Taxi of tomorrow: The space-age pods with no driver, no fumes... Just a 'start' button*. Retrieved November 26, 2009, from http://www.dailymail.co.uk/ sciencetech/ article-1205970/ Taxi-future-Space-age-pods

De Bruijn, H., & Veeneman, W. (2009). Decision-making for light rail. *Transportation Research Part A, Policy and Practice*, *43*(4), 349–359. doi:10.1016/j.tra.2008.11.003

Debrezion, G., Pels, E., & Rietveld, P. (2009). Modelling the joint access mode and railway station choice. *Transportation Research Part E, Logistics and Transportation Review*, *45*(1), 270–283. doi:10.1016/j.tre.2008.07.001

Derrible, S., & Kennedy, C. (2010). Characterizing metro networks: State, form, and structure. *Transportation*, *37*(2), 275–297. doi:10.1007/s11116-009-9227-7

Domanski, R. (1979). Accessibility, efficiency and spatial organization. *Environment & Planning A*, *11*, 1189–1206. doi:10.1068/a111189

E.C. (2001). *White paper European transport policy for 2010: Time to decide*. Brussels, Belgium: Energy and Transport DG.

E.C. (2005). *Trans-European transport network: TEN-T priority axes and projects 2005*. Brussels, Belgium: Energy and Transport DG.

E.C. (2008). *TEN-T trans-European transport network: Implementation of the priority projects progress report*. Brussels, Belgium: Energy and Transport DG.

ESPON. (2006). *Territory matters for competitiveness and cohesion*. Luxembourg: Ministry of Interior and Spatial Development.

European Union. (2004). Trans-European transpot network (TEN-T). Priority axes and projects. Retrieved January 14, 2011, from http://ec.europa.eu/ten/ transport/maps/doc/ axes/pp00.pdf

Freilich, R. H. (1998). The land-use implications of transit-oriented development: Controlling the demand side of transportation congestion and urban sprawl. *Urban Lawyer*, *30*(3), 547–572.

Fröidh, O., & Nelldal, B.-L. (2008). Regional high-speed trains on the svealand line: Evaluation of effects. In Bruinsma, F., Pels, E., Rietveld, P., Priemus, H., & van Wee, B. (Eds.), *Railway development: Impacts on urban dynamics* (pp. 295–314). Heidelberg, Germany: Physica-Verlag.

Fryer, C. E. J. (1997). *A history of slipping and slip carriages*. Oxford, UK: The Oakwood Press.

Garmendia, M., Urena, J. M., Ribalaygua, C., Leal, J., & Coronado, J. M. (2008). Urban residential development in isolated small cities that are partially integrated in metropolitan areas by high speed train. *European Urban and Regional Studies*, *15*(3), 249–264. doi:10.1177/0969776408090415

Garmendia, M., Urena, J. M., Rivas, A., Coronado, J. M., Menendez, M., Gallego, J. I., & Romero, V. (2009). *High speed rail, a new mode of suburban metropolitan transport*. Paper presented at the XV Urban Transport Conference.

Geurs, K. T., van Wee, B., & Rietveld, P. (2006). Accessibility appraisal of integrated land-use-transport strategies: Methodology and case study for the netherlands randstad area. *Environment and Planning. B, Planning & Design*, *33*(5), 639–660. doi:10.1068/b31099

Givoni, M. (2006). Development and impact of the modern high-speed train: A review. *Transport Reviews*, *26*(5), 593–611. doi:10.1080/01441640600589319

Givoni, M., & Rietveld, P. (2007). The access journey to the railway station and its role in passengers' satisfaction with rail travel. *Transport Policy*, *14*(5), 357–365. doi:10.1016/j.tranpol.2007.04.004

Gordon, C. (2010). *Density and LRT: The case of Canberra, Australia*. Paper presented at the 89th Transportation Research Board Annual Meeting.

Greenair. (2008). *Air France plans to enter high-speed rail venture with Veolia on routes to London and Amsterdam*. Retrieved July 29, 2010, from http://www.greenaironline.com/ news.php?viewStory=253

Guihéry, L. (2002). The future of railway transport in europe. In Black, W. R., & Nijkamp, P. (Eds.), *Social change and sustaianble transport* (pp. 241–247). Bloomington, IN: Indiana University Press.

Gutiérrez, J. (2001). Location, economic potential and daily accessibility: An analysis of the accessibility impact of the high-speed line Madrid-Barcelona-French border. *Journal of Transport Geography*, *9*(4), 229–242. doi:10.1016/S0966-6923(01)00017-5

Gutiérrez, J., Condeco-Melhorado, A., & Martín, J. C. (2010). Using accessibility indicators and gis to assess spatial spillovers of transport infrastructure investment. *Journal of Transport Geography*, *18*(1), 141–152. doi:10.1016/j.jtrangeo.2008.12.003

Gutiérrez, J., Monzón, A., & Piñero, J. M. (1998). Accessibility, network efficiency, and transport infrastructure planning. *Environment & Planning A*, *30*(8), 1337–1350. doi:10.1068/a301337

Hall, P. (1999). The european high-speed train and urban development: Experiences in 14 European regions. *Progress in Human Geography*, *23*(4), 670–671.

Handy, S., & Niemeier, D. A. (1997). Measuring accessibility: An exploration of issues and alternatives. *Environment & Planning A*, *29*(7), 1175–1194. doi:10.1068/a291175

Hansen, W. G. (1959). How accessibility shapes land-use. *Journal of the American Institute of Planners*, *25*, 73–76. doi:10.1080/01944365908978307

Harman, R. (2006). *High speed trains and the development and regeneration of cities*. Retrieved January 12, 2010, from http://www.greengauge21.net/

Haynes, K. E. (1997). Labor markets and regional transportation improvements: The case of high-speed trains - an introduction and review. *The Annals of Regional Science*, *31*(1), 57–76. doi:10.1007/s001680050039

Haywood, R. (2009). *Railways, urban development and town planning in Britain: 1948-2008*. Aldershot, UK: Ashgate.

ISTAT. (2010). *Principali indicatori demografici della popolazione residente*. Retrieved March 26, 2010, from http://demo.istat.it/

Knapp, W., Kunzmann, K. R., & Schmitt, P. (2004). A cooperative spatial future for Rheinruhr. *European Planning Studies*, *12*(3), 323–349. doi:10.1080/0965431042000195047

Krugman, P. (1979). Increasing returns, monopolistic competition, and international trade. *Journal of International Economics*, *9*(4), 469–479. doi:10.1016/0022-1996(79)90017-5

Krugman, P. (1994). Competitiveness: A dangerous obsession. *Foreign Affairs (Council on Foreign Relations)*, *73*(2), 28. doi:10.2307/20045917

Kuby, M., Barranda, A., & Upchurch, C. (2004). Factors influencing light-rail station boardings in the United States. *Transportation Research Part A, Policy and Practice*, *38*(3), 223–247. doi:10.1016/j.tra.2003.10.006

Kwan, M. P., & Weber, J. (2008). Scale and accessibility: Implications for the analysis of land use-travel interaction. *Applied Geography (Sevenoaks, England)*, *28*(2), 110–123. doi:10.1016/j.apgeog.2007.07.002

Laird, J. J., Nellthorp, J., & Mackie, P. J. (2005). Network effects and total economic impact in transport appraisal. *Transport Policy*, *12*(6), 537–544. doi:10.1016/j.tranpol.2005.07.003

Lane, B. W. (2008). Significant characteristics of the urban rail renaissance in the United States: A discriminant analysis. *Transportation Research Part A, Policy and Practice*, *42*(2), 279–295. doi:10.1016/j.tra.2007.10.001

Latour, B. (1993). *Aramis, ou l'amour des techniques*. Paris, France: Editions de la Découverte.

Leonardi, G. (1978). Optimum facility location by accessibility maximizing. *Environment & Planning A*, *10*, 1287–1305. doi:10.1068/a101287

Linneker, B., & Spence, N. (1996). Road transport infrastructure and regional economic development. The regional development effects of the m25 London Orbital Motorway. *Journal of Transport Geography*, *4*(2), 77–92. doi:10.1016/0966-6923(96)00001-4

Linstone, H., & Turoff, M. (2002). The Delphi method. In M. Turoff, & H. Linstone (Eds.), *Techniques and applications*. Retreived November 26, 2009 from http://www.is.njit.edu/ pubs/delphibook/

Litman, T. (2007). Evaluating rail transit benefits: A comment. *Transport Policy*, *14*(1), 94–97. doi:10.1016/j.tranpol.2006.09.003

López, E., Gutiérrez, J., & Gómez, G. (2008). Measuring regional cohesion effects of large-scale transport infrastructure investments: An accessibility approach. *European Planning Studies*, *16*(2), 277–301. doi:10.1080/09654310701814629

López, E., Monzon, A., Ortega, E., & Mancebo, S. (2008). *Strategic assessment of transport infrastructure plans on European integration: Application for the Spanish strategic transport and infrastructure plan 2005-2020*. Paper presented at the 87th Transportation Research Board Annual Meeting.

López, E., Monzón, A., Ortega, E., & Mancebo, S. (2009). Assessment of cross-border spillover effects of national transport infrastructure plans: An accessibility approach. *Transport Reviews: A Transnational Transdisciplinary Journal, 29*(4), 515 - 536.

Maggi, R., Masser, I., & Nijkamp, P. (1992). Missing networks in european transport and communications. *Transport Reviews: A Transnational Transdisciplinary Journal, 12*(4), 311 - 321.

Marshall, A. (1920). *Principles of economics* (revised edition). London, UK: Macmillan.

Marshall, S. (2001). The challenge of sustainable transport. In Layard, A., Davoudi, S., & Batty, S. (Eds.), *Planning for a Sustainable Future* (pp. 131–147). London, UK: Spon.

Martin, F. (1997). Justifying a high-speed rail project: Social value vs. Regional growth. *The Annals of Regional Science*, *31*(2), 155–174. doi:10.1007/s001680050043

Martín, J. C., Gutiérrez, J., & Román, C. (2004). Data envelopment analysis (dea) index to measure the accessibility impacts of new infrastructure investments: The case of the high-speed train corridor Madrid-Barcelona-French border. *Regional Studies*, *38*(6), 697–712. doi:10.1080/003434042000240987

Martín, J. C., & Reggiani, A. (2007). Recent methodological developments to measure spatial interaction: Synthetic accessibility indices applied to high-speed train investments. *Transport Reviews*, *27*(5), 551–571. doi:10.1080/01441640701322610

Menerault, P. (1998). Processus de territorialisation des réseaux: Analyse de la grande vitesse ferroviaire à l'échelle régionale. *Networks and communication studies, 12*(1-3), 161-184.

Meyer, D., & Miller, E. J. (2001). *Urban transportation planning* (2nd ed.). New York, NY: McGraw-Hill.

Migliorini, F. (2007). *Un corridoio tutto da inventare: L'alta velocità per far crescere città e distretti*. Venezia, Italy: Marsilio.

Munnell, A. H. (1992). Policy watch infrastructure investment and economic growth. *The Journal of Economic Perspectives*, *6*(4), 189.

Nelson, A. C., & Niles, J. (1999). *Essentials for transit-oriented development planning: Analysis of non-work activity patterns and a method for predicting success*. Paper presented at the 7th Transportation Research Board Conference on the Application of Transportation Planning Methods.

Nijkamp, P., & Van Geenhuizen, M. (1997). European transport: Challenges and opportunities for future research and policies. *Journal of Transport Geography*, *5*(1), 4–11. doi:10.1016/S0966-6923(96)00041-5

OpenStreetMap. (2011). *Street map of North-Eastern Italy*. Retrieved January 14, 2011, from http://www.openstreetmap.org/?lat=45.89&lon=12.25& zoom=8&layers=M

Page, M., Kelly, C., May, A., Jones, P., & Forrester, J. (2009). Enhancing appraisal methods to support sustainable transport and land use policies. *European Journal of Transport and Infrastructure Research*, *9*(3), 296–313.

Pazour, J. A., Meller, R. D., & Pohl, L. M. (2010). A model to design a national high-speed rail network for freight distribution. *Transportation Research Part A, Policy and Practice*, *44*(3), 119–135. doi:10.1016/j.tra.2009.11.006

Pereira, A. M., & Roca-Sagalés, O. (2003). Spillover effects of public capital formation: Evidence from the Spanish regions. *Journal of Urban Economics*, *53*(2), 238–256. doi:10.1016/S0094-1190(02)00517-X

Plassard, F. (1992). L'impact territorial des liaisons à grande vitesse. In Derycke, P. H. (Ed.), *Espace et dynamiques territoriales* (pp. 243–262). Paris, France: Economica.

Plotkin, D. (1997). Carrying freight on high-speed rail lines. *Journal of Transportation Engineering*, *123*(3), 199–201. doi:10.1061/(ASCE)0733-947X(1997)123:3(199)

PRC Ministry of Railways. (2010). *Planning of the National Railway Network* (in Chinese). Retrieved July 13, 2010, from http://www.china-mor.gov.cn/

Priemus, H., & Konings, R. (2001). Light rail in urban regions: What Dutch policymakers could learn from experiences in France, Germany And Japan. *Journal of Transport Geography*, *9*(3), 187–198. doi:10.1016/S0966-6923(01)00008-4

Priemus, H., Nijkamp, P., & Banister, D. (2001). Mobility and spatial dynamics: An uneasy relationship. *Journal of Transport Geography*, *9*(3), 167–171. doi:10.1016/S0966-6923(01)00007-2

Reggiani, A. (1998). Accessibility, trade and locational behaviour: An introduction. In Reggiani, A. (Ed.), *Accessibility, trade and locational behaviour* (pp. 1–14). Aldershot, UK: Ashgate.

Regione del Veneto. (2001). *Il sistema ferroviario metropolitano regionale dell'area centrale veneta*. Este, Italy: Grafica Atestina.

Regione del Veneto. (2004). *Piano regionale dei trasporti del veneto: Quaderno di sintesi*. Venezia, Italy: Segreteria Regionale Infrastrutture e Mobilitá.

Resor, R., Blaze, J., & Morlok, E. (2004). Short-haul rail intermodal: Can it compete with trucks? *Transportation Research Record*, *1873*, 45–52. doi:10.3141/1873-06

Reusser, D. E., Loukopoulos, P., Stauffacher, M., & Scholz, R. W. (2008). Classifying railway stations for sustainable transitions - Balancing node and place functions. *Journal of Transport Geography*, *16*(3), 191–202. doi:10.1016/j.jtrangeo.2007.05.004

Rietveld, P., & Bruinsma, F. (1998). *Is transport infrastructure effective?*Berlin, Germany: Springer-Verlag.

Robinson, J. (2003). Future subjunctive: Backcasting as social learning. *Futures*, *35*(8), 839–856. doi:10.1016/S0016-3287(03)00039-9

Rowe, G., & Wright, G. (1999). The delphi technique as a forecasting tool: Issues and analysis. *International Journal of Forecasting*, *15*(4), 353–375. doi:10.1016/S0169-2070(99)00018-7

Roy, R. (2003). European versus national-level evaluation: The case of the PBKAL high-speed rail project. In Pearman, A., Mackie, P., & Nellthorp, J. (Eds.), *Transport projects, programmes and policies: Evaluation needs and capabilities* (pp. 125–138). Aldershot, UK: Ashgate.

Sasaki, K., Ohashi, T., & Ando, A. (1997). High-speed rail transit impact on regional systems: Does the shinkansen contribute to dispersion? *The Annals of Regional Science*, *31*(1), 77–98. doi:10.1007/s001680050040

Schneider, J. B. (1993). Selecting and evaluating intermodal stations for intercity high-speed ground transportation. *Transportation Quarterly*, *47*(2), 221–245.

Scott, A. J., & Storper, M. (Eds.). (1992). *Pathways to industrialization and regional development*. Oxford, UK: Routledge.

Sivakumaran, K., Lu, X. L., & Hanson, M. (2010). *The use of passenger transit infrastructure for goods movement: A bay area economic feasibility study*. Paper presented at the 89th Transportation Research Board Annual Meeting.

Spiekermann, K., & Wegener, M. (1994). The shrinking continent - New time-space maps of Europe. *Environment and Planning. B, Planning & Design*, *21*(6), 653–673. doi:10.1068/b210653

Straatemeier, T. (2008). How to plan for regional accessibility. *Transport Policy*, *15*(2), 127–137. doi:10.1016/j.tranpol.2007.10.002

Tapiador, F. J., Burckhart, K., & Marti-Henneberg, J. (2009). Characterizing European high speed train stations using intermodal time and entropy metrics. *Transportation Research Part A, Policy and Practice*, *43*(2), 197–208. doi:10.1016/j.tra.2008.10.001

Thompson, E. R. (2003). A grounded approach to identifying national competitive advantage: A preliminary exploration. *Environment & Planning A*, *35*, 631–657. doi:10.1068/a35110

Torchin, F., Grilly, D., Combes, S., Hasiak, S., & Menerault, P. (2008). *High speed rail for regional transport: Case studies in European countries*. Paper presented at the European Transport Conference.

Troin, J. F. (1995). *Rail et aménagement du territoire, des héritages aux nouveaux défis*. Aix-en-Provence: Edisud.

Urena, J. M., Menerault, P., & Garmendia, M. (2009). The high-speed rail challenge for big intermediate cities: A national, regional and local perspective. *Cities (London, England)*, *26*(5), 266–279. doi:10.1016/j.cities.2009.07.001

van Exel, J., Rienstra, S., Gommers, M., Pearman, A., & Tsamboulas, D. (2002). EU involvement in TEN development: Network effects and European value added. *Transport Policy*, *9*(4), 299–311. doi:10.1016/S0967-070X(02)00016-1

van Ham, J., & Muilerman, G. (2002). *Opportunities for time-sensitive intermodal rail freight*. Paper presented at the International congress on Freight Transport Automation and Multimodality.

Vickerman, R. (1997). High-speed rail in Europe: Experience and issues for future development. *The Annals of Regional Science*, *31*(1), 21–38. doi:10.1007/s001680050037

Vickerman, R. (2008). Transit investment and economic development. *Research in Transportation Economics*, *23*(1), 107–115. doi:10.1016/j.retrec.2008.10.007

Vickerman, R., Spiekermann, K., & Wegener, M. (1999). Accessibility and economic development in europe. *Regional Studies*, *33*(1), 1–15. doi:10.1080/00343409950118878

Voskuhl, D. E. F. (1995). Interlinking the region with its centre: The example of the Karlsruhe region in Germany. *Journal of Transport Geography*, *3*(4), 281–285. doi:10.1016/0966-6923(95)00026-7

Vuchic, R. V., & Casello, J. M. (2002). An evaluation of maglev technology and its comparison with high speed rail. *Transportation Quarterly*, *56*(2), 33–49.

Wardman, M., & Tyler, J. (2000). Rail network accessibility and the demand for inter-urban rail travel. *Transport Reviews*, *20*(1), 3–24. doi:10.1080/014416400295310

Wegener, M., & Fürst, F. (1999). *Land-use transport interaction: State of the art*. Dortmund, Germany: Institut für Raumplanung.

Weisbrod, G. (2008). Models to predict the economic development impact of transportation projects: Historical experience and new applications. *The Annals of Regional Science*, *42*(3), 519–543. doi:10.1007/s00168-007-0184-9

Weisbrod, G., & Treyz, F. (1998). Productivity and accessibility: Bridging project specific and macroeconomic analyses of transportation investments. *Journal of Transportation and Statistics*, *1*(3), 65–79.

Willigers, J., & Floor, H. (2007). Accessibility indicators for location choices of offices: An application to the intraregional distributive effects of high-speed rail in the Netherlands. *Environment & Planning A*, *39*(9), 2086–2098. doi:10.1068/a3913

Yang, C. W., & Sung, Y. C. (2010). Constructing a mixed-logit model with market positioning to analyze the effects of new mode introduction. *Journal of Transport Geography*, *18*, 175–182. doi:10.1016/j.jtrangeo.2009.01.005

ADDITIONAL READING

Aschauer, D. A. (1989). Is public expenditure productive? *Journal of Monetary Economics*, *23*(2), 177–200. doi:10.1016/0304-3932(89)90047-0

Batey, P., & Friedrich, P. (Eds.). (2000). *Regional Competition*. Heidelberg, Germany: Springer-Vrelag.

Belli, A., De Luca, G., Fabbro, S., Mesolella, A., Ombuen, S., & Properzi, P. (2008). *Territori regionali e infrastrutture. La possibile alleanza*. Milano, Italy: Franco Angeli.

Cheshire, P., & Gordon, I. (1998). Territorial competition: Some lessons for policy. *The Annals of Regional Science*, *32*(3), 321–346. doi:10.1007/s001680050077

Coutard, O., Hanley, R., & Zimmerman, R. (2004). *Sustaining Urban Networks: The Social Diffusion of Large Technical Systems* (p. 239). Oxford, UK: Routledge.

Doucet, P. (2006). Territorial Cohesion of Tomorrow: A Path to Cooperation or Competition? *European Planning Studies*, *14*(10), 1473–1485. doi:10.1080/09654310600852449

Fabbro, S., & Mesolella, A. (2010). Multilevel Spatial Visions and Territorial Cohesion: Italian Regional Planning between the TEN-T corridors, ESDP polycentrism and Governmental 'Strategic Platforms'. *Planning Practice and Research*, *25*(1), 25–48. doi:10.1080/02697451003625364

Karlsson, C., Johansson, B., & Stough, R. (Eds.). (2009). *Innovation, Agglomeration And Regional Competition*. Cheltenham, UK: Edward Elgar.

Olson, M. (1971). *The Logic of Collective Action: Public Goods and the Theory of Groups (Revised edition ed.)*. Cambridge, MA: Harvard University Press.

Rullani, E. (2006). *Dove và il Nordest*. Venezia, Italy: Marsilio.

KEY TERMS AND DEFINITIONS

Accessibility: Often referred to as 'easiness of connection', accessibility brings activities, people, areas, systems or networks in correlation both in time and space. Thus, it associates the concepts of mobility and sustainability in providing access to spatially and temporally dispersed resources. In the field of public transport, it could be interpreted as a combination of infrastructure connectedness (e.g. existence of links and stops) and service availability (e.g. service quality and frequency at those stops).

Competitive Cooperation: Group collaboration of cities and regions to work synergistically toward a common goal of growth, valuing relative capabilities and resources.

High Speed Rail (HSR): Railway systems planned to transport passengers at the interregional level. Usually defined by technical parameters, running trains at top speeds of 250 - 350km/h, HSR has also an economic definition based on the infrastructure cost that depends essentially by the operating model applied (Campos & de Rus, 2009) with a range between 20 to 60 million €/km.

Network Effects: Accessibility spillovers from one mode, system or network into neighbouring ones deriving from increases in efficiency, densities or economies of scale in the former.

Regional Competitiveness: Strategic repositioning of a territory, intended as a geopolitical environment where to build competitive capabilities, both in terms of real infrastructures and cultural development.

Transit Strategy: System planned to transport passengers at the urban and regional levels as strategy to overcome tunnel effects (see below). While technical parameters may differ for each specific system, in this study only rail based technology has been considered for comparison.

Tunnel Effect: Effect impacting on the distribution of benefits from non-users to users as if infrastructures were running underground. This effect is often met with drainage effects, when resources shift where benefits could be perceived. In 1992 Plassard recognised that HSR projects present tunnel effects when crossing territories without exchanges aside from rare station locations.

Chapter 10
Urban Competitiveness, U-City Strategies and the Development of Technological Niches in Songdo, South Korea

Luís Carvalho
Erasmus University Rotterdam, The Netherlands

ABSTRACT

This chapter is concerned with the recursive conflation between the concept of city competitiveness in ICT and different versions of the "knowledge city" concept based on ICT and digitalization, often responsible for ambiguous political discourses and unclear local economic development strategies. To overcome this problem, the chapter distinguishes both concepts, indentifies links between them and illustrates a way through which national and local polices can support ICT-related city competitiveness and knowledge city strategies: the development of innovation arenas. The chapter illustrates these notions with the case of the on-going development of the Songdo district in Incheon (South Korea), its competences in ubiquitous computing and the connection between this technological prowess with the ambitions of creating one of the first and most advanced "u-cities" in the world.

DOI: 10.4018/978-1-61350-174-0.ch010

INTRODUCTION

Cities need to be competitive, it seems. The idea that cities are competing with each other to attract and retain a number of functions and "costumers" – e.g. talent, inhabitants, companies, investments, visitors (van den Berg & Braun, 1999; Malecki, 2007) – has been gaining ground in urban practice and planning. Indeed, fueled by a rampant number of academic literatures and policy documents, during the last decade-and-a-half the notion(s) of *competitiveness* steadily entered the policy jargon and the agendas of many urban policymakers worldwide (Begg, 1999; Bristow, 2005).

There have been several types of celebrated, and sometimes conflated, policy initiatives to play the "territorial competitiveness game": the support of Porter's clusters (Porter, 1990, 2000), the steering of innovation systems (Cooke, 2001), the promotion of creative cities (Laudry, 2000; Florida, 2002), the engineering of intelligent cities (Komninos, 2002) and knowledge-based urban development (Yigitancalar et al., 2008), and so on. This phenomenon has been intimately linked with considerable fuss around the development of several urban and regional plans and strategies in which knowledge, innovation and creativity became unavoidable buzzwords. Unfortunately, the meaning of these concepts is far from stable and shared between different scholars, policy makers and stakeholders. Suffering from this conceptual conflation, lack of shared meaning and the intrinsic elusiveness of some of those concepts (Kitson et al., 2004), evidence so far shows disappointing results of territorial competitiveness and innovation strategies.

This chapter focuses in one of those arenas of conceptual and policy conflation: the role and relevance of Information and Communication Technologies (ICT) fostering urban competitiveness. The rationale here is to show that recent ICT breakthroughs – namely the contemporary developments in ubiquitous computing technologies (Weiser, 1991; Jeon et al., 2007) – can play a significant role in urban competitiveness, but this may happen through different mechanisms, which are related, but should be disentangled if we are to better inform policy making. To do so, the chapter distinguishes between the notion of *city competitiveness in ICT*, focused on the development of internationally competitive and territorially anchored clusters of economic activities, and *city competitiveness through the use of ICT*, concerned with the use of ICT in cities as a tool to foster e.g. efficient urban services, facilitate the integration of urban functions, social inclusion and participative democracy.

After making this conceptual frame clear, the chapter explores these two dimensions through the case of the pioneer *(u)biquitous-city strategy* and sensor technology development taking place in the new Songdo district (hereafter Songdo) in the Korean city of Incheon, 30 Km away from Seoul. Songdo anchors several national and local policy initiatives towards the development of an international cutting-edge sensor and u-technology cluster, and hosts unique knowledge-supportive services, as well as many national and foreign companies. Simultaneously, the on-going spatial planning and infrastructure development prospects in the area are directed to provide fertile ground for the experimentation of futuristic u-city concepts and services in homes and public space, targeting demanding city users (e.g. expatriate workers). Korean government envisages a network of u-cities throughout the country, and, simultaneously, to endow national firms with first-mover competitive advantages.

The chapter proceeds by arguing that these two dimensions both relate with the district and the city's competitiveness prospects, though in different ways. Borrowing and introducing the notion of *technological niche management* from social studies of innovation (Hoogma et al., 2002; Truffer et al., 2002), the chapter argues that these two dimensions may come virtuously together in Songdo, namely as the district plays the role of an "innovation laboratory" for exploring, testing and early exploiting u-technologies and its applications, while supporting its embeddedness in daily city life. It argues that the distinctive role of Songdo concerning the link ICT-competitiveness unfolds by providing ground for nurturing these two dimensions separately, but also by fostering the permanent interaction between the technological development of u-solutions (inside companies and R&D centers) and its real-life city applications toward an attractive, efficient and livable city. The chapter concludes by pointing out policy recommendations and opening latent research questions.

The evidence presented here has been collected and analyzed under the setting of an international comparative project on the development of new "knowledge locations" in European and Asian cities (van Winden et al., 2011). The empirics for this case rely on desk research and on 17 in-depth interviews with key stakeholders, namely companies, knowledge institutes and policymakers involved in the u-city strategies for Songdo, locally conducted during April 2010.

BACKGROUND: TERRITORIAL COMPETITIVENESS AND ICT

Do Cities and Regions Compete?

The concept of territorial competitiveness has been extensively discussed in the literature during the last decade. International journals such as Urban Studies (1999) and Regional Studies (2004) published special issues with fundamental reviews and advance on this matter (see also, e.g. Bristow, 2005; Harrison, 2007). From a policy-oriented perspective, seminal publications are the ones by the OECD (2006) on *Competitive Cities in the Global Economy* and the European Commission report on *Factors of Regional Competitiveness* (2003), produced by Cambridge Econometrics and ECORYS consulting. This section does not claim to be comprehensive or review these excellent publications, but introduces some key points at stake when trying to understand the links between ICT and urban competitiveness, and namely on how the former may impact on the latter.

The widespread interest on the link between competitiveness and the long-term performance of territories was ignited by the research of the Management & Strategy scholar Michael Porter (1990). His influential work stressed the fact that progress on the living standards of nations had been intimately dependent on the capacity of those nations and its firms to increase productivity levels overtime, i.e. on the efficiency in resource deployment, e.g. labour. Porter was thus primarily interested in the competitive advantage of countries or *nations*. However, a central contribution of his work, far from evident by the time, was that the *holy grail* of competitive advantage and productivity increases was "...created and sustained through a highly localized process" (Porter, 1990, p. 19), namely through clusters of economic activities, i.e. "...geographic concentrations of interconnected companies, specialized suppliers, service providers, firms in related industries and associated institutions (e.g. universities, standards agencies, trade associations) in a particular field that compete but also co-operate" (Porter, 2000, p. 15).

These localized clusters, it is now well acknowledged, may provide a set of important competitive advantages for firms that locate "within": easier and flexible access to specialized labour and infrastructures; facilitated "access" to rumors, peer pressure and privileged information about the industry, the competitors and innovations in the form of knowledge spillovers; presence of lower entry barriers, common business culture, specialized venture capital and social bounds[1]. Probably the most iconic case where these mechanisms have been identified over and over again is, of course, Silicon Valley, whose firms have been enjoying strong *competitive advantages* in ICT-related industries already for four decades now (e.g. Kenney and Patton, 2006). During these decades, companies, talent and investment moved from other US locations and overseas to the Valley in order to tap into this favorable "cultural and competitive infrastructure" (Saxenian, 1990, 2007). The Valley's dynamics, with impacts on their firm's innovations and competitiveness, have been having a relevant contribution for the US´ productivity and exports of highly valued new ICT goods and services, and, thus, on national competitiveness[2].

Notwithstanding the shortcomings of Porter's work (see, e.g. Martin & Sunley, 2003; Van Winden & Carvalho, 2009), the conceptual problems with the "competitiveness" notion early arrived when the concept moved from the initial units of analysis - firms and nations (see Cellini & Soci, 2002, referenced in Kitson et al., 2004) – to meso levels like cities and regions. Again, Saxenian (1996) was among the first to explicitly suggest that the specific "contexts" of Silicon Valley's were an important source of *regional competitive advantage* vis-à-vis other locations in the US with ICT industries. Some authors reacted arguing that the concepts of "competition" and "competitiveness" were inadequate to speak about cities and regions since, unlike firms, those cannot become insolvent and their "competitiveness" relies primarily on the sum of their firm's internal capabilities (e.g. Krugman, 1997).

However, others stressed Saxenian's perspective that the aggregate competitive performance of firms depend indeed on place-specific assets (e.g. Camagni, 2002; Boschma, 2004); moreover, cities increasingly compete with each other to attract companies, talent, residents, tourists, investments and events (van den Berg and Braun, 1999). It is out of this line of reasoning that we arrive to the notion that to be "competitive" in a world of increasing digitalization and flows (Castells, 1996), the development of clusters, innovation systems and other creative, learning-based and "intelligent" urban strategies are pivotal to cities struggling to keep and improve their competitive edge, anchor knowledge and innovation capabilities of existent and new economic activities, and, thus, maintain and increase its well being overtime.

All in all, Kitson et al. (2004) argue that the concept of competitiveness has meaning and value for cities and regions, but as a richer and more complex notion than what can be informed by the firm-centric logics of competition. Thus, it should focus more on studying "…the determinants and dynamics of a region's (or city's) long-run prosperity than on more restrictive notions of competing over *shares* [emphasis added] of markets and resources. (…) [Competitiveness] is [a concept] that recognizes that, ultimately, competitive regions and cities are places where both companies and people want to locate and invest in" (Kitson et al., 2004, p. 997). From this perspective, many studies put forward a number of structural assets for a city or region's long-term development and, thus, its competitiveness outcomes overtime. From an urban lens, van Winden (2006) mentions "urban specific factors" like governance and structural assets, to know: knowledge base, economic base, quality of life, accessibility, diversity and social equity. These assets have been pointed as critical conditioning the potentials of urban competitiveness and progress in the knowledge economy, namely a city's capacity to create and apply new knowledge, attract and retain talent.

Being pervasive to all societal domains, ICT have (and are likely to keep having) impacts in many of these urban competitiveness determinants. However, it is precisely here that the confusion arises on what it is meant by the link between ICT and urban competitiveness: the impacts may be many, in many realms, through many mechanisms. In this vein, a central issue in this paper is to understand how and to which extend ICT make cities competitive. In the remaining of this section I present a heuristic to disentangle this link in two simplified dimensions, not mutually exclusive but also not overlapping: competitiveness *in* ICT and competitiveness *through the use of* ICT.

Competitiveness in ICT

The link ICT-city competitiveness can be seen through the optic of the development of internationally distinctive clusters of companies, knowledge institutes and organizations with competences in ICT research, development, funding and commercialization. The exploration and mechanism through which this link takes places directs us to the literatures on economic clustering and innovation, well developed in the discipline of economic geography (see, e.g. Silicon Valley's example above).

Ron Boschma (2004) refers that "it is meaningful to talk about regional competitiveness when *the region* affects the performances of local firms to a considerable degree. This is especially true when the competitiveness of a region depends on intangible, non-tradable assets based on a knowledge and competence base embedded in a particular institutional setting that are reproduced and modified through the actions and repeated interactions of actors" (Boschma, 2004, p. 1001, emphasis added). Despite the possibilities to use the same knowledge and competence base in different industries, namely when they are related to each other (Asheim et al, 2008), those assets are specific to a certain type of economic and innovation activities (e.g. ICT-related), developed overtime through the concentration of related companies, supportive organizations, demanding clients, social links and aligned institutions (Maskell & Malmberg, 2007; Braunerhjelm & Feldman, 2006).

Once these distinctive clustering dynamics are at work, they contribute to city competitiveness by strengthening the export base of the city *per se*, by penetrating old industrial sectors with transversal innovations (e.g. machinery) or by bridging and finding cross fertilization links with other related fields (e.g. media, audiovisual, green tech). From this perspective, urban competiveness is nurtured though the retention and attraction of companies, investments and talent to the region (e.g. through increasingly aligned government funding and venture capital, foreign investors), but also through the continuous development of indigenous companies potential. This is *competitiveness in ICT*.

Some often mentioned knowledge cities and their regions are particularly competitive in ICT, e.g. San Francisco/Silicon Valley, London/Cambridge, Stockholm/Kista, Helsinki, Bangalore, etc. The fact that some of these ICT clusters develop inside or in the proximity of large urban agglomerations is not by chance. The development of new technologies often rely on qualified graduates from renowned universities; moreover, a large part of the "creative" ICT work evolves through project-based networks and informal cooperation user-producer, requiring proximity and frequent interaction (Trippl et al., 2009). Many of these users locate in large urban areas and national capitals, e.g. in the media and other creative industries (Scott, 1999). This is not to say that the vibrancy, "quality of life" and "buzz" ascribed to large metropolis are necessary or sufficient fostering competitive ICT clusters. Other smaller technological cities with limited vibrancy and amenities (e.g. Eindhoven in the Netherlands, Oulu in Finland) also developed excellence ICT clusters, based on large companies (Philips, Nokia), their spin-offs, dynamic technology suppliers and a renowned technical university. What seems to be central is indeed the possibility to develop stable contacts and project-based work through inter-firm cooperation and user-client interactions (Asheim et al., 2007). Technological procurement of large clients is often important to spur innovation. For example, the procurement of communication solutions by the US defense industry was determinant in the emergence and development of many leading ICT firms in San José and Palo Alto, at the core of Silicon Valley (e.g. Markusen, 2003).

Hence, it seems that the development of ICT new knowledge and marketable innovations - feeding territorial competitiveness - rely to an important extent on networks between territorially co-located players (firms, users, universities, venture capitalists, etc), despite the global communication possibilities opened by those same ICT (Malmberg & Maskell, 2006). Despite the well-known off shoring of back office functions and software development to cities like Bangalore, facilitated by ICTs, there is nowadays plenty of evidence that digitalization didn't fundamentally change the way knowledge – the key driver of *competitiveness in ICT* – is (co-)created and transmitted to others (Malecki, 2010). Although ICT definitely facilitates the access to information, knowledge spillovers still result from complex human and organizational processes for which there is not yet a close to perfect substitute to face-to-face interaction between culturally and socially close individuals (Storper & Venables, 2004), whether in a spatially permanent or temporay fashion (Asheim et al., 2007).

To foster competitiveness in ICT, policies often focus on incentives to create cluster platforms, support research consortia and networks, incentives to entrepreneurship, access to venture capital and development of science parks. Other more general incentives have been focusing in attracting ICT talent to cities through several urban regeneration schemes and amenity-based redevelopment. It is behind the scope of this chapter to discuss the effectiveness of these instruments, but it is important to stress that this policy toolkit is hardly in the hands of visionary mayors or collective action of urban actors, as sometimes suggested in the literature. Local governments often have limited degrees of freedom and capacity to implement these policies and much less to develop this type of cluster-based competitiveness from scratch, i.e. when missing the institutional settings and the critical mass of competences to do so (Acs et al., 2008).

Competitiveness Through the Use of ICT

The relation between ICT and city competitiveness can also be seen through the optic of the benefits for urban communities accruing from the use of ICT and digitalization. Literatures on knowledge-based urban development in the information era have precisely been stressing this dimension, as well as the profound changes that ICT may cause in city planning and the options opened towards (hopefully) more attractive, livable and cohesive cities (e.g. Yigitcanlar et al., 2008). From this perspective, contributions to urban competitiveness are also associated with the possibilities opened by ICT fostering an integral approach to urban development and management.

Recalling van Winden´s (2006) structural assets, we can understand *competitiveness through the use of ICT* primarily by the impact of ICT on the urban dimensions of "quality of life", "accessibility" and "social inclusion"[3]. On the one hand, competitiveness-enhancing effects are likely to take place through the attraction of new residents and companies to a high quality space, but (mainly) through the increasing satisfaction of indigenous populations in their communities, e.g. in the access to services and urban functions in an integrated fashion, or through the increasing participation in civic life. For example, current ICT may smooth the functioning and openness of public administration, an urban asset valued by knowledge workers, at least in the US (Florida, 2008). On the other hand, the cost-effective availability of digital infrastructure (e.g. optic fiber networks) may also be a distinctive competitiveness factor for some specific types of businesses and knowledge activities requiring rather large broadband and ICT processing capacity, like new media. Simultaneously, ICT and digitalization provoke impacts in the functioning patterns of neighborhoods, public space and building designs, imposing challenges to urban design (Mitchell, 2000).

It is important at this point to stress that this perspective on the link ICT-urban competitiveness considers city as a physical, socially and economically constructed reality, where ICT play a role on how people and activities interact with each other and with artifacts in a given location. In this vein, cities are places where "digitality" is supportive of "…new ways in which infrastructure and behaviour are being augmented by hardware, software, and data in terms of how these functions are being embodied in smart highways and intelligent buildings as well as in new forms of electronic service delivery."(Batty, 2001, p. 479). This perspective differs from more metaphorical versions of "digital cities" (e.g. Ishida, 2000), i.e. non-physical spaces where people can plug in, e.g. as an avatar, or digital web platforms in which the word "city" isn't but an allegory. The perspective of Batty (2001) seems more appropriate to relate ICT with urban competitiveness in the way this chapter defines it, i.e. as a means to generate urban advantages and attractiveness for a still largely non-digital urban reality.

In a similar fashion, Komninos (2006) speaks about "digital city" as community-based digital spaces at the service of the activities taking place at the physical city, "…a wide range of digital networks and software applications facilitating multiple aspects of the social and economic life of cities: commerce, transactions, security, health, education, work, leisure, transport, and others (p.3)". Translating it to action and policy, Komninos is speaking about the types of digital investments that took place in many cities worldwide in the last decade in order to improve urban organization under the heading of "digitality": development of broadband infrastructure, several types of digital platforms, e-yellow pages, telematics, virtual sight tours, traffic and parking web platforms, e-government portals, etc. It is important to note that the types of "networks" facilitated by these platforms are, generally, supportive of spreading information rather than co-generating new complex types of knowledge, prone to be translated into innovations in the way those are mentioned in the literatures, e.g. of clusters or innovation systems. There is plenty of evidence that innovation in many of the so-called "knowledge industries" like ICT, media or biotechnology rely on much more dense, trust based, socially constructed and structured networks than what can be expected to emerge out of these "digital city" type of ICT platforms. This is a central issue deserving clarification in the "knowledge city" and "digital city" literature.

At this point, it should be clear that the two previous dimensions of the link between ICT and urban competitiveness – urban competitiveness *in* and *through the use of* ICT – don't necessarily overlap. Very proactive cities in applying digital solutions in the design of their public space, or in the digital involvement of citizen in e-government platforms, don't necessarily need to have strong industrial and knowledge *competitiveness in ICT*, or the other way around. However, this is not to say that there are no links between these two dimensions. Rather the opposite. They might happen and indeed drive the development of these two dimensions in an integrated and virtuous fashion. For example, Finnish cities like Helsinki or Oulu have strong competences in ICT, headquartering Nokia and several highly competitive ICT and new media companies. Simultaneously these cities pioneer the development and implementation of digital community platforms for the integration of deprived residents, steer new mobility patterns of elderly residents based on ICT, and champion user-centric development of new ICT applications through "living lab" methodologies (Carvalho et al., 2009). New ICT products and services are linked with new social needs of the local residents, and some of the most relevant innovations have been happening along this interface. These links may emerge rather spontaneously, but may also be nudged through selective policy interventions. One way to do this is through the development of experimentation arenas or locally emergent "technological niches". I will turn to this issue in the next section, exploring how the Korean Government's strategy to develop strong competences in ubiquitous computing is linked with the planning of new futurist versions of a "ubiquitous-city" in a new sea-reclaimed district close to Seoul.

URBAN COMPETITIVENESS AND U-CITY STRATEGIES IN SONGDO, SOUTH KOREA

The New "Big Thing" in ICT: Ubiquitous Computing

The trend towards *ubiquitous computing* is being seen as the next ICT breakthrough with pervasive impacts in society, and, thus, also on urban planning and city competitiveness (Lee et al., 2008). The term *ubiquitous computing* was firstly coined by Mark Weiser, at the Xerox Palo Alto Research Center, referring to a new computing paradigm in which "...technologies weave themselves into the fabric of everyday life until they are indistinguishable from it". Following previous stages of ICT revolutions ("many people, one computer"; "one person, one computer"), ubiquitous computing refers to the fact that in a not-so-distant future, "many computers share each of us" (Weiser, 1996).

Jeon and Leem (2007) compared many definitions of ubiquitous computing in order to depict its spectrum, referring to "invisible computing" or "everyware". The most encompassing one and probably more widely used has been proposed by Ken Sakamura, who refers to ubiquitous computing as the possibility of being able to use computers *anywhere, at any time*. In practice, it refers to the possibility of augmenting and connecting "objects" (e.g. buildings, cars, pieces of paper, even persons) and digital technologies in an undistinguishable fashion. Ubiquitous computing will probably become like electricity: spread everywhere (Almeida, 2009). Human beings and objects will be permanently connected, receiving and releasing information from one another.

This trend raises important innovation challenges in ICT. From the perspective of computing systems and engineering, Almeida (2009) illustrates this future in a rather vivid way:

The paper we write on, the table, window, wall, or any other object will recognize and allow us to interact. Even our clothes will be enhanced, with safety features or simply informative ones: location, our public profile, their life-cycle, price and where to buy. Objects will be self-aware and will be able to communicate between themselves without human intervention, saving our ever more precious time (...) In a wider scope, not only individual buildings, but existing cities will be augmented with features and services presently unheard of. Some will even be built from scratch to properly accommodate ubiquitous computing. With ubiquitous computing our lives will be filled with new and immersive experiences. (Almeida, 2009, p. 2)

Songdo's U-City Strategy and Urban Competitiveness

One of these projected "u-cities", pioneer and literally built from scratch is the new Songdo district. It is envisaged, among other objectives, to become a u-city where people can be connected to a large number of digital, web services anytime, anywhere. Songdo locates at the edge of Incheon (the third largest South Korean city), only 30 Km from Seoul (see Figure 1). It presently hosts the largest private real estate development in the world, in 53 Km^2 of new sea-reclaimed land in the Yellow Sea (roughly 60% the size of Manhattan). The district is one of the three "new cities" envisaged under the Korean Government's strategy to endow the country with Free Economic Zones to *compete* with other Asian hot-spots (e.g. Singapore, Beijing, Hong-Kong, Shanghai) for the attraction of foreign investment, business, talent and tourists. The first of these Korean Free

Economic Zones locates in Incheon and is called IFEZ (Incheon Free Economic Zone), composed by three different sea-reclaimed poles/cities, with distinct profiles.

Songdo has been developed and occupied since 2003. Its expected completion year is 2020, with 250 thousand new inhabitants; it is planned to become a fully fledged international business and innovation hub in Northeast Asia. Songdo's master plan includes a well balanced functional mix of housing, business and industrial space, as well as e.g. universities and research institutes and a vast array of urban functions and amenities, like a concert hall, urban parks, commerce, convention centre, high-rise landmarks and golf courses. Songdo can be seen as the urban "face" of a number of converging national and local strategies to create distinctive competitive advantages for South Korea, Seoul and Incheon vis-à-vis other Asian business locations ("competitors"). Among those is the development of more flexible and business friendly legislation to attract foreign investors and companies, improve Korea's international image, develop new high-tech growth sectors and decentralize "knowledge activities" from Seoul. The vision of Incheon Metropolitan City is to develop new knowledge growth poles in the form of new "knowledge districts", claiming higher functional independence from the capital Seoul (Ducruet, 2007). This has been facilitated by recent development of excellent accessibility infrastructure, like a brand-new international airport in Yeongjong (also in see reclaimed land), now connected to Songdo through one of the largest bridges in the world

Initially inspired by Dubai's experience and by the urban features of well know European and North American metropolis (see e.g. Segel, 2005), the development of Songdo has been orchestrated by the Metropolitan City of Incheon with the strong commitment of the Korean Government. Besides tax cuts and other financial incentives to attract foreign business, large sums of money are channelled to fund science, technology and inno-

Figure 1. South Korea and Incheon. Source: Worldtravels.com (2010). © Globe Media Ltd. Used with permission.

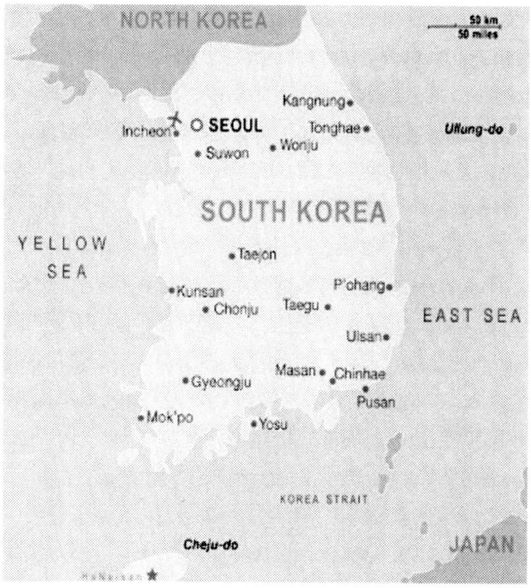

vation efforts of Korean firms and R&D institutes re-locating and investing in Songdo. Together with Bio and Nano-Technology, ICT is one of the flagship sectors in which Songdo plans to excel and "compete" worldwide (Ministry of the Knowledge Economy, 2009), namely on the development of ubiquitous computing technologies.

Already in 2004, the Korean Government launched IT389, a plan to transform Korea into a "ubiquitous society", through the development of a network of u-cities. Motivated by this National level initiative, Songdo targeted the development of an international cutting-edge cluster of sensors and u-technologies, providing state-of-the-art services and conditions for indigenous and foreign firms. Moreover, the spatial planning and on-going infrastructure development in Songdo envisage providing fertile ground for the experimentation of futuristic u-city concepts and services in homes and public space, making use of u-technologies like radio-frequency identification systems, telematics, sensor networks, smart card applications and so on. In Songdo, all the information systems in, e.g., buildings, business, office space, medical services and transport will share data. Together with the development of the first network of u-cities in the world, the Korean government wants to simultaneously endow national firms with technological prowess and first-mover *competitive advantages*.

At this point, and informed by the previous distinction, how can we understand the links between ICT and the competitiveness of Songdo?

First, the strategy for Songdo foresees the development of specific location advantages to make the area highly *competitive in ICT*. The ultimate objective is to lead the technological edge in u-computing together if not outperforming the US, Japan, Germany and Singapore. Incheon and Seoul's region have important assets in this field, accumulated during the last decades. Digital infrastructure – namely optical fiber – is excellent and is often ranked as the world's best. Broadband penetration is close to 100% (Florida, 2005; Kelly et al., 2003). This makes the country, and Seoul/ Incheon in particular perfect test-beds for new ICT related innovations.

Seoul is an important ICT region. It hosts the headquarters of world leaders in ICT and electronic apparel - LG and Samsung - as well as highly qualified engineers, top universities and technical schools with aligned ICT courses and curriculum, and one of the world's largest number of ICT-related patents (Florida et al., 2008). It concentrates the core of high-tech activities and industries in the country, requiring advanced ICT in their activities (e.g. machineries, new media, green tech, gaming, defense industry). Moreover, it hosts a plethora of ICT companies and supportive sector organizations for small and medium sized companies (e.g. the IITPA – Incheon Information Technology Industry Promotion Agency). Government funding is sound and aligned at different levels to support ICT-related R&D, innovation and entrepreneurship. The branching from these ICT activities towards ubiquitous computing is thus smoothly happening with strong support from the government, accommodated by the region's competences. Besides planning to become world leader in this set of technologies, Korean Government plans to make technological cross fertilization with green technologies, as well as other established industries like automotive, ships and electronics

At the time of this writing, Songdo attracted already significant public and private investments in the development of u-technologies. More than 20 high-tech small and medium sized companies (Korean, American, Finnish) working with u-technologies re-located operations and staff to Songdo; moreover, the two of the largest world players in associated ICT systems and technologies (e.g. surveillance and network systems) – CISCO and IBM – already invested heavily in Songdo, with own premises to conduct R&D and tap in Songdo's "u-developments". Simultaneously, the Korean Government and Incheon Metropolitan City invested heavily in the so called RFID/USN and MEMS technology centre (Radio Frequency Identification /Ubiquitous Service Discovery; Micro-Electro-Mechanical Systems), running since 2006. This center hosts more 14 smaller companies inside its premises and provides exclusive services and dedicated facilities for companies developing related technologies. Among those are different testing and engineering labs, pilot facilities, training rooms as well as brokerage services, participation in events and many amenities. Several new ICT innovations are under way as well as the related patenting processes.

Second, national and local Governments have ambitions to turn Songdo into a globally competitive city *through the use of ICT*, transforming it in the world's first fully integrated u-city. IFEZ authority coordinates this encompassing project. This is expected to make Songdo a very efficient and user-friendly place, where urban functions and services are integrated and connected with each other, appealing to new business, residents and visitors. According with IFEZ (2010), the u-city "space" to unfold in Songdo should "create new urban value" and urban attractiveness. It will contribute to spread public information and services among citizens, fostering communication networks connecting humans, objects, nature and space towards more comfort, efficiency, generating a sense of "pride" and "uniqueness".

The full implementation of u-technologies has deep implications in the planning and functioning of the urban space, as well as its management (Lee et al., 2008). Trough e.g. video networking, Wi-Fi, radio frequency technology, sensors, information systems and special management software, several activities in a city can be connected to and coordinated by a common network, e.g. safety and crime prevention, transportation, health care, environment, etc. "U" concepts are planned to be implemented in Songdo in different urban amenities and facilities (e.g. u-plaza; u-mall); moreover, the development of an "urban integrated operation centre" is planned, connecting and controlling all infrastructures and services in Songdo, like transport, health, safety, etc.

The development and implantation of a u-city in Songdo's daily life is still in its infancy, despite the international media attention and the strong promotional efforts of the IFEZ authority. Nowadays, u-concepts and pilots can only be experienced in a permanent exposition/venue called "Tomorrow City" in Songdo. Thus, so far, the link between ICT and competitiveness is primarily through the extent to which Songdo has been developing a number of sound ubiquitous computing technologies, while attracting the (Korean and international) players and developing the institutions to do so; much less has to do (yet) with contributions to enhance inclusion, efficiency, quality of life, etc, though this might also become the case in the future.

Songdo as a U-Protected Space

Notwithstanding this fact, closer observation reveals that the two dimensions are related, and it is that relationship that explains an important part of the u-innovation efforts going on in Songdo, which so far guarantee *competitiveness in (u-) ICT*. As previously mentioned, by tapping on past competences of the entire region, it is the explicit objective of the Korean and Incheon Government to support the development of technological prowess in ICT that will feed the pioneer implementation of u-city concepts in Songdo. To this effect, the IFEZ authority is creating the incentives for companies and other organizations to develop u-technologies and u-concepts that otherwise would not be demanded by the markets or society, and thus would probably exist only as political discourses and foresight exercises[4]. Examples of those incentives are e.g. the procurement of the new technologies, testing and prototyping that will feed the visionary u-city (or the competitiveness of Songdo *through the use of ICT*), the availability of sound R&D funding, attraction of global industrial players to Songdo and the development of supportive, "anchor" infrastructures (e.g. the RFID/USN centre).

In the jargon of Socio-Technical Studies of Innovation (e.g. Kemp et al., 1998), ICT companies and Government are strategically managing "technological niches" (Hoogma et al., 2002). Technological niches are "*protected spaces* created by specific actors – be they industry managers, policy-makers or citizen groups – with the strategic aim to test and develop a technology and to prepare it for further diffusion" (Truffer et al., 2002, p.113, emphasis added). Contrarily to market niches, the user advantages and producers′ interest in the technology development are unclear. Thus, stakeholders interested in the technological potential (in this case, industrial players and Government) will deploy temporary incentives and actions to nudge potential users in the technology's experimentation, as well as supporting producers in fine-tuning and developing technologies further. This is of critical relevance in order to overcome market and social barriers in the development and implementation of technologies and concepts, in this case "u-related" (e.g.

lack of demand, privacy constrains, "big brother" fear, legitimation, etc).

According with Truffer et al. (2002), the nurturing of technological niches, which can eventually be scaled up in the future (e.g. implementation of u-city models worldwide and diffusion of associated u-technologies) primarily rely on real world experiments that can nurture two fundamental processes: leaning and societal embedding[5]. Learning is associated with the development of new knowledge about technologies and its functioning in real-life situations. Songdo provides fertile ground to this aim. First, the R&D challenges and technological requirements of a u-city require the exploration of new products, processes and technologies. Songdo's incentives facilitate the development of localized platforms for knowledge creation and learning, e.g. through interactive learning and joint problem-solving with the demanding u-technology users, often responding to locally accessible financial resources. Moreover, Songdo's opportunities around the development of u-technologies facilitate anchoring of external knowledge (e.g. CISCO, IBM, foreign companies in the RFID/USN centre), its contextualization and diffusion in the district (e.g. through transnational corporate knowledge networks, new academic research, spin-offs, development of academic curriculum, e.g. in the Songdo Campus of the University of Incheon). Second, as recognised by the IFEZ and Songdo authorities, the futuristic u-city visions for Songdo endow the involved companies with first-mover advantages. The interest of these companies is not to develop the "u-technology for Songdo", but to develop the killer applications for the future global market. To this aim, Songdo is "the place to be". Moreover, city officials become ambassadors of the technology themselves, intertwining the u-city developments of Songdo with concrete u-technology providers, implicitly giving these companies competitive advantage. For example, IFEZ authority plans to demonstrate some of the new u-services in the forthcoming 2014 Asian Games in Incheon, with high media exposure.

Challenges Ahead: Societal Embedding

The second process involved with the development of technological niches is societal embedding. Truffer et al. (2002) associate it with three interrelated processes: i) network management, i.e. the creation of supporters of the technologies and its uses; ii) infrastructure matching, i.e. the adjustment of the new technologies in an existing socio-technical environment of regulations, standards and physical artefacts, e.g. an urban structure, and; iii) expectation building, i.e. the development of favourable expectations about the advantages of the new technologies and processes. In the medium and long-run, the success (or competitiveness in ICT) of Songdo and its companies in u-technologies depends on the extent to which u-concepts become embedded and legitimated in society (Bergek, 2008), i.e. the extent to which a u-city becomes a functioning and accepted model of city, effectively desired, validated and regulated by society. Songdo's competitiveness *through the use of ICT* is very much contingent upon this validation.

Korea has social and cultural advantages to test and first implement u-city concepts vis-à-vis other western countries. Again, the use of internet and broadband is almost universal in Korea (e.g. Almeida, 2009). Moreover, privacy expectations are lower and the society is highly collectivistic, giving the Government power to implement u-city concepts and standards. However, recent research (Shin, 2009) is less optimistic about the society absorption of the u-city technological apparatus. Despite the huge government investments in this strategy throughout Korea (e.g. 300 million USD in the RFID centre in Songdo; 60 Billion USD in the first stage for all the Korean u-city pilots), Shin argues that infrastructure provision, technological and corporate interests have dominated the project so far. Korea and Songdo in particular developed a sound number of competitive u-related technologies, but with very limited involvement of the citizens, potentially hampering the effective integration and use of these new technologies in society.

Ours and Shin's research in the recent progress of u-city strategies in Songdo reveal that none of the key involved actors – Ministry of Knowledge Economy, Industry Conglomerates and Local Government – have promoted a more inclusive and "bottom-up" involvement of the citizens and have been focusing instead, respectively, on infrastructure development, technological capacity, cluster promotion, image and regional economic diversification. However, as mentioned, besides the deployment of u-technologies in the city's physical artefacts, u-concepts require integration, or embeddedness, in the cultural and social apparatus of cities.

This poses important challenges and opportunities for Songdo. So far the u-developments in Songdo have indeed been very much oriented by a "technology-push" and high-tech clustering vision; simultaneously, the social and cultural fabric of Songdo, at this stage, is still practically inexistent (occupation of residents and companies started roughly 5 years ago); therefore, the interaction between technology and the social system would be in any case barely impossible to attain due to the "virgin land" character of Songdo district. However, as Songdo starts to be occupied and new companies and residents move in, there are emerging opportunities to effectively test some of these new concepts and spur innovation. And here relies probably one of the great advantages of Songdo – it may well play the role of a protected and experimentation arena to foster further learning and societal embedding. Moreover, the types of users in Songdo (e.g. expats, qualified technicians) are likely to be demanding IT users, thus constituting an important test to the potential applicability and implementation of the concept. Songdo is flexible and "new" enough to accommodate the infrastructure, although it is still to be seen whether the development of the u-city concept will clash or not with the planned urban design nowadays. In any case, these efforts require rather permanent evaluation and monitoring.

FUTURE RESEARCH DIRECTIONS

The development of Songdo as a city, in general, and as a u-city, in particular, is still in its infancy. Thus, the interplay between *competitiveness in ICT* and *through the use of ICT* has a lot of room to develop, and these dimensions deserve to be researched further as the implementation of the concepts evolves. Can the first progress without the latter? How will this relation evolve over time? Which governance mechanisms are involved, and how can they be managed? At which spatial dimensions is this interplay likely to occur (local, regional, national, networked globally)? These are questions that may contribute to further explorations in the interface of urban and regional studies and information science studies interested in the competitiveness of cities and the use of ICT, as well as on clearer policy recommendations to foster both.

As a world pioneer in this respect (application of ubiquitous computing in city planning and urban functions), Songdo represents a critical case study, calling for intensive, in-depth research methodologies. In this sense, the evolution of the area represents an outstanding arena to research these phenomena further. This research should to a large extend be process oriented, closely monitoring policy processes and outcomes (Rotmans, 2005).

However, in order to better inform us on the mechanism and processes involved in the development of technological niches, other cases with apparently similar evolving experiences should also be taken into analysis. International comparative analysis would be particularly interesting, in order to grasp how different cultural features and political institutions may shape this interaction. The Korean, and to a large extend the Asian model of strong State intervention in industrial policy issues and societal collectivism may be hardly reproducible in other western contexts. Which are the implications for the development of technological niches and implantation of u-city models in other world regions? And, focusing on the Asian context, is the model of Developmental State and national industrial policy of the 70s and 80s (Wade, 1990) fit to intervene and manage cities in such complex and integration-intensive issues?

Finally, at the interface between information science, urban studies and social studies of technological transitions, an interesting research direction should focus on the implementation of futuristic ICT-city concepts in society. On the one hand, the study of socio-technical transitions and development of technological niches has focused on mobility and energy transition topics, and could benefit from refreshing views on other types of societal transitions. On the other hand, studies that look into cities and ICT have been putting its focus on the technology side and relatively less on the processes of larger societal absorption of the concepts – input from the socio-technical transitions literature may complement this approach further.

CONCLUSION

The role of on-going ICT (r) evolutions is often mentioned as one of the most important fundamental changes with manifold impacts in "urban knowledge developments" and "city competitiveness". Examples are the discussions around the potentials for concentration versus de-concentration of economic activity, innovation, creativity and knowledge transmission (Malecki, 2010), digital accessibility and quality of life (e.g. van den Berg et al., 1999), social equity, civic involvement and urban governance (Komninos, 2002). Obviously, ICT has become more and more pervasive, permeating practically all fields of human life. Not surprisingly, ICT have impacts in cities, in their socio-economic development and the way cities manage to keep attractive and competitive overtime. Simultaneously, ICT became a growing knowledge-based and exporting industry in its own right, clustered in specific locations, also with impacts in the diversification and upgrading of many other industries in cities and regions. Thus, several policies and cluster initiatives worldwide target the localized development of ICT growth clusters, in isolations or in articulation with other cluster industrial platforms.

If we acknowledge the previous, we should also acknowledge that it is critical to qualify and clearly specify what we mean when we speak about the links between ICT and urban competitiveness. So-called urban competitiveness strategies have proliferated widely in the last decade. However, due to the elusiveness of the concept (Kitson et al., 2004), its uses have often been extended to the minimum common denominator in urban policy discourses, associated with buzzwords like "creative", "intelligent" and, more recently, "sustainable" types of city visions. It is thus not surprising that in highly dynamic regions like East Asia, the urban competitiveness discourse – originally developed in western economies - became widespread.

In line with the central topic of this book, this chapter tried to make sense of the notion of urban competitiveness and its relation with ICT, namely the on-going developments in ubiquitous computing. These topics were explored with reference to the on-going development in the implementation of a u-city in Songdo, South Korea. Particular emphasis was put on how different dimensions of urban competitiveness may come together through the deliberate management of ICT technological niches (u-computing). By doing so, dimensions of technological learning and societal embedding were put together, stressing the determinant role of urban environments and urban policy in fostering technological innovations but also in testing and embedding them in the socio-economic fabric while supporting a place's competitiveness in different fronts. Strategic approaches to urban development and knowledge may benefit from taking this more nuanced view and tools on board in their efforts to steer urban competitiveness overtime.

I am grateful to IHS (Institute for Housing and Urban Development Studies) for the financial support associated with this study. Kindly acknowledged is also the support of the Portuguese Foundation for Science and Technology (FCT - doctoral grant).

REFERENCES

Acs, Z. J., Glaeser, E., Litan, R., Fleming, L., Goetz, S. J., Kerr, W., & Strange, W. (2008). *Entrepreneurship and urban success: Toward a policy consensus*. Kansas City, MO: Ewing Marion Kauffman Foundation. Retrieved from http://ssrn.com/ abstract=1092493

Almeida, M. (2009). *Ubiquitous computing and natural interfaces for environmental information*. Lisbon, Portugal: Faculdade de Ciências e Tecnologia da Universidade Nova de Lisboa.

Asheim, B., Boschma, R., & Cooke, P. (2008). *Constructing regional advantage: Platform policies based on related variety and differentiated knowledge bases*. Unpublished manuscript, Utrecht, The Netherlands.

Asheim, B., Coenen, L., & Vang, J. (2007). Face-to-face, buzz, and knowledge bases: Sociospatial implications for learning, innovation, and innovation policy. *Environment and Planning. C, Government & Policy*, *25*(5), 655–670. doi:10.1068/c0648

Bathelt, H., Malmberg, A., & Maskell, P. (2004). Clusters and knowledge: Local buzz, global pipelines and the process of knowledge creation. *Progress in Human Geography*, *28*(1), 31–56. doi:10.1191/0309132504ph469oa

Batty, M. (2001). Contradictions and conceptions of the digital city. *Environment and Planning. B, Planning & Design*, *28*(4), 479–480. doi:10.1068/b2804ed

Begg, I. (1999). Cities and competitiveness. *Urban Studies (Edinburgh, Scotland)*, *36*(5), 795–809. doi:10.1080/0042098993222

Benneworth, P., Coenen, L., & Truffer, B. (2010). *Back to the future? The geographical dilemma of contemporary transition studies*. Paper presented at 2010 AAG Annual Meeting, Washington, DC.

Bergek, A., Jacobsson, S., Carlsson, B., Lindmark, S., & Rickne, A. (2008). Analyzing the functional dynamics of technological innovation systems: A scheme of analysis. *Research Policy*, *37*(3), 407–429.

Boschma, R. (2004). Competitiveness of regions from an evolutionary perspective. *Regional Studies*, *38*(9), 1001–1014. doi:10.1080/0034340042000292601

Braunerhjelm, P., & Feldman, M. P. (2006). *Cluster genesis: Technology-based industrial development*. New York, NY: Oxford University Press.

Bristow, G. (2005). Everyone's a'winner': Problematising the discourse of regional competitiveness. *Journal of Economic Geography*, *5*(3), 285. doi:10.1093/jeg/lbh063

Camagni, R. (2002). On the concept of territorial competitiveness: Sound or misleading? *Urban Studies (Edinburgh, Scotland)*, *39*(13), 2395 doi:10.1080/0042098022000027022

Carvalho, L., Mingardo, G., & Van Haaren, J. (2010). *Green urban transport policies and cleantech innovations. Evidence from Curitiba, Göteborg and Hamburg*. Paper presented at 2010 AAG Annual Meeting, Washington, DC.

Carvalho, L., van Tuijl, E., & van Haaren, J. (2009). *Developing locations in the knowledge economy: Arabianranta, Helsinki*. Rotterdam, The Netherlands: Eramus University Rotterdam.

Castells, M. (1996). *The rise of the network society*. Malden, MA: Wiley-Blackwell.

Cellini, R., & Soci, A. (2002). Pop competitiveness. *Quarterly Review. Indexes - Banca Nazionale del Lavoro*, *55*, 71–101.

Cooke, P. (2001). Regional innovation systems, clusters, and the knowledge economy. *Industrial and Corporate Change*, *10*(4), 945. doi:10.1093/icc/10.4.945

Ducruet, C. (2007). Incheon, vitrine de la Corée du Sud et masque de Séoul. *M@appemonde, 85*(1).

European Commission. (2003). *A study on the factors of regional competitiveness*. Brussels, Belgium: Directorate-General Regional Policy.

Fernald, J. G., & Ramnath, S. (2004). The Acceleration in the US total factor productivity after 1995: The role of Information Technology. *Economic Perspectives*, *28*(1), 52–68.

Florida, R. (2005). *The flight of the creative class: The new global competition for talent*. New York, NY: HarperBusiness.

Florida, R. (2008). *Who's your city?*New York, NY: Basic Books.

Florida, R. L. (2002). *The rise of the creative class: And how it's transforming work, leisure, community and everyday life*. New York, NY: Basic Civitas Books.

Harrison, J. (2007). From competitive regions to competitive city-regions: A new orthodoxy, but some old mistakes. *Journal of Economic Geography*, *7*(3), 311. doi:10.1093/jeg/lbm005

Hoogma, R., Kemp, R., Schot, J., & Truffer, B. (2002). *Experimenting for sustainable transport. The approach of strategic niche management*. London: Spon Press.

IFEZ. (2010). *Business hub of Asia–Incheon free economic zone*. Incheon, South Korea: IFEZ Public Relations Office.

Ishida, T. (2000). Understanding digital cities. *Digital Cities,* 7-17.

Jeon, N. J., Leem, C. S., Kim, M. H., & Shin, H. G. (2007). A taxonomy of ubiquitous computing applications. *Wireless Personal Communications*, *43*(4), 1229–1239. doi:10.1007/s11277-007-9297-9

Kelly, T., Gray, V., & Minges, M. (2003). *Broadband Korea: Internet case study*. Seoul, South Korea: International Telecommunication Union.

Kemp, R., Schot, J., & Hoogma, R. (1998). Regime shifts to sustainability through processes of niche formation: The approach of strategic niche management. *Technology Analysis and Strategic Management*, *10*(2), 175–198. doi:10.1080/09537329808524310

Kenney, M., & Patton, D. (2006). The co-evolution of technologies and institutions: Sillicon Valley as the iconic high-technology cluster. In Braunerhjelm, P., & Feldman, M. P. (Eds.), *Cluster genesis: Technology-based industrial development*. New York, NY: Oxford University Press.

Kitson, M., Martin, R., & Tyler, P. (2004). Regional competitiveness: An elusive yet key concept? *Regional Studies*, *38*(9), 991–999. doi:10.1080/0034340042000320816

Komninos, N. (2002). *Intelligent cities: Innovation, knowledge systems, and digital spaces*. London, UK: Spon Press.

Komninos, N. (2006). The architecture of intelligent cities. *Intelligent Environments*, *6*, 53–61.

Krugman, P. (1997). *Pop internationalism*. Cambridge, MA: MIT Press.

Landry, C. (2000). *The creative city: A toolkit for urban innovators*. London, UK: Earthscan.

Lee, S. H., Yigitcanlar, T., Han, J. H., & Leem, Y. T. (2008). Ubiquitous urban infrastructure: Infrastructure planning and development in Korea. *Innovation: Management. Policy & Practice*, *10*(2-3), 282–292. doi:10.5172/impp.453.10.2-3.282

Malecki, E. J. (2007). Cities and regions competing in the global economy: Knowledge and local development policies. *Environment and Planning. C, Government & Policy*, *25*(5), 638–654. doi:10.1068/c0645

Malecki, E. J. (2010). Everywhere? The geography of knowledge. *Journal of Regional Science*, *50*(1), 493–513. doi:10.1111/j.1467-9787.2009.00640.x

Malmberg, A., & Maskell, P. (2006). Localized learning revisited. *Growth and Change*, *37*(1), 1–18. doi:10.1111/j.1468-2257.2006.00302.x

Markusen, A. (2003). Fuzzy concepts, scanty evidence, policy distance: The case for rigour and policy relevance in critical regional studies. *Regional Studies*, *37*(6), 701–717. doi:10.1080/0034340032000108796

Martin, R., & Sunley, P. (2003). Deconstructing clusters: Chaotic concept or policy panacea? *Journal of Economic Geography*, *3*(1), 5. doi:10.1093/jeg/3.1.5

Maskell, P., & Malmberg, A. (1999). The competitiveness of firms and regions: 'Ubiquitification' and the importance of localized learning. *European Urban and Regional Studies*, *6*(1), 9–25. doi:10.1177/096977649900600102

Maskell, P., & Malmberg, A. (2007). Myopia, knowledge development and cluster evolution. *Journal of Economic Geography*, *7*(5), 603–618. doi:10.1093/jeg/lbm020

Mitchell, W. J. (2000). *E-topia: "Urban life, Jim-but not as we know it*. Cambridge, MA: MIT Press.

OECD. (2006). *Competitive cities in the global economy*. Paris, France: OECD.

Porter, M. (1990). *The competitive advantage of nations*. New York, NY: Free Press.

Porter, M. E. (2000). Location, competition, and economic development: Local clusters in a global economy. *Economic Development Quarterly*, *14*(1), 15–34. doi:10.1177/089124240001400105

Regional Studies. (2004). Special issue on regional competitiveness. *Regional Studies, 38/39*.

Rotmans, J. (2005). *Societal innovation: Between dream and reality lies complexity*. Rotterdam, The Netherlands: Erasmus University Rotterdam.

Saxenian, A. L. (1990). Regional networks and the resurgence of Silicon Valley. *California Management Review*, *33*(1), 89–112.

Saxenian, A. L. (1996). *Regional advantage: Culture and competition in Silicon Valley and Route 128*. Cambridge, MA: Harvard University Press.

Saxenian, A. L. (2007). *The new Argonauts: Regional advantage in a global economy*. Cambridge, MA: Harvard University Press.

Scott, A. (1999). The cultural economy: Geography and the creative field. *Media Culture & Society*, *21*, 807–817. doi:10.1177/016344399021006006

Segel, A. I. (2005). *New Songdo City* (Harvard Business School case no. 9-206-019). Boston, MA: Harvard Business School Publishing.

Shin, D. H. (2009). Ubiquitous city: Urban technologies, urban infrastructure and urban informatics. *Journal of Information Science*, *35*(5), 515. doi:10.1177/0165551509100832

Storper, M., & Venables, A. J. (2004). Buzz: Face-to-face contact and the urban economy. *Journal of Economic Geography*, *4*(4), 351–370.

Studies, U. (1999). Special issue on competitive cities. *Urban Studies (Edinburgh, Scotland)*, *36*(5-6).

Trippl, M., Tödtling, F., & Lengauer, L. (2009). Knowledge sourcing beyond buzz and pipelines: Evidence from the Vienna software sector. *Economic Geography*, *85*(4), 443–462. doi:10.1111/j.1944-8287.2009.01047.x

Truffer, B., Metzner, A., & Hoogma, R. (2002). The coupling of viewing and doing: Strategic niche management and the electrification of individual transport. *Greener Management International*, *37*, 111–124.

Van den Berg, L., & Braun, E. (1999). Urban competitiveness, marketing and the need for organising capacity. *Urban Studies (Edinburgh, Scotland)*, *36*(5-6), 987. doi:10.1080/0042098993312

Van Winden, W. (2006). Globalisation and urban competitiveness: Challenges for different types of urban regions. In OECD (Ed.), *Competitive cities in the global economy*. Paris, France: OECD.

Van Winden, W., & Carvalho, L. (2009). Exploration and exploitation networks in space: The case of the shipbuilding cluster of Turku, Finland. In Mahnken, G., & Matthiesen, U. (Eds.), *Das Wissen der Städte*. Wiesbaden, Germany: VS Verlag. doi:10.1007/978-3-531-91648-4_25

Van Winden, W., Carvalho, L., Van Tuijl, E., Van Haaren, J., & Van den Berg, L. (2011). *Creating Knowledge Locations in Cities*. Rotterdam, The Netherlands: Erasmus University Rotterdam, Euricur.

Van Winden, W., Van den Berg, L., & Pol, P. (2007). European cities in the knowledge economy: Towards a typology. *Urban Studies (Edinburgh, Scotland)*, *44*(3), 525–549. doi:10.1080/00420980601131886

Wade, R. (1990). *Governing the market: Economic theory and the role of government in East Asian industrialization*. Princeton, NJ: Princeton University Press.

Weiser, M. (1991). The computer for the 21st century. *Scientific American*, *265*(3), 94–104. doi:10.1038/scientificamerican0991-94

Worldtravels.com. (2010). *South Korea map*. Retrieved August 12, 2010, from http://www.wordtravels.com/ Travelguide/Countries/South+Korea/Map

Yigitcanlar, T., Velibeyoglu, K., & Baum, S. (2008). *Knowledge-based urban development: Planning and applications in the information era*. Hershey, PA: Information Science Reference. doi:10.4018/978-1-59904-720-1

ADDITIONAL READING

Crevoisier, O., & Jeannerat, H. (2009). Territorial knowledge dynamics: From the proximity paradigm to multi-location milieus. *European Planning Studies*, *17*(8), 1223–1241. doi:10.1080/09654310902978231

Ducruet, C. (2007). *Incheon, Port City of South Korea-The Oxford Encyclopedia of Maritime History* (*Vol. 2*, pp. 186–187). Oxford, UK: Oxford University Press.

Edler, J., & Georghiou, L. (2007). Public procurement and innovation—Resurrecting the demand side. *Research Policy*, *36*(7), 949–963. doi:10.1016/j.respol.2007.03.003

Giuliani, E. (2005). Cluster absorptive capacity: Why do some clusters forge ahead and others lag behind? *European Urban and Regional Studies*, *12*(3), 269–288. doi:10.1177/0969776405056593

Kenney, M., & Von Burg, U. (1999). Technology, entrepreneurship and path dependence: Industrial clustering in Silicon Valley and Route 128. *Industrial and Corporate Change*, *8*(1), 67. doi:10.1093/icc/8.1.67

Lee, S. H., Yigitcanlar, T., Han, J. H., & Leem, Y. T. (2008). Ubiquitous urban infrastructure: Infrastructure planning and development in Korea. *Innovation: Management. Policy & Practice*, *10*(2-3), 282–292. doi:10.5172/impp.453.10.2-3.282

Lee, Y. I. (2004). MNCS' perception on the feasibility of South Korea as a business hub of Northeast Asia. *Asian Academy of Management Journal*, *9*(2), 97–112.

Lee, Y. I., & Hobday, M. (2003). Korea's new globalization strategy: Can Korea become a business hub in Northeast Asia? *Management Decision*, *41*(5),498–510.doi:10.1108/00251740310479340

McKinsey&Company. (2010). *South Korea: Finding its place on the world stage: Five essays from leading thinkers explore the country's present and future.*

Mortice, Z. (2008). New Songdo city looks back at the new world for older urban models. *The news of the American Community of Architects.*

OECD. (2010). OECD Economic Surveys [-*overview*. Paris, France: OECD.]. *Korea & World Affairs*, (June): 2010.

Stubbs, R. (2009). Whatever happened to the East Asian Developmental State? The unfolding debate. *The Pacific Review*, *22*(1), 1–22. doi:10.1080/09512740802650971

Van Winden, W., & Carvalho, L. (2008). Urban competitiveness in the knowledge economy: Competitiveness paths of the Portuguese metropolises. In Yigitcanlar, T., Velibeyoglu, K., & Baum, S. (Eds.), *Knowledge-based urban development: planning and applications in the information era*. Hershey, PA: Information Science Reference. doi:10.4018/978-1-59904-720-1.ch012

Whitman, C. T., Reid, C., von Klemperer, J., Radoff, J., & Roy, A. (2008). *New Songdo city: The making of a new green city.* Paper presented at the Conference on Tall Buildings and Urban Habitat (CTBUH) 8th World Congress 2008, Dubai, United Arab Emirates.

Wolfe, D., & Gertler, M. S. (2006). Local antecedents and trigger events: Policy implications of path dependence for cluster formation. In Braunerhjelm, P., & Feldman, M. P. (Eds.), *Cluster genesis: Technology-based industrial development*. New York, NY: Oxford University Press.

Yigitcanlar, T., & Velibeyoglu, K. (2008). Knowledge-based urban development: The local economic development path of Brisbane, Australia. *Local Economy*, *23*(3), 195–207. doi:10.1080/02690940802197358

KEY TERMS AND DEFINITIONS

Clusters: In its simplest form, refers to a number of geographically concentrated companies operating in the same or related sectors. Other definitions add to this one the nature of the connections between these companies (e.g. competition, cooperation, knowledge exchange), and includes other agents besides firms, like universities, venture capitalists, consultants and manifold associations.

Competitiveness: Comparative concept referring to the ability of a company to supply (internationally tradable) goods and/or services in an open market, against other potential suppliers. Widely used in the disciplines of economics and business.

Free Economic Zones: Specific areas planned by Governmental authorities where (usually) foreign companies access to several benefits (usually tax cuts) in order to motivate their settlement and investment in those locations.

Technological Niches: Protected innovation arenas supported by Government authorities and other stakeholders in order to develop and experiment a specific technology or group of technologies not yet validated by society, and for which there exists no immediate demand, but with promising foreseen applications.

Territorial Competitiveness: Extension of the previous concept, developed during the last decades, to speak about territorial entities such as countries, regions or cities. It often refers to the dynamic capacity of these territories to retain, attract and nurture globally mobile factors like talent, investment, businesses, as well as residents and tourists.

Ubiquitous City: Is a city endowed with ubiquitous computing technologies, where information subsystems and urban functions are linked through several digital infrastructures and technologies, for example traffic, health care, crime prevention, etc.

Ubiquitous Computing: Also called "invisible computing" or "everyware", it is a new computing paradigm in which ICTs blend themselves in people's everyday life, without being noticed. In this paradigm, information and its transmission is integrated in objects, artefacts and persons. In practice, it refers to the possibility of augment and connecting "objects" (e.g. buildings, cars, pieces of paper, even persons) and digital technologies, in an undistinguishable fashion. Human beings and objects will be permanently connected, receiving and releasing information from one to the other.

ENDNOTES

1 For recent advances and more nuanced views on these mechanisms, namely on the genesis, functioning and evolution of clusters, please see the "additional reading" section at the end of this chapter.

2 Moreover, new ICT technologies have been responsible for significant increases in total factor productivity, namely in the service sectors, just see the case of the logistic and distribution innovations pioneered, e.g. by Wal-Mart (Fernald et al, 2004)

3 Just the way *competitiveness in ICT* was primarily linked to the "knowledge base and economic base" and to some degree to the economic "diversity" of the urban area.

4 This is a similar situation to what we can find nowadays with the scenarios for the consolidation of new clean energy sources based on electricity and hydrogen for urban mobility.

5 Recent research argues that these processes, underlying the development of technological niches, start in a rather localized fashion, and are eventually articulated in complex geographies between concrete places, e.g. cities, endowed with the right assets and institutional infrastructure (e.g. Benneworth et al, 2010; Carvalho et al, 2010). This has very much to do with the localized nature of learning and its cognitive, cultural and social dimensions (Maskell and Malmberg, 1999; Bathelt et al, 2004).

Chapter 11
Personal Smart Cards:
From Transportation to a City Smart Card—The Database Integration of Public Services in Curitiba

Gustavo Taniguchi
URBTEC, Brazil & Pontifical Catholic University of Paraná, Brazil

Fábio Duarte
Pontifical Catholic University of Paraná, Brazil

ABSTRACT

An effective information management is one of the most valuable assets of any organization, including public administration. Database integration from various municipal Information Systems is an essential tool for urban and strategic planning. The city of Curitiba, Brazil, has implemented several Information Systems to enhance its public services, from education to health care—most of these systems use several non-integrated personal smart cards to provide access to services. In this chapter we analyze some public Information Systems in use in Curitiba, and their inputs and outputs. We also advocate that the personal smart card already used in the metropolitan transportation system only as a fare collection card, is, among all others, the best option, from a technological and administrative standpoint, to integrate all municipal Information Systems, improving the effectiveness of public services and assisting for a comprehensive planning process.

DOI: 10.4018/978-1-61350-174-0.ch011

INTRODUCTION[1]

Information is a main key to understand and manage many aspects of contemporary society. As we are immersed in a global network of ubiquitous and pervasive technologies, most exchanges of data that permeate our lives are invisible, but represent, at the same time, the very essence of today's life. Both private and public sectors base the provision of products and services on an information exchange with their clients – consumers or citizens. From granting access to some places, to the individualization of services provided for specific clients, the information transactions among consumers and companies involve the exchange of personal data. The same occurs with the access to public facilities and services.

Information technology systems (ITS) allow this data exchange. Therefore, an effective information management is one of the most valuable assets of any organization, including public administration. The city of Curitiba, Brazil, has been using ITS to enhance different public services, as to control the access to public schools or to follow the medical history of patients in health care facilities. Most of these services in Curitiba are mediated by a personal smart card, which is the main focus of this chapter.

The transportation system of the Metropolitan Region of Curitiba is one of the biggest public services that has adopted the personal smart card in the last decade. Implemented in 2005, the use of the transport card (*cartão transporte*) is not mandatory, but it has been increasing over the years – in 2006, 44 percent of the passengers were using the transport card as a mean of payment (more than twice the percentage of usage just after its implementation).

Like other smart cards adopted by the public administration in Curitiba, the transport card holds personal information about its users. However, each public service has its own card, holding only a few data, which differs from the information held by other services' smart cards. It means that, to make use of more than one kind of public service, a citizen must carry different smart cards together and, consequently that, the databases of different public services are not integrated – what questions the *smart* side of this technology.

The main argument of this chapter is that from a strategic point of view, the adoption of a single personal smart card for all municipal services could help building a comprehensive citizen's database that could enhance public services and assist a comprehensive planning process. We also advocate that the personal smart card already used in the metropolitan transportation system only as a fare collection card, is, among all others, the best option, from a technological and administrative standpoint, to integrate all municipal information systems, improving the effectiveness of public services.

We begin by briefly discussing the use of ITS in a context where ubiquitous and pervasive technologies are an inherent part of many mundane activities of our daily life. As the exchange of individual data is becoming omnipresent for the delivery of high quality public services, the adoption of personal smart cards (PSC) spreads worldwide in the municipal level. Technologically, PSC is an appropriate identification tool, but some concerns of possible misuses of private data are a key issue and will also be discussed here.

The use of PSC in public transportation is seen in this chapter as a viable strategy for integrating different public databases, as it allows the understanding of some urban dislocation patterns over time on a specific territory – what can be a valuable asset for many other public policies. Therefore, an overview of PSC adoption in transportation will be presented. The case study discussed here is the city of Curitiba, in the South of Brazil. Curitiba, a 1.8 million people city, has adopted PSC in different public services and has a metropolitan transportation system well structured that guides the distribution of public facilities.

Based on ITS and PSC already in use, it would be possible to develop a model of a municipal or metropolitan Integrated Information System. This system would assist the planning processes, directing public resources to better serve local and regional demands of the population.

Public Sector and the Information Environment

In the 1960s, Marshall McLuhan had already predicted the emergence of a society where most of the social, economic and cultural exchanges between organizations and people would be mediated by information and communication technologies (ICT). But at that time, each media had its own language, and an expensive and slow translation between them was yet to take place. Digital media were responsible for organizing any information using only one code, making registration, storage and transmission of data easier, cheaper and faster than ever.

ICTs became broadly used in the 1990s. Following a general tendency, the adoption of ICTs in public services increased, and has been advocated from both a managerial and a democratic point of view. From the managerial perspective, it brings more effectiveness to the public sector (Heintze & Bretschneider, 2000); and from the democratic perspective, it empowers people to participate more directly in governance (Kauchakje *et al.*, 2006).

A vast array of ICTs is currently deployed in virtually all activities of our daily life. Many scholars (Aurigi & De Cindio, 2008; Firmino *et* al., 2010) argue that we live in an augmented reality, with technologies infiltrated everywhere. And as digital media are becoming omnipresent, an increasing amount of personal and organizational relations depend on what has been called invisible ubiquitous computing (Weiser, 1991).

Media devices are seen as key enablers of these interactions. On the one hand, such devices should allow multiple interactions; on the other hand, they must keep the individuality of their holder safe. Therefore, smart cards are privileged media, as they host "security algorithms and functions" in order to store data and participate in automated electronic transactions (Mayes, 2008, p. 2). Credit cards, access of personal bank accounts and employee's identification are among the possible uses for PSC (Rankl & Effing, 2002).

Nevertheless, privacy is a critical issue. David Lyon (2002) puts that surveillance is an everyday experience, built in a rhizomatic form that may not be perceived by the "surveilled". Lyon (2007, p. 14) clarifies that surveillance is a "focused, systematic and routine attention to personal details for purposes of influence, management, protection or direction". Gathering information is not an intrinsically hideous action, and different levels of today's activities and transactions are structured upon the manipulation of coded information that puts in contact people and organizations. But some societies seem to be "over coded".

In Brazil there is an overlapping of systems of identification and, therefore, identification documents. Virtually any interaction between public and private agents with their public (consumers and citizens) is mediated by a personal ID. As Murakami Wood and Firmino (2010, p. 302) put it, it is so "embedded in the way of life that people do not even think about it very much", at a point that "the presentation of either identification numbers or tokens of identification, and often more than one on each occasion".

In public transportation, the adoption of PSC "is raising major privacy concerns on the part of the users (…) because the operator has a better knowledge of the location of users" – and, for example, it can be used as an evidence in police investigation (Pelletier *et al*, 2009, p. 6).

In spite of a possible control over individual attitudes that may be exercised by those who master the information systems, personal smart card may also bring benefits to their users. As David Lyon (2002) well asserted it, surveillance is ambiguous, meaning that the very use of surveillance systems is uncertain and depends on the social embedment of such technologies.

Curitiba, as many other cities around the world, has adopted PSC in its public health care system. Enhancing privacy and authentication, PSC can store "critical emergency data" (Smart Card Alliance, 2006, p. 13), and "enable convenient access to patient data and that supports new applications that deliver both clinical and administrative benefits" (*Idem,* p. 4).

In public transportation, smart cards have become a safe solution to improve security in fare collection. Hong Kong's *Octopus* card is one of the first multimodal smart cards ever implemented, and presently has more than 9 million contactless smartcards used in all transport modes (Sambandaraksa, 2006). One year after its implementation, cards had ran out of stock, as users exceeded distribution expectations (Ho, 2001, p. 24). It was an instant success: it was a convenience to all users and operators, it improved transportation efficiency, and eliminated cash as a form of payment in the system. All data are collected and processed in a central database that can be integrated with other services such as parking, commerce, and telecom.

Octopus is basically a prepaid transport card, but it is also accepted in retail stores. The risk, as argue Fleishman *et al* (1998, p. 31), is that "because there is no requirement that a multiple-use card be used for transit, an employer providing vouchers or actual fare cards could be effectively subsidizing retail purchases or telephone calls rather than transit use for some employees". Other cities, like New York, have adopted a multimodal card, used on buses, metro, trains and boats. But both New York's and Hong Kong's cards are "anonymous" smart cards, as named by Bagchia (2005, p. 243), who states that with this, transport agencies are putting aside a potential "to engage in direct marketing with those individuals".

Agencies with or fearing high levels of fare fraud tend to prioritize SC – it is the case in the United States, as shows a survey conducted by Iseki *et al.* (2006). In Curitiba, the transportation agency had repeatedly reported high rates of frauds before implementing the transport smart card technology. Apart from the benefits in fare collection, transport planners also advocate that smart cards may reduce boarding time, increasing the overall performance of the transportation system.

Even if the subject of this chapter were analyzing SC only as an automatic and safe fare collection system, we can take a step forward by looking into other kinds of technological interactions between databases and the possibilities to improve services. Teleplay is a mobile payment system customized for transport systems. In this systems, transport users can get travel information, buy tickets and check-in using mobile phones, "using SMS, WAP and short range communication technologies" (Teleplay, 2002, p. 9). Ticket value is charged on the customer's phone bill, what gives to this technology an additional security advantage: users do not need to carry a prepaid card along with them, as it is the case with most of the transport smart cards. The system has been tested in some European cities, and as Jakubauskas (2006, p. 258) puts it, "these practical examples revealed persuasive advantages of e-ticketing: seamless interoperable ticketing, increased revenue due to lower costs for transport and infrastructure operators, less time wasted queuing by passengers, better functioning multimodal urban public transport systems, among others."

However, ensuring safe fare collection methods, for the benefit of both the transport agency and passengers, is just one aspect of the information system mediated by smart cards. Several authors have discussed the advantages of using PSC in public transportation in a broader way, rather than just as a safe fare collection system. Torres, Izquierdo & Sierra (2007, p. 2249) see that "a new trend indicates a shift towards more enhanced cards with networking capabilities. Personal smart card is the technology that could improve public facilities management".

PSC used in public transportation can provide a large quantity of data, in a daily basis. For Bagchia and White (2005, p. 464), there is a perception that "such systems will produce high quality data on journeys undertaken on public transport services and the customers undertaking those journeys"; and that, in an improved information system, "such data [could] be linked back to the individual card".

More recently the combination of PSC technology with the automatic vehicle location (AVL), integrated with geographic information systems (GIS), has opened up "new opportunities to collect comprehensive transit travel data at practically no additional cost" (Chapleau, Trépanier & Chu, 2008, p. 2). To the authors, this system "behaves like an automated continuous travel survey" (Chapleau, Trépanier & Chu, 2008, p. 23); and with some statistical inferences and specific algorithms, Chapleau, Trépanier and Chu (2008, p. 23) have demonstrated that the combination of such technologies helps constructing a precise origin-destination matrix – essential for public transportation planning. Mapping trips patterns in a daily basis (and virtually in an individual basis), has an enormous potential to improve transportation management.

In the following sections we will present how the city of Curitiba has adopted information systems based on PSC in different public services; how the municipality organizes all these data; and how the transport card is, among all others, the best option to integrate all the municipal information systems, improving the effectiveness of public services and assisting for a comprehensive planning process.

BACKGROUND

Smart Cards: An Overview

The improvement of electronic data processing in the 1970s made smart cards a viable device (Rankl & Effing, 2000). Even though banking applications were already commonly used, it was only in the early 1980s that France's Public Telephone and Telegraph Systems launched the first successful commercial deployment of smart cards. They were loaded with pre-purchased values in order to create a coinless public telephone system.

International standardization is regulated by the International Organization of Standardization (ISO), in co-operation with the International Electrotechnical Commission (IEC) and the European Standardization Committee (CEN, for *Comité Européen de Normalisation*).

Smart cards embedded with integrated circuits replaced the magnetic-strip cards, with the advantages of data storage, processing and transmission – and, just as important, the virtual elimination of frauds.

Smart Cards can be sorted by the way they transmit data, as microprocessor cards, contactless smart cards, and hybrid cards.

Microprocessor Cards

Smart Cards that transfer data to a Point of Sale (POS) or Smart Card Reader through physical contact have a chip with a microprocessor unit, RAM (Random Access Memory), ROM (Read Only Memory) and a data storage unit – usually an EEPROM (Electrically Erasable Programmable Read-Only Memory).

Contacts remain one of the most frequent sources of failure in electromechanical systems, despite the fact that failure rate is below than one in a thousand (Rankl & Effing, 2000).

Contactless Smart Cards

The contactless smart card has a discrete way of exchanging data between the card and reader hardware, opening a wide variety of applications like transport and access controls.

In addition to the microcontroller unit, there is a radio frequency interface activated by proximity to a contactless device. The electromagnetic field provided by an external device supply the power needed to exchange data between the card and external hardware. Field experiences show that this kind of transaction is viable in distances up to 50 centimeters (20 inches) between the card and the reader.

Hybrid Cards

Combining both data transfer technology, hybrid or "combi" cards can also use a common area that is accessible either way, contact or contactless. As technology keeps advancing, smart card capacity in kilobytes grows continuously, allowing multiple application development and more data storage.

Smart Cards in Brazil: A Short Story

Most of smart cards in use in Brazil are credit cards, or those used in mobile phones and public transportation. They were introduced in the 1990s. Along the years, other public services began using this technology, such as health systems, education, human resources management and commerce.

From parking systems to bicycle renting, there are smart card solutions for every kind of transport system. They were implemented to gain boarding agility, to control the fare collection and to prevent fraud by reducing the use of cash in those systems.

Automated Fare Collection Systems (AFCS) were firstly used in subway and rail systems in the 1970s. In the 1980s this system migrated to buses and Bus Rapid Transit (BRT), and by the end of the 1990s they were all around the country, tackling mass transit problems as fare evasion, fraud and operational issues.

A survey conducted by the Brazilian National Urban Transport's Company Association (NTU), in several cities that were implementing Automatic Fare Collection Systems, showed that more than 40% of the biggest cities were already using smart cards for mass transit systems, and only 14% had not had plans to use such a technology (NTU, 2003). Among the cities using smart cards, 69% were using contactless smart cards.

Beyond security issues, modal integration and the use of different fare zones were the main focus for AFCS implementation.

MAIN FOCUS OF THE CHAPTER

In this section, we present an overview of the information systems already in use in Curitiba. We begin with the public transportation, both because it was one of the first public services to adopt ITS and smart cards in particular, and because its use is the broadest in terms of reaching citizens from a variety of socioeconomic profiles – which is a major advantage in favor of this card for integrating different databases. We also discuss how public services can be integrated through the use of personal smart cards in order to optimize their management.

Public Transportation in Curitiba

Curitiba is known worldwide for its urban planning experiences, especially concerning public transportation systems. From the 1970s, the city implemented a comprehensive urban planning articulating land use, a hierarchical road system and a modern public transportation based on buses. Under this principle, high densities were allowed almost exclusively along the transit main structures (designed as the main development axes for the city), in order to stimulate the use of public transportation and optimize public urban infrastructure.

Over the years, the transportation system of Curitiba has become what was named Bus Rapid Transit (BRT). BRT is defined as an "integrated system of facilities, equipment, services, and amenities that improves the speed, reliability, and identity of bus transit" (TCRP, 2007, p. 7). Or, as described by ITDP (2007, p. 15), "the BRT concept invokes a range of both quantitative and qualitative attributes that together help to create

Figure 1. Aerial view of a BRT corridor in Curitiba, Brazil. Source: © 2009, Irã Taborda Duduque. Used with permission.

a high-quality transit experience for the customer".

The system's good reputation depends on some technical characteristics, that are now, common among BRT projects around the world: segregated bus lanes; the use of a network of integrated bus routes; terminals to handle feeder-to-trunk transfers; a fast boarding system – with on level access between the bus floor and platforms; a single fare collection system, and prepaid fares. Curitiba is recognized for pioneering the BRT system, and as an example of a "full" BRT implementation (as it makes use of almost all BRT technical characteristics).

The city has implemented and has been improving its BRT over the last 30 years in a continuous process of gradual upgrades and advances – for instance, even with the segregated bus lanes in place since the first years of the system, only in 2009 the city built the first corridor with overtaking lanes for express buses.

Besides these kind of adaptations in existing infrastructures, the implementation of information technology systems (ITS) to the planning and management of the system is the most recent and incomplete input of Curitiba's BRT. From the very beginning, in the 1970s, Curitiba has tried to adopt a prepaid fare collection system. But only recently the city adopted personal smart cards in public transportation – but without exploring all the potential of this technology. Transport cards, as we will argue, could go beyond fare collection to become an information hub for the municipal ITS.

Automatic Fare Collection in Curitiba

The first attempts to implement an AFCS in Curitiba go back to 1974, when the operation of the

BRT was just beginning. Using a new technology for that time, with a graphite paper ticket loaded with 10 trips and an automatic cutting and reader device attached to buses turnstiles, it quickly proved to be easily copied, and was abandoned.

By the mid 1980s, tokens replaced money in the mass transit system. Notwithstanding, tokens still had value, what created a parallel market for them. At that time the Brazilian economy was fragile, with the inflation getting over 750% a year (1990-1995). In this context, tokens became a good "investment" and soon counterfeit tokens were inserted in the system.

This situation urged a smart solution, and finally in 2003 tokens were discontinued and AFCS was fully implemented, with more than 500 thousand contactless smart cards distributed in several fare classes. More than 1,800 card readers were installed in buses and transport terminals allowing the municipal transportation management company (URBS) to prevent monthly debts by counterfeits.

When smart cards were implemented, more than 22 million tokens were in use in Curitiba – but more than 10 million were bogus cards. Several card classes were implemented to control transport access, gratuities and discounts, mostly for students and elderly people.

All access data, like card number, card type, date and hours of use, are stored in card readers and transferred to a central database. Everyday a list of stolen and lost smart cards is uploaded in every reader, diminishing wrong use and abuse. Transportation credits can be bought on the internet or from card chargers throughout the city.

Health System in Curitiba

Another important use for smart cards in Curitiba is in the public health care system. Called Quality Health Card (*Cartão Qualidade Saúde*), it is one of the most accessed public services in Curitiba.

Technologically, it is a simple magnetic stripe ID card. More than 1.3 million cards have been issued and can be used in more than 138 Health Units. The transport card can also be used to get access to health services, but not all Health Units are equipped with smart card readers.

This system gathers all health information about the patient and her or his family, controlling medicine dispensed by the public pharmacies, where patients get appropriate medicine for free. All the information is stored in a datacenter in the Municipality Health Department and mirrored in another datacenter at the Information Institute of Curitiba (ICI). Information is used to plan health investments and promote campaigns for the whole city or specific neighborhoods.

Education System in Curitiba

In 2003 a specific smart card was introduced in the municipal education system. The main focus was to control the access of students to schools. About 90 thousand contactless smart cards were issued and are used in all municipal schools.

Similar to the bus access, turnstiles are used to restrain access for those not allowed inside the educational environment. Security is a main concern in this case, leaving unwanted people (drug dealers i.e.) out of the schools.

Information collected by this system also makes easier the planning of schools' purchases, such as food and stationary, reducing leftovers.

School staff can also control student's attendance to classes, intervening or calling up the families. The education smart card also gives access both the student and her or his family to the municipal libraries, called Lighthouses of Wisdom. Each book borrowed by a student in these libraries has its information sent to her or his school's database.

There are also a few other advantages, like drugstores that offer discounts for students holding the card when shopping, and football teams that away tickets for those students with highest frequency in classes.

Table 1. Comparison between Curitiba and Hong Kong: services and users

Service \ City	Curitiba	Hong Kong
Culture		
Health	X	
Education	X	
Transport	X	X
Multi-modal Transport		X
Parking		X
Commerce	X	X
Number of users	1,000,000	9,000,000

Telephone Information System 156

Even with no use of smart cards, the 156 telephone information system is an important element for Curitiba on its way to become a whole informational city.

At the beginning of its implementation, the system was meant to be just a receptive phone number, taking notes of citizens' concerns and demands, but this service soon proved to be a great tool for planning and interacting with people.

Workflow of activities and services were developed to make phone calls faster, and now 3,200 standard services can be activated directly by operators after a citizen's call. From garbage collection to street lamp replacement, when a service is demanded it is uploaded in the system. After a call, a demand form is automatic generated and a service order is sent to the responsible department in the city administration. If the system does not receive a message from this department it automatically alerts higher hierarchical stances in the administration, until it reaches the mayor. When the problem is solved an informative phone call or email is sent to the citizen that demanded it.

All services requests are tracked and can be summarized by region, type and volume. This information is used to next year's budget planning and investments.

Service demanding represents 50% of all calls, while others are informative calls only, like bus lines and time tables inquires. From all services requested in the municipality, 70% are made by phone, 12% on the internet and 18% are personal requests.

Quality Card

Working as an identification badge while at work, Quality Card is mainly used by city government employees since 1997. It uses a contactless technology to control access in public buildings. Besides, over 30,000 employees can shop in general commerce, get bank loans and use the card as a debt card.

For a brief comparison, Table 1 highlights the number of users, and the public services involved in the smart cards systems currently in use in Curitiba and Hong Kong.

Despite being based in different technologies, smart cards in Curitiba are used in more public services than in Hong Kong, where it is mainly as a debt card, accepted in general commerce and also in public transportation. Therefore, this situation seems to bring to Curitiba an opportunity to integrate different public services through personal smart cards, allowing the municipality to better manage investments and planning decisions.

Situation Room: Business Intelligence

The so-called "situation room" is currently the main IT project in Curitiba, coordinated by the Curitiba Information Institute (ICI or *Instituto Curitiba de Informática*). ICI is a social organization, legally similar to a Non-Governmental Organization, with a nonprofit regime, working under contract for the city. Its main task is to operate and maintain all IT systems, software and hardware of Curitiba's City Hall.

The situation room is based on a Business Intelligence (BI) technology to deal with all information generated by the various municipal information systems.

Implementing BI had some strategic goals:

- Standardize all information collected by various systems, allowing a common information core;
- Establish a social and economic standard for municipal services users, with demographic, educational, occupation and families incomes, familiar structure and home conditions;
- Build an integrated database to feedback results about social politics and help the municipality in planning social programs;
- Evaluate public investments through indexes constructed from an integrated database;
- Subsidize cost-benefit studies of service and programs offered by the municipality;

Operational goals were:

- Give instant feedback about the municipal service users to the attendant, allowing access to information about familiar and social status, giving the attendant the opportunity to give a personal help to the citizen;
- Build priority criteria easier for attendants;
- Implement database integration, allowing multi sector access by multiple services.

Curitiba's ITS Analysis

The integration of Geographic Information System (GIS) with the BI tool generates a multitude of cross information that became fully available only in 2010.

Health, education, tributes, call centers and many other systems are already providing information to BI technology. This information is used to plan field actions in various municipal sectors, giving strategic advice to all decision makers. Although transportation data are not yet integrated, information has been stored for future integration by the ICI.

Pelletier et al (2009) have summarized several works about smart card data used in public transit planning in strategic, tactic and operational levels. Strategic and long-term planning are focused in user characterization and classification, without the use of personal information. In tactical planning service adjustment is the main concern. And in the operational level performance indicators and statistic data are the most important information used. Smart card brings a myriad of information for transportation and transit systems, like trip habits, demand management, real-time itinerary design, evaluation of actual and planned conditions in the network, creation of future demand matrix, public transport network planning, customer-loyalty scheme, transfer journeys, user patterns vs. fare structure, transit user behavior and many others.

Here lies the main concern of this chapter: how can a PSC add substantial value to the information already collected by different systems used by the municipality? With the implementation of GPS technology on the buses, and a PSC embedded with a Radio-frequency identification technology, a link between card usage and geographic coordinates can provide an Origin-Destination map in real time. It can help municipality to recognize

Figure 2. Business Intelligence Workflow and Systems Integration. Source: © 2010, Gustavo Taniguchi. Used with permission.

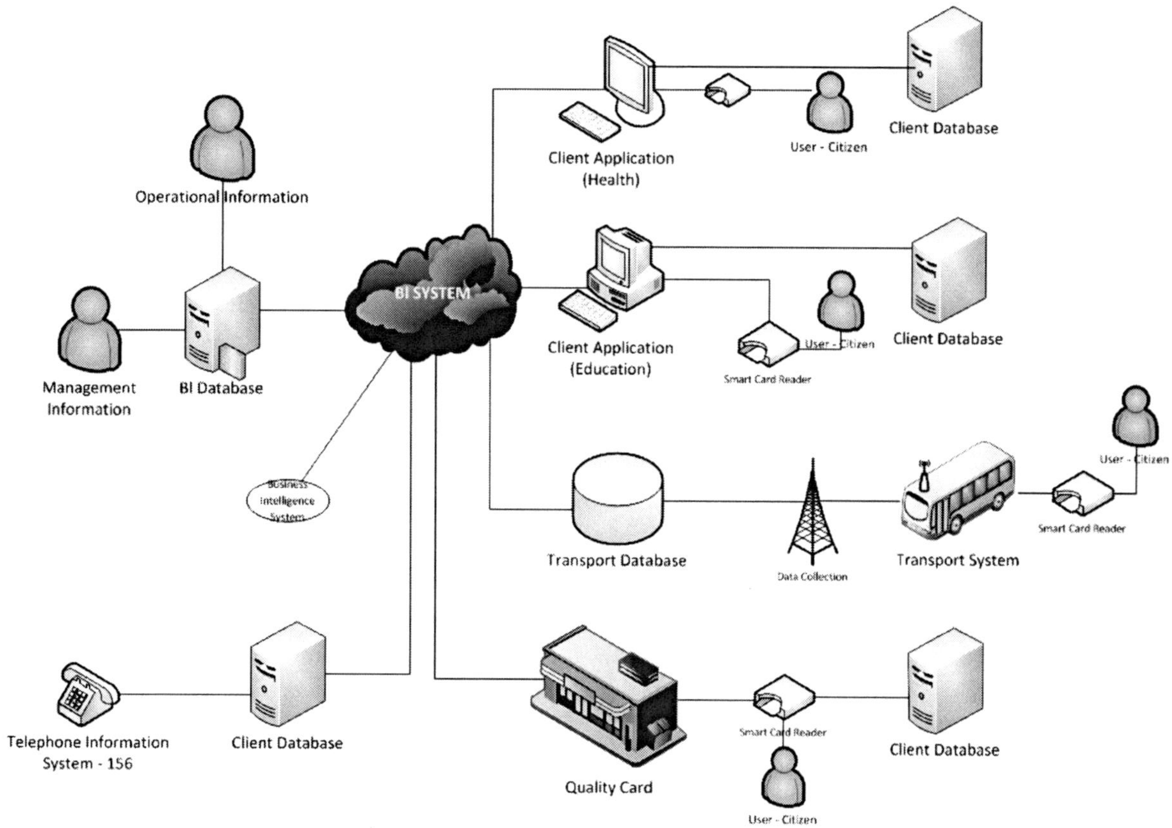

patterns of urban dislocations over time and space. With this information, the city administration can optimize investments and management in the provision and maintenance of public facilities and services.

Therefore, a PSC can play a key role in city planning. Containing personal and public data, it makes virtually possible the planning of personal costumer assistance to every citizen in most public areas such as health, transport and education. For example, with such this integration in place, decision about the location of a new hospital, based on the demands of actual health facilities and the dislocations for this specific use, can be turned into a viable reality.

Certainly, personal data security and privacy shall be a concern, but security algorithms can technically ensure data integrity.

Table 2 shows data collected in 4 different PSC systems used in municipal services in Curitiba.

Currently, the so-called health card and education card are the ones that collect more data from their users, despite data not being the same. Health card has an additional technical problem, which is the difficulty of inputting new data, because of the limitations of the magnetic stripe technology. Another important issue is that, in general, they are used by different people, as education card is only used by students in municipal public schools, while the health card is more often mostly used by adults and children. In this context, transportation card could have an important role. It is the

Table 2. Data collected through Personal Smart Cards in Curitiba

Data Collected	Transport Card	Quality Card	Education	Health Card
Technology	*Smart card*	*Smart card*	*Smart Card*	*Magnetic Stripe*
Name	X	X	X	X
Mother's name	X		X	X
ID	X		X	X
Birth Date	X	X	X	X
Gender	X			X
Public Service ID		X		
Nationality				X
City of Birth				X
Marriage Status				X
Scholarity				X
Scholar Number			X	
School name			X	
Profession				X
Address			X	X
City / State			X	X
Zip Code			X	X
Picture		X	X	

most widely used for different social groups; and it can carry on the very important information represented by the pattern of displacement of people. Linked with its use in educational equipments (schools, but also public libraries), it can also provide the pattern of displacement by purpose.

In Curitiba, urban facilities, like health units and public schools, are distributed throughout the city. That makes transport smart card a better way to permeate city's tissue and provide information about all regions.

Strategic planning is easier when all city data is accessible through one single tool. A single ID for public services could reduce duplicated data in several systems.

Among strategic information that could be provided by database integration are:

- Epidemiologic watching vs. infectious diseases in schools;
- Regular nutrition vs. health unit usage, by verifying public supermarkets usage and its effectiveness;
- Students that have fare benefits vs. distance between home and school, as this is a requisite to gain transport fare benefits;
- Health conditions vs. school, identifying how health and home conditions can affect student's grades and performance;
- Travel patterns vs. urban infrastructure, verifying the way citizens move across the city and urban services usage, making transportation planning easier.

Eguchi et al (2007) propose some policy issues for transportation smart card systems that can be also important to multifunction systems. For the authors, municipalities should:

- Choose a technology standard that allows for integration with other smart card systems;
- Offer two types of cards – one with embedded personal identification data, and another without;
- Consider a distance-based fare structure;
- Continue to respect different pricing structures for different social groups;
- Provide financial incentives to encourage adoption;
- Provide backup functionality in case of central system failure;
- Routinely screen system for software failures;
- Provide parental controls in cases where smart card functionality expands.

Nevertheless, some hindrances can difficult the using of smart cards. Institutional arrangements are crucial to the viability of this integration and the city's choice for an open system that can improve its functionalities. Transport agencies that use closed systems and issue its own smart cards make transport and service integration more difficult.

CONCLUSION

Database integration can play a key role in providing good services to the public, optimizing investments.

Using transport information in database integration can boost Business Intelligence. Transportation always had a key role in urban development in Curitiba. Therefore, it is not the case of only improving the PSC used in transportation, but rather use it as a basis to create a kind of a City Smart Card (CSC). This CSC can be based on the transportation card because it potentially holds important information, related to dislocations' patterns in town. The awareness of these patterns over space and time can be a turning point in planning public services and facilities for the city.

The use of CSC as a feedback tool is also important. Every time a citizen is informed that a public service is available and she or he has an opportunity to evaluate it, the city "gets" more information. This database has the potential to keep growing endlessly providing information to decision makers. It can also keep an administration running with less political interference, showing the mayor the biggest and more demanded problems. The more transparent this information become, the easier is to work on it. The more transparent and accountable an administration is, the greater is the chance of having potential investors considering businesses or investing in the cities.

REFERENCES

Aurigi, A., & De Cindio, F. (Eds.). (2008). *Augmented urban spaces*. Hampshire, UK: Ashgate.

Bagchia, M., & White, P. R. (2005). The potential of public transport smart card data. *Transport Policy*, *12*, 464–474. doi:10.1016/j.tranpol.2005.06.008

Baldassi, V. (2000). *The Venice card: Our city at our fingers*. In Major cities of Europe: ITusers group. Retrieved from http://www.majorcities.org/pics/medien/1_1102426321 /Baldassi_eins.pdf

Chapleau, R., Trépanier, M., & Chu, K. K. (2008). The ultimate survey for transit planning: Complete information with smart card data and GIS. In *8th International Conference on Survey Methods in Transport: Harmonisation and Data Comparability*, Annecy, France [Proceedings] (pp. 1-25).

Eguchi, M., Fredholm, S., Liu, S., Ponce de León Baridó, P., & Ye, J. (2007). *Policy issues in implementing smart cards in urban public transit systems*. Cambridge, MA: Massachusetts Institute of Technology, Technology and Policy Program.

Firmino, R., Duarte, F., & Ultramari, C. (Eds.). (2010). *ICTs for mobile and ubiquitous urban infrastructures: Surveillance, locative media and global networks*. Hershey, PA: IGI Global. doi:10.4018/978-1-60960-051-8

Fleishman, D., Schweiger, C., Lott, D., & Pierlott, G. (1998). *Multipurpose transit payment media* (TCRP Report 32, pp. 1-132). Washington, DC: National Academy Press.

Heintze, T., & Bretschneider, S. (2000). Information Technology and restructuring in public organizations: Does adoption of Information Technology affect organizational structures, communications, and decision making? *Journal of Public Administration: Research and Theory*, *10*(4), 801–830.

Ho, K. (2001). Hong Kong – Moving into the 21st century. In Gar-On Yeh, A., Hills, P. R., & Ka-Wing Ng, S. (Eds.), *Modern transport in Hong Kong for the 21st century* (p. 283). Hong Kong: Centre of Urban Planning and Environmental Management, University of Hong Kong.

Iseki, H., Yoh, A. C., & Taylor, B. D. (2006). *Survey on status of knowledge and interest of smartcard fare collection systems among US transit agencies* (pp. 1–43). Los Angeles, CA: California PATH Research Report, University of California, Los Angeles.

ITDP. (2007). *Bus rapid transit planning guide*. New York, NY: Institute for Transportation & Development Policy.

Jakubauskas, G. (2006). Improvement of urban passenger transport ticketing systems by deploying intelligent transport systems. *Transport*, *21*(4), 252–259.

Kauchakje, S., Penna Neto, M. C., Frey, K., & Duarte, F. (2006). Redes socio-técnicas y participación ciudadana: Propuestas conceptuales y analíticas para el uso de las TICs. *REDES. Revista Hispana para el Análisis de Redes Sociales*, *11*, 1–26.

Lyon, D. (2002). Everyday surveillance: Personal data and social classifications. *Information Communication and Society*, *5*(2), 242–257. doi:10.1080/13691180210130806

Lyon, D. (2007). *Surveillance studies*. Cambridge, UK: Polity Press.

Mayes, K. (2008). *Smart cards, tokens, security and applications*. New York, NY: Springer. doi:10.1007/978-0-387-72198-9

Murakami Wood, D., & Firmino, R. (2010). Empowerment or repression? Opening up questions of identification and surveillance in Brazil through a case of identity fraud. *Identity in the Information Society*, *2*(3), 297–317. doi:10.1007/s12394-010-0038-y

Ntu-Petry, R., & Cruz, D. (1999). *Sistemas automatizados de arrecadação tarifária para o transporte urbano*. Retrieved from http://www.ntu.org.br/ frame_publicacoes.htm

Pelletier, M. P., Trépanier, M., & Morency, C. (2009). Smart card data in public transit planning: A review. *Cirrelt*, *46*, 1–19.

Rankl, W., & Effing, W. (2002). *Smart cards handbook*. West Sussex, UK: Wiley.

Rezende, D. (2005). *Planejamento de Informações Públicas Municipais: Guia para Planejar Sistemas de Informação, Informática e Governo Eletrônico nas Prefeituras e Cidades* (1st ed.). São Paulo, Brazil: Atlas.

Robinson, J. (2005). *Eurocult21 Integrated Report. Helsinki: Book on Demand, Lasipalatsi Media Centre* (p. 114). Retrieved from http://eurocul21.org

Russo, A. P. (1999). Venice: Coping with culture vultures. *The UNESCO Courier*. *Ano*, *57*, 42–43.

TCRP. (2007). *Bus Rapid Transit Practitioner's Guide (TCRP Rep. 18)*. Washington, DC: Transportation Research Board.

Teleplay. (2002). *Telepayment system for multimodal transport services using portable phones*. Telepay Conference, Lyon, France. Retrieved from www.ertico.com/download/ telepay_documents/ 2_TELEPAY.pdf

Torres, J., Izquierdo, A., & Sierra, J. M. (2007). Advances in network smart cards authentication. *Computer Networks*, *51*, 2249–2261. doi:10.1016/j.comnet.2007.01.010

Weiser, M. (1991). The computer for the 21st century. *Scientific American*, *265*(3), 94–104. doi:10.1038/scientificamerican0991-94

ADDITIONAL READING

ANTP/BNDES. (2003). *Bilhetagem Automática e Gestão nos Transportes Públicos. Série Cadernos Técnicos* (*Vol. 1*). São Paulo: ANTP.

Atkins, W. (2004). *The smart card report*. Oxford, UK: Elsevier.

Aurigi, A. (2005). *Making the digital city*. Hampshire, UK: Ashgate.

Bekkers, V. J. J. M., van Duivenboden, H., & Thaens, M. (2006). *Information and communication technology and public innovation*. Amsterdam, The Netherlands: IOS Press.

Beuscart, R., Renard, J.-M., Delerue, D., & Souf, A. (1999). Telecommunication in healthcare for a better coordination between hospitals and GP's: Routine application of the "ISAR-Telematics" project. *Information Technology in Biomedicine*, *3*(2), 101–108. doi:10.1109/4233.767085

Cetin, M., & List, G. (2006). Integrated modeling of information and physical flows in transportation systems. *Transportation Research Part C, Emerging Technologies*, *14*(2), 139–156. doi:10.1016/j.trc.2006.06.003

Chandramouli, R., & Lee, P. (2007). Infrastructure standards for smart ID card deployment. *Security & Privacy*, *5*(2), 92–96. doi:10.1109/MSP.2007.34

Choi, E. H. C., Taib, R., Shi, Y., & Chen, F. (2007). Multimodal user interface for traffic incident management in control room. *Intelligent Transport Systems*, *1*(1), 27–36. doi:10.1049/iet-its:20060038

Curitiba. (2001). *Lei Municipal n. 10.333*, de 11 de dezembro de 2001. Dispõe sobre a instalação de equipamentos de bilhetagem eletrônica no transporte coletivo público de Curitiba. Curitiba: Prefeitura Municipal de Curitiba.

Falletti, E., Letizia Lo Presti, Sellone, F. (2006). SAM LOST smart antennas-based movable localization system. *Vehicular Technology*, *55*(1), 25–42. doi:10.1109/TVT.2005.858169

Firmino, R. (2005). Planning the unplannable: How local authorities integrate urban and ICT policy making. *Journal of Urban Technology*, *12*(2), 49–69. doi:10.1080/10630730500307979

Firmino, R., Duarte, F., & Ultramari, C. (Eds.). (2010). *ICTs for mobile and ubiquitous urban infrastructures: Surveillance, locative media and global networks*. Hershey, PA: IGI Global. doi:10.4018/978-1-60960-051-8

Frey, K. (2005). ICT-enforced community networks for sustainable development and social inclusion. In Albrechts, L., & Mandelbaum, S. J. (Eds.), *The network society: A new context for planning* (pp. 183–196). London, UK: Routledge.

Germani, E. (2003). *Tecnologia da Bilhetagem Automática: Evolução e Perspectivas. Bilhetagem Automática e Gestão nos Transportes Públicos* (1st ed., p. 120). São Paulo, Brazil: ANTP – Associação Nacional de Transportes Públicos.

Greisler, D., & Stupak, R. (Eds.). (2002). *Handbook of technology management in public administration*. Boca Raton, FL: Francis & Taylor.

Husemann, D. (1999). The smart card: Don't leave home without it. *Concurrency (Chichester, England)*, *7*(2), 24–27.

IPPUC. (2004). *Plano diretor: O Planejamento Urbano de Curitiba*. Curitiba: IPPUC.

Lazar, J. (2002). *Managing IT: Community partnerships in the 21st century*. Hershey, PA: IGI Global.

Pelletier, M., Trépanier, M., & Morency, C. (2009). *Smart card data in public transit planning: A Review*. CIRRELT – Centre Interuniversitaire de recherché sur les réseaux d'entreprise, la logistique et le transport Quebec and Montreal, Canada.

Reddick, C. G. (2010). *Citizens and e-government: Evaluating policy and management*. Hershey, PA: IGI Global. doi:10.4018/978-1-61520-931-6

Rezende, D. (2008). *Tecnologia da informação e planejamento estratégico: Alinhamento e integração com a inteligência organizacional privada ou pública*. Rio de Janeiro, Brazil: Brasport.

Silva, D. (2000). *Sistemas Inteligentes no Transporte Público Coletivo por Ônibus. Dissertação de Mestrado. Programa de pós-graduação em Engenharia de Produção*. Porto Alegre: UFRGS – Universidade Federal do Rio Grande do Sul.

Symonds, J. (2010). *Ubiquitous and pervasive computing: Concepts, methodologies, tools, and applications*. Hershey, PA: IGI Global.

Targowski, A. (2009). *Information Technology and societal development*. Hershey, PA: IGI Global.

The World Bank Group. (2006). Hong Kong Smart Card System. Information on the Hong Kong Smart Card System provided by AESPRODATA, the Railroad-Technology Journal and MTR Corporation. Retrieved April 20th, 2006, from http:lnweb18.worldbank.org/External/lac/lac.nsf/Sectors/ Transport/D5A576A039A80

KEY TERMS AND DEFINITIONS

Bus Rapid Transit: Mass transportation system based on buses. Its main characteristics are segregated bus corridors, physical fare integration of different bus lines, fare payment before boarding, boarding platform and bus floor on the same level, use of information technologies.

Business Intelligence: Refers to computer-based techniques used in spotting, digging-out, and analyzing business data. Usually it refers to the information that is available for the enterprise to make decisions. A data warehousing (or data mart) system is the backend, or the infrastructural component for achieving business intelligence. Business intelligence also includes the insight gained from doing data mining analysis, as well as unstructured data (thus the need for content management systems).

Information and Communication Technologies: Information and communication technologies combined. All information is stored and distributed as digital data, which can be manipulated by computers.

Personal Smart Card: It hosts "security algorithms and functions" in order to store personal data and participate in automated electronic transactions.

Ubiquitous Pervasive Technologies: Information and communication technologies embedded in "ordinary" devices, such as cell phones, clothes, watches, which are able to receive, process, and distribute digital data.

ENDNOTE

1 We would like to thank Fundação Araucária and CNPq for funding this research. We would also like to thank Rodrigo Firmino for his critical revision of this chapter

Chapter 12
The Effect of Google Data Centers on City Competitiveness

Jacqueline Bueno Sousa
Florida International University, USA

ABSTRACT

The decision by Google several years ago to begin building major data center campuses prompted economic-development officials nationwide to vie to have their communities chosen as a Google Data Center site. They hoped the facilities would bring hundreds of jobs, lure other technology companies and help create technology hubs in their areas. However, this study of three of the communities where Google built major data centers -- Goose Creek, South Carolina; Lenoir, North Carolina; and Council Bluffs, Iowa—shows that the centers haven't always lived up to expectations. The author interviews economic development officials in each of the three communities.

INTRODUCTION

Google data centers are a source of fascination both in and out of the tech environment. Technology industry professionals for years have wondered about the contents and operations of the centers, which are reputed for their efficiency and which Google believes is key to the company's competitive advantage. Google's secrecy about the location and make up of its centers has served to fuel the aura of mystery surrounding the facilities.

Among those who have been fascinated by the centers are economic-development officials and city leaders, many of whom believe that Google's brand presence and technological benefits can give them a competitive advantage over other cities. For example, in early 2010, when Google announced it would select an American city where

DOI: 10.4018/978-1-61350-174-0.ch012

it would build an ultra-high-speed broadband network, city leaders throughout the country vied to have their community selected. To stand out in the competition for Google's attention, city leaders turned to outlandish stunts. The mayor of Duluth, Minnesota, submerged himself in the icy waters of Lake Superior. The mayor of Sarasota, Florida, swam in a tank – with sharks. The mayor of Topeka, Kansas, renamed the city Google for a month (Helft, 2010).

The competition to be chosen as the site for a Google data center hasn't been as sensational, but many city leaders have pursued the data centers with substantial vigor. Some have been willing to offer millions of dollars in incentives to lure the facilities. They've done so with the hope and expectation that the facilities would directly create hundreds of jobs, help lure other tech businesses to the area and transform their communities into high-tech hubs.

The interest in luring Google data centers is part of a broader interest by many cities to upgrade their connection to the information highway in order to gain a competitive edge over other communities. Many community leaders now believe that broadband and information connectivity is the essential utility of our times, on par with having energy sources and good roads. This interest comes from the realization that broadband is not only creating new companies, but also new industries. By getting on the information highway, companies can export their knowledge and skills around the world, employers can seek out labor throughout the globe, and industries can utilize Web-based tools and services (Bell, 2008).

Luring a name-brand technology company such as Google, some economic-development officials believe, brings the double benefit of having a trophy company located in one's city as well as the possibility of having that company enhance the area's broadband infrastructure – which, in turn, could lure even more technology companies (J. Sousa, personal communication, April 13, 2010).

However, a study of three communities where Google built three of its largest and most recent data centers shows that those hopes and expectations have been mostly unfulfilled. The effect of Google Data Centers on city competitiveness, as measured by the number of jobs they created, ability to lure other technology companies to the area, and impact on local technology education and training, has been mixed. Two of the three communities – Goose Creek, South Carolina; Lenoir, North Caronlina; and Council Bluffs, Iowa – instead have found that economic recessions impact even emerging industries, such as technology. And unlike industries, such as manufacturing, that help spring new businesses because of the need for suppliers and large amounts of labor, the Google data centers so far have done little to bring about an economic-domino effect. The exception is Lenoir, NC, where the presence of a Google data center, combined with an aggressive approach to luring new companies, is producing results.

THE DATA CENTER

Think of the early years of computing and many people imagine large, cable-filled rooms full of complex machinery and devices – all designed to help computers do their jobs properly. Data centers are the modern-day equivalent of those rooms. During the dot-com bubble of the 1990s, large numbers of businesses began relying on Internet connectivity, which prompted the need for large data facilities. The larger businesses could afford to develop their own Internet data centers, which required specialized staffs, large amounts of space and sophisticated equipment. However, smaller businesses could not afford the expense of having their own data centers. As a result, private data centers were created to meet the needs of smaller businesses. Today large data centers may exist as a private enterprise that provides access to smaller businesses or they may exist as the property of a large organization, for the purpose of fulfilling the technological needs of that organization.

Like those early computer rooms, modern data centers are controlled environments that are best kept in dedicated rooms because they require high levels of security and – since they house equipment that generates high levels of heat – so that they can be maintained at very cool temperatures. Containing hundreds -- and sometimes thousands -- of servers, they are huge energy consumers. Much of the literature on data centers focuses on their energy use as they now consume approximately 61 billion kilowatts hours or 1.5 percent of all U.S. electric consumption nationwide (Anderson, 2010).

This high level of energy consumption has become a key factor in the site-selection process since an area's local energy costs will impact the overall cost of maintaining a data center. As a result, much of the focus when selecting a data center site has turned to areas where energy is plentiful and inexpensive. For example, Sabey Corp., a Seattle, Washington-based developer of data centers cited energy cost as being among the key criteria when selecting sites, including the selection of rural Douglas County, Washington, as the site for a $100 million data center (Starner, 2006). As an energy and cost-saving measure, the company is considering ways to recapture and reuse the heat generated by the equipment housed in the data centers. Another factor that impacts site selection includes a geographic area's impact on overall security, such as the absence of earthquake fault lines and high winds (Anderson, 2010).

In part because of their specialized energy, air-conditioning and security requirements, data centers also have developed their own design and construction standards, another area of focus for much of the literature. Data centers often are out-of-the way, nondescript buildings designed to draw little attention. Their specialized design and construction needs take into account that data centers must operate 24 hours a day, seven days a week and 365 days a year (Kurkjian, 2007).

The highly complex and technical nature of their construction results in data center development representing large and significant capital expenditures, often in the hundreds of millions of dollars. As a result, many communities have been willing to provide tax incentives to lure data center development to their areas. In 2007, when Microsoft chose San Antonio, Texas, as the site for a new $550 million data center, the company received ten-year exemptions on personal and real property. In addition, the city, along with Bexar County, funded improvements to the 40-acre site (Bruns, 2007).

Like the technology it houses, the physical needs of data centers undergo constant changes. Among the changes being foreseen in the industry is the unification of data center systems. Under such a system, new technology by companies like Cisco, Hewlett-Packard, IBM and Dell allows everything from servers to storage to networking services to be consolidated into server blades, making it possible for data centers to use less space (George, 2009). For example, Tudor Perrini, a large construction company, recently was able to "collapse" its needs from 400 servers to 36 Cisco server blades. At the same time, it reduced its need for data center space, consolidating from five data centers into one (George, 2009).

Nonetheless, for companies, such as Google, that must store, process and transfer massive amounts of data and information, data centers continue to require large space needs and represent significant investments.

HISTORY OF GOOGLE

Founded in 1998 by Larry Page and Sergey Brin, Google has become one of the world's better known companies in a relatively short period of time. A provider of Internet-based products and services, the company, best known for its popular search engine, has stated that its mission is "to organize the world's information and make it universally accessible and useful" (Google, n.d.).

Almost from its inception, Google has shown an unusually casual approach to conducting business. The company's corporate philosophies, for example, include such statements as "you can be serious without a suit." The company encourages its engineers to spend 20 percent of their time on independent projects that they find interesting. Marissa Mayer, Google's Vice President of Search Products and User Experience, has estimated that about half of the company's new products have come from ideas developed from those independent projects (Mayer, 2006).

Over the years, the company has developed or acquired a variety of well-known products and services, including Google Maps and AdWords. The company's advertising programs account for nearly all of Google's revenue, which topped $23 billion in 2009, with profits of $6.5 billion (Cendrowski, 2010). However, Google continues to be best known for a service it provides for free: its search engine. The system allows a user to enter key words into a search field, and then the engine searches billions of web pages for those words. ComScore recently ranked Google as the top search engine in the United States, involved in more than 65 percent of the online searches in the country (ComScore, 2009).

What is a Google Data Center & Where Do You Find One?

The company's menu of technical services and products requires that it also operate the technical equipment to process and store immense amounts of information. As a result, Google operates more than a million servers in data centers throughout the world ("Google: One Million", 2007).

With the exception of a few well-publicized facilities, the company has historically declined to disclose the locations of its data centers, claiming that details about the facilities – their configuration and operations – help give the company an advantage over its competitors. For example, the centers are said to be configured in way that is highly efficient, giving the company a cost advantage on energy use (Miller, 2008).

The high amount of interest in Google's data centers prompted DataCenterKnowledge.com two years ago to compile a list of Google data centers throughout the world. At the time, DataCenterKnowledge.com counted at least 12 "significant" centers in the United States, not including three major centers that were under construction at the time. In addition, Google leases space at data centers owned by other parties. In all, DataCenterKnowledge.com listed seventeen U.S. data center facilities for Google (Miller, 2008). Outside of the United States, the list included another seventeen centers, including two that were under development at the time (Miller, 2008) [see Google Data Center Locations list in Table 1].

CASE STUDIES: LENOIR, NORTH CAROLINA; GOOSE CREEK, SOUTH CAROLINA; AND COUNCIL BLUFFS, IOWA

In recent years, Google has shown a preference for building its own data centers rather than leasing space at shared facilities. In 2007, the company began developing three major centers -- in Lenoir, North Carolina; Goose Creek, South Carolina; and Council Bluffs, Iowa. Despite the company's preference for keeping quiet about the location of its centers, much publicity surrounded the development of these facilities as the communities vied to be selected as the data-center sites.

The facilities entailed significant investments in the communities, including the acquisition of hundreds of acres for each center. In several media stories and press announcements, Google estimated its investment in Lenoir, North Carolina, and Goose Creek, South Carolina, at $600 million for each community. Furthermore, the centers were part of a new trend in data center development,

Table 1. Google Data Center Locations

UNITED STATES
Mountain View, California
Pleasanton, California
San Jose, California
Los Angeles, California
Palo Alto, California
Seattle, Washington
Portland, Oregon
The Dalles, Oregon
Chicago, Illinois
Atlanta, Ga. (two sites)
Reston, Virginia
Ashburn, Virginia
Virginia Beach, Virginia
Houston, Texas
Miami, Florida
***Lenoir, North Carolina**
***Goose Creek, South Carolina**
***Pryor, Oklahoma**
***Council Bluffs, Iowa**
INTERNATIONAL
Toronto, Canada
Berlin, Germany
Frankfurt, Germany
Munich, Germany
Zurich, Switzerland
Groningen, Netherlands
***Mons, Belgium**
Eemshaven, Netherlands
Paris
London
Dublin, Ireland
Milan, Italy
Moscow, Russia
Sao Paolo, Brazil
Tokyo
Hong Kong
Beijing

Source: DataCenterKnowledge.com
*Under construction at the time the list was developed in 2008
Major centers in bold

which traditionally had been located in major business hubs such as New York and Silicon Valley. Beginning in about 2006, technology companies such as Yahoo! and Microsoft were announcing major "server farms" in little-known towns such as Quincy, Washington (BusinessWeek, 2006). Driving the trend toward smaller, out-of-the way towns was the availability of cheap land, inexpensive energy and the increased security that comes from remote locations (BusinessWeek, 2006).

For local-government officials, the centers came with the promise of hundreds of new jobs and the hope of luring other technology companies. It enticed North Carolina, for example, to provide Google with a \$4.7 million state grant for the Lenoir project, part of an incentive package that was estimated at more than \$200 million (Dewan, 2007). The North Carolina project was one of four similar facilities Google announced in 2007.

Three years later, based on the three case studies below, the projects have not turned out to be the boon to job creation some expected and have had mixed results in meeting economic-development expectations.

Goose Creek, South Carolina

In early 2007, South Carolina announced that Google would build a data center in the town of Goose Creek, a bedroom community near Charleston. The state projected the data center would employ about 200 workers and represent a \$600 million investment by Google. To show Google just how much the state wanted the data center, South Carolina agreed to help pay for some of the site preparation costs associated with developing the 520-acre site.

"Given the stature of this company and the magnitude of this investment, this is a real win for South Carolina and will have a tremendous impact on the local and state economy," Gov. Mark Sanford said in a prepared statement at the time (Miller, 2007a).

But three years later, the 200 jobs that were expected haven't materialized. In March of 2010, Google informed the state that it would not claim a \$2.5 million grant that was conditioned on the 200 jobs. Furthermore, only one of the two facilities planned at the campus has been built and is operational. The shell of the second building is up but has shown little construction activity (Miller, 2010).

Google Spokeswoman Emily Wood told *The State* newspaper: "Given the current economy and the difficulty in forecasting our business climate, we do not believe that the (job tax credits) would be a wise investment for either Google or South Carolina at this time" (Miller, 2010).

In fact, officials don't know how many jobs have actually been created at the center, says Gene Butler, Director of Economic Development for Berkeley County, which houses the town of Goose Creek. "There were job numbers thrown out there," he said. "But how many people they have working there? We don't know, and they [Google] won't tell you" (J. Sousa, personal communication, April 13, 2010).

Aside from the jobs that have been created, has Butler seen any kind of economic impact from having the center? "Not really," he said. "It's a storage kind of situation. They're so secretive and quiet. They don't even want people to know what power usage they have. It's fenced in and tightly secured. You need to do a background check just to get in there" (J. Sousa, personal communication, April 13, 2010).

Butler said it is prestigious for the community to have a global company such as Google. However, he said, "I can't say that I can put a dollar number on that or say that a company has come here because of Google" (J. Sousa, personal communication, April 13, 2010).

Butler noted that there has been a recent increase in interest by developers to build data centers in the area. He believe that interest is part of a data-center development trend that is catching on, and not particularly a result of the presence of the Google data center (J. Sousa, personal communication, August 4, 2010).

Lenoir, North Carolina

At around the same time it was announcing its plans for South Carolina, Google called on the Appalachian town of Lenoir, North Carolina. The community was striving to survive the gradual decimation of its long-standing furniture industry. For years, furniture companies in the area had been closing down or moving offshore, leavings thousands of people without jobs in a community of about 17,000 residents. It was a community where people had enjoyed a long tradition of craftsmanship and where skills were passed down from generation to generation. When Google purchased 215 acres that once included an old lumberyard, it promised to bring something the community desperately wanted: jobs. Google announced that it would employ about 250 people, with an average salary of $48,000, at two facilities on the data center campus (Dewan, 2007).

To make the deal happen, town leaders helped assemble the land, going so far as to paying for a divorce to free up land and finding long-forgotten heirs (Straight Up Search, 2008).

It didn't matter that many of the town's workers lacked the technical background and education normally associated with data-center employees. Google's Director of Global Operations, Lloyd Taylor, said at the time that the company didn't need "a team of rocket scientists holed up in a room." Google began working with the local community college to create a technology training program (Dewan, 2007).

Nonetheless, not everyone was happy about Google's arrival; the local media criticized the package of incentives the state offered Google. The state passed a law exempting certain technology companies – Google among them -- from having to pay sales taxes on electricity. Because the data centers consume huge amounts of power, the cost of electricity has been a key factor in the site selection process.

Furthermore, the state promised to forgive all of the company's personal property taxes and the bulk (80 percent) of its real estate taxes. The exclusions would be in place for 30 years, so long as the facilities were operational (Dewan, 2007).

The facility opened in 2008 but has yet to create the 200-plus jobs projected. In 2009, Google informed the state that it would not meet the job-creation criteria to quality for the $4.7 million state grant, which required 200 new jobs in four years (Miller, 2009).

Harry Whalen, Executive Director of the Caldwell County Economic Development Commission, which includes the town of Lenoir, believes the town still has gained from Google's data center (J. Sousa, personal communication, April 13, 2010). Whalen acknowledges that the center has not created the hundreds of jobs expected, nor the thousands of jobs some locals thought would come as a result of having a global organization that could attract other companies. Still, jobs aside, Whalen notes that the Google data center brought about IT training programs at its local schools and the extensive addition of fiber-optic cable to the area (J. Sousa, personal communication, August 4, 2010).

Indeed, when it comes to attracting technology companies, this region of North Carolina has done better than many areas. In 2009, Apple Inc. announced it would create a $1 billion data storage site in nearby Maiden, North Carolina. Around the same time, five counties announced they had formed an alliance to draw more data centers to the area (Fuller, 2009). In the summer of 2010, that effort bore fruit when Dacentec, a cloud-computing company, announced it would locate a data center in Lenoir (Teague, 2010). The center's manager, William Groh, told the local media that the company chose Lenoir after considering five other cities. The key factors in choosing Lenoir, he said, were assistance from local governments, existing power infrastructure, real estate cost and Google's presence in the area.

"We really wouldn't have come to Lenoir if that seed had not been planted," he said (Teague, 2010).

Whalen notes that other data center operators are looking in the area. He can't say with certainty that all those companies have been attracted to the area specifically because of Google. However, he says, he has no doubt that Google's presence has helped. "The main benefit," Whalen said, "is the branding of the community. Having that presence in the community represents a level of attainment" (J. Sousa, personal communication, April 13, 2010).

Council Bluffs, Iowa

In mid 2007, one of the world's most well-known corporations announced it would make a major investment in one of the world's lesser-known communities: Council Bluffs, Iowa. Located in the western part of the state, the town had been eagerly awaiting for months to learn if Google would choose it as the site of a major data center. The announcement claimed that a $600 million data center would bring 200 jobs to the area. Governor Chet Culver stated that the project "will have a tremendous impact on Council Bluffs, western Iowa and the entire state" (Miller, 2007b).

Why Council Bluffs? The area generates a large amount of renewable energy, thanks to its local utility, MidAmerican Energy, one of the nation's biggest generators of wind energy. Officials with the utility company projected that about 18 percent of the energy consumed by the Google data center would be wind generated or derive from other renewable sources (Miller, 2007b).

The Council Bluffs facility would include one of the largest land holdings yet for Google – 1,185 acres. Officials estimated that Google would pay $65 million in property taxes over the next 15 years, in addition to $6 million in sales taxes over two years as a result of construction-related purchases (Miller, 2007b).

It's still too early to measure the revenue benefits the state and area will derive from having the Google data center, says Greg Halverson, Vice President of Economic Development for the town of Council Bluffs. Halverson says the economic benefit from having the center is based primarily on the presence of having a major new building in the area. He doesn't know how many people are employed at the center (J. Sousa, personal communication, April 21, 2010).

"Google keeps a lot of the information close to the vest," Halverson said. "There's a new building, but there are no spin-off businesses or anything like that."

The community hasn't seen many secondary jobs created as a result of the center, he said. "With data centers," he said, "you don't' see a lot with the supply side. It's just a room full of servers."

However, Halverson said Google's presence has benefited in other ways, which may pay off in the future. For example, he said, Google is providing grants to the community to develop WiFi networks. "They're very supportive of the community with regard to technology," he said (J. Sousa, personal communication, August 3, 2010).

ANALYZING THE CASE STUDIES

Any analysis of the impact of the Google data centers in Goose Creek, South Carolina; Lenoir, North Carolina; and Council Bluffs, Iowa, would be incomplete without acknowledging the time frame in which the research takes place. All three data centers represent relatively new facilities, in operation only two to three years. As a result, it is too soon to determine their long-term impact on the respective communities (see Table 2).

Likewise, more time needs to pass before a thorough cost-benefit analysis can be conducted in the communities that paid incentives to help lure the Google data centers. Such an analysis at this time is restricted not only by the recentness of the projects but also by Google's reluctance to provide information such as job creation numbers. Future incentive qualifications, such as property tax benefits, will become more apparent in time and will require disclosure of information on the part of the company to determine whether it qualifies for the incentives.

In the short term, however, it has become clear that the expectations that surrounded the announcements of the centers have not been realized in the time frame some expected. Google itself has acknowledged as much by informing the states of North Carolina and South Carolina that it would not qualify for incentives in those states. In addition, a fourth data center that Google also announced in 2007 – in Oklahoma – has been placed on indefinite hold (Miller, 2009).

Nonetheless, having a Google data center is still an attractive proposition for many communities, assuming they don't offer incentives that outweigh the benefits of a center. Economic-development officials at all three communities in the case studies said that the construction of the centers temporarily employed more than a hundred construction workers and resulted in sales to local, pre-existing businesses that supplied construction and other materials as part of the development. None, however, could provide hard numbers on the value of that benefit.

CONCLUSION

Google data centers have not created the hundreds of jobs officials expected. A typical Google data center employs about 45 people, and this is true even of the new, larger facilities Google built in recent years (Miller, 2008). Furthermore, economic-development officials report mixed results in the Google data centers' ability to draw other businesses to an area. While Google's presence may have indirectly helped to draw other technology companies in North Carolina, it has done little to bring ancillary businesses such as suppliers to South Carolina.

Nonetheless, none of the communities is sorry to be housing a Google data center. All three of the communities examined were little-known towns that hoped the presence of a global company such as Google and would bring about hundreds of jobs and catapult their communities into major centers in the technology industry. While that hasn't happened to the degree they hoped,

Table 2. Google Data Centers: Economic Development Impact

	Goose Creek, SC	Lenoir, NC	Council Bluffs, IA
Jobs			
Google Jobs Projected	200	250	200
Projection Met?	No (total not disclosed)	No (total not disclosed)	No (total not disclosed)
Business Recruitment			
Other Technology Companies Drawn to the Area by Google's Presence	None	Apple (2009) Dacentec (2010)	None
Education			
Impact on Local Technology Education/Training	None	Google working with local college to create technology training program	Google is providing grants to the community to develop WiFi networks

they say they believe they now have some kind of presence, however small, in the technology world. In short, they say they would rather have the centers than not. The key to being satisfied with the impact of having a center, it seems, is having realistic expectations about their power to transform a community.

REFERENCES

Anderson, S. (2010). Improving data center efficiency. *Energy Engineering, 107*(5), 42–63.

Bell, R. (2008). Five ways for communities to shape their broadband futures. *Economic Development Journal, 7*(3), 13–19.

Bruns, A. (2007, September). Tech Magnet. *Site Selection, 52*(5), 744–759.

BusinessWeek. (2006, June 12). *Servers as high as an elephant's eye*. Retrieved from http://www.businessweek.com/ magazine/content/ 06_24/ b3988087

Cendrowski, S. (2010). *20 most profitable companies*. Retrieved from http://money.cnn.com

ComScore. (2009). *2009 US search engine rankings*. Retrieved from http://www.comscore.com/ Press_Events/Press_Releases/ 2009/3/ US_Search_Engine _Ranking

Dewan, S. (2007, March 15). *Google is reviving hopes for ex-furniture makers*. Retrieved from http://www.nytimes.com/ 2007/03/15/us/ 15google.html?_r= 2&0ref=slogin

Fuller, B. (2009, October 30). *Iredell one of five counties to form NC data center corridor*. Retrieved from http://www2.mooresville tribune.com

Google. (n.d.). *Homepage*. Retrieved from www.google.com

Helft, M. (2010, March 22). *Hoping for gift from Google? Go jump in the lake*. Retrieved from http://www.nytimes.com/2010 /03/22/technology/ 22stunts.html

Kurkjian, C., & Glass, J. (2007, February). Meeting the needs of 24/7 data centers. *ASHRAE Journal, 49*(2), 24–35.

Mayer, M. (2006, May 17). MS&E 472 course: Entrepreneurial thought leaders seminar series. *ETL Seminar Series*. Stanford, CA: Stanford University.

Miller, R. (2007a, April 4). *Google announces $600M SC project. Data Center Knowledge.* Retrieved from http://www.datacenter knowledge.com/ archives/2007/04/04/ google-announces-600m-so-car-project/

Miller, R. (2007b, June 19). *Google confirms data center in Council Bluffs, Iowa. Data Center Knowledge.* Retrieved from http://www.datacenter knowledge.com/ archives/2007/06/19/ google-confirms-data-center -in-council-buffs-iowa/

Miller, R. (2008, March 27). *Google Data Center FAQ. Data Center Knowledge.* Retrieved from http://www.datacenter knowledge.com/archives /2008/03/27/ google-data-center-faq/

Miller, R. (October 9, 2009). *When incentives don't work. Data Center Knowledge.* Retrieved from http://www.datacenter knowledge.com/archives /2009/10/09/ when-incentives-dont-work-out/

Miller, R. (2010, March 17). *Google won't claim SC incentives. Data Center Knowledge.* Retrieved from http://www.datacenter knowledge.com/ archives /2010/03/17/ google-wont-claim-s-c-incentives/

Pandia. (2007, July 2) Google: One million servers and counting. *Pandia Search Engine News.* Retrieved from http://www.pandia.com

Starner, R. (2006). Energizing the Internet. *Site Selection, 51*(5), 674.

Straight Up Search. (2008, April 29). *Google builds data center in small town, North Carolina.* Retrieved from http://www.straightup search.com/ archives /2008/04/ google_builds_d.html

Teague, P. (2010, July 30). *Dacentec makes official announcement of new data center.* Retrieved from http://www.newstopic.net/ printer_friendly/ 8933471

Chapter 13
Urban Geospatial Management System

Rudj Maria Todaro
Insula S.p.A., Italy

Ayça Cangel
International Competitiveness Research Institute, Turkey

ABSTRACT

A city is a complex system of dynamic elements, including streets, buildings, subways, and sewers, that are subject to constant changes. For the last decade, Information Technology has played a vital role in planning and management of such complexity through creating integration, flexibility and homogeneity in terms of data and processing. While an urban Information System aims to serve and support everyday tasks of people who are responsible for daily operations of the city, such a focus often involves top-down approach. A system that involves contributions of data from many users may provide better administrative level tools for correct decision-making, and urban planning literature presents many computerized tools and applications which could be applicable to more interactive real-life processes. This chapter presents a geospatial maintenance system of Venice that has been developed to include the contributions of users in different levels.

The City of Venice has been a unique tourism destination for decades and has attracted thousands of visitors every year coming for various purposes (leisure, restoration, business, etc.). Particularly important to Venice is its reputation for being the "City of Water," but this legacy also involves challenges. The city is constantly under the threat of degradation caused by salty lagoon water that flood into the city in certain times of the year. At this point, urban maintenance of city elements is essential to ensure minimum functional levels of city infrastructure. The system introduced in this chapter is an Urban Geospatial Maintenance System designed for Venice that focuses on planning and execution of maintenance work and is based on the logic of integration of classical management concepts and Geospatial Information Systems (GIS) concepts.

DOI: 10.4018/978-1-61350-174-0.ch013

INTRODUCTION

The rapid development of Information and Communication Technologies plays a major role in the improvement of planning processes and utilization of resources in urban infrastructures management (Huang, 2003; Hamilton, Burns, et al., 2005; Hamilton, Wang, et al., 2005; Wang etal., 2007). However, the integration of new technologies stands as a big challenge for urban managers and engineers (Huber, 1990; Pierre et al., 2001a, 2001b; Quintero et al., 2005). Because the planning, design and operation of a complex urban system such as a city requires a complex information system in order to perform these operations with prospective success level. Today, public administrations have to integrate a wide range of data such as design data, maintenance data or geographic information from different sources and utilize these data in an operation of an urban infrastructure with minimum level of data loss or duplicity (Quintero et al., 2005) and maximum level of precision, consistency and integrity (Michele & Daniela, 2011). In addition to data, the service quality, financial status, time management, feasibility are some of the other aspects that have to be considered while realizing daily operations of an urban territory.

The system to be planned and managed is a combination of different subsystems that have mutual relationships and multiple functions as well as different user types that interact with the system and different parties undertaking planning and maintenance activities. Since the introduction of new technologies, functions such as feedback data management, the planning maintenance management or the works management have been developed to assist public administrations during the management process (Michele & Daniela, 2011). Rubenstein-Montano (2000) states that all of the computerized tools that have been introduced in the urban planning literature, somehow provide some type of decision support and classifies urban information systems according to their decision support nature as: decision support systems, group decision support systems, spatial decision support systems, expert systems, and knowledge-based decision support systems. There are various studies in the literature presenting different applications of information systems used in urban management. Hadzilacos et al. (2000) mentions a decision-support system for rehabilitation planning and optimization of the maintenance of underground pipe networks of water utilities while Ghaemi et al. (2009) presents a web-based platform in their study that enables interactive environmental planning, which aims to provide tools for municipalities and community-based groups to identify candidate project sites. Pederson et al. (2004) describes a GIS-based management tool for conversation and management of biodiversity and the process of implementation in Oslo. The multi-agency system presented enable local authorities to find out any conflicts between conversation interests and other types of land use and acts as an information source for the public, and consultants who are related to environmental impact assessments. GIS serves as a facilitator for spatial decision-making and planning processes (Boroushaki & Malczewski, 2010). GIS based technologies have become widely used especially since 90's in urban infrastructure management especially within public administrations and have been combined with other specialist systems in order to fulfill the needs of advanced applications (Abel et al., 1992; Armstrong et al., 1991; Chang et al., 1997; Kim et al., 2000; Miyamoto et al., 1997; Parrott & Stutz, 1991; Pior & Osman, 1997; Sanchez, 1993; Siderelis, 1991; Yeh & Chow, 1996; as cited in Pior & Shimizu, 2001). In a GIS, any kind of data of an infrastructure is in relation with the geographical position of the on the territory map (Michele & Daniela, 2011). Planner's mode of work and their way of decision-making have evolved with the assistance of GIS since these systems have led to improvement of organizational efficiency and effectiveness (Leonard-Barton & Kraus, 1985; Aldosary & Zaheer, 1996). In the last decade as a GIS based technology, the development of spatial decision support systems (SDSS) enabled multiple users to gain access to territorial data together with inventory data at different levels of decision-making processes. Improvements in geo-spatial technologies have paved the way for public and private urban decision-makers to develop tools and applications for better management, maintenance and planning of urban infrastructures. For instance, Coutinho -Rodrigues et al. (2010) presents an SDSS prototype that enables users such as government and municipal agencies to evaluate, plan and analyze urban infrastructure investments based on formal multiple criteria methodologies. The system is depicted through a case study of an urban infrastructure investment regarding the expansion of water supply network in City of Coimbra, Portugal. The system is employed to evaluate 4 investment options based on ten criteria. On the other hand, Pior and Shimizu (2001) presents another aspect of GIS-infrastructure relationship and introduces a GIS-aided evaluation system for infrastructure improvements. An urban infrastructure investment requires large amount of capital, therefore investments actualized are expected to last long term. On the other hand, as cities are enlarging, the need for more complex urban infrastructure investments arises which eventually leads to more complex problems regarding urban systems maintenance and trouble-shooting procedures. As it is the case in every system, management and operation of urban infrastructures hold a number of problems too. According to Leem et al. (2007) making the maintenance and management plan of urban infrastructures during planning and designing, avoiding the planning and management of facilities by an individual maintenance body, and establishing a common maintenance plan that are fully compatible with developed technologies such as GIS are some of the strategies that can be followed in order to avoid basic maintenance problems. It is undoubtedly obvious that the development of IT systems for computer-aided monitoring of real-time processing and infrastructure management has facilitated the work of urban specialists and managers (Berner et al., 1995) because a computerized management system enables users to retrieve the information as up-to-date and complete. Abudayyeh et al. (2005) summarizes the objectives of a computerized maintenance management system as follows:

- Improve maintenance efficiency
- Reduce maintenance costs
- Reduce asset downtime by scheduling preventive maintenance
- Increase the design life of an asset
- Provide historical records to assist in maintenance planning and budgeting
- Provide maintenance reports in a format that is required by the user

Most Computerized Maintenance Management Systems provide aforementioned objectives. Nonetheless, there are many software tools that serve to different parts of the infrastructure management process. At this point, interoperability, integration and efficient sharing of infrastructure data come forward as strategically important terms for a well-designed infrastructure management system (Halfawy et al., 2002; Michele & Daniela, 2011).

Often, those who work on territorial information systems (both solution providers and technical referrals on the user's side) are not concerned with management needs and problems of workflow. When one is faced with the need to organize a management process for an activity on the territory, the solution is generally to provide the management system with a link to a point on a map. However, this leads to a series of flags on the map, but adds no value either to the map or to the process. On the other hand the managers of a city need a body of cartographic information to sustain decision-making systems; the norm is to generate measurement and digitalization campaigns that gather the data required for the specific need which is somehow rarely updated. For example in managing underground utility systems, there is a general recourse to comprehensive surveys, sometimes with geo-radars, whereas it is rare to see management processes of installation jobs that bother to collect the data as they work by taking adequate "as built" measurements.

Another factor that is decisive for a successful outcome of a project is the fact that these systems often "come down from the top". Which means: an analysis of needs is drafted, with the purpose of understanding all the aspects that need to be solved and the solutions that could be implemented, on the basis of requests by the top-management. The system is then designed in all its complexity and the tool is created, concentrating all the efforts in order to put the system in production within a fixed deadline. In general the users, the people who must deal with the system daily, will not be introduced to it until it is finished. This not only creates the risk of actually missing the target of the project, but it also makes it harder to share the solution.

In order to manage an urban territory, it is necessary to be able to access all the components of it when needed and to manage its governing and production processes. The system that will be introduced in this chapter is a geographical management system that helps run and control the maintenance processes. The objective here is to highlight the importance of a start with operative processes and everyday works when the point at issue is "city" because a city primarily thrives on operative processes including all the minor activities that must be regulated and managed daily.

It is of prime importance that system is not only a container for registry (even for geometric or geo-referential data) but the system also must become actual management systems based on geospatial objects. On the way to create the system required, the first thing to do is to define the typologies of data masters that the objects should be divided into. For instance, for each individual master object the discriminating factor must be defined. The object "street" has different information needs depending on the operator who must use them, and there are many details that have to be considered such as whether the street name describes a street or the uniqueness of its path or whether a different master for each lane and for each sidewalk should be defined. In addition, interoperability is also a critical factor that has to be considered since the system should adopt shared data models that enable interaction with other subjects. For this reason, agencies such as INSPIRE [1] pay attention to establish the definition of the data structures of these masters.

BACKGROUND

Italy's City of Water, Venice is among the UNESCO World Heritage sites, and faces "Acqua alta" (high water) and floods in certain times of every year. This phenomenon serves as a unique tourist attraction but also a challenging factor for the urban managers of the city because the city faces a serious and constant threat against its preservation due to erosive salty water. This threat has two-sided consequences; one is for residents of the city in particular, and the other is for residents of the entire world since the city possesses a unique tourism attraction, which must be preserved for the future generations. Uniqueness does not only come from its historical places and monuments but also from streets, squares, canals and piazzas. Drained canals, raised streets, pedestrian paths are all specific to City of Venice and this heritage must be maintained properly as heretofore. On the other side, in the face of increasing international competition among tourism destinations, the city has to be maintained regularly in order to remain competitive; canals have to be dredged so that waterways remain navigable, bridges should be restored and pavements should be raised to allow pedestrian passage even at high tide, embankments should be repaired to ensure the stability of buildings, the sewer system should be kept running to provide the best possible sanitary conditions and the underground grids (water, electricity, gas and telephones) should be renewed and complete.

In order to run and control all these maintenance tasks as well as possible, geographical information system has been utilized by Insula Spa, a public limited company set up by the Venice City Council in 1997. This system, which is called Urban Maintenance System (SMU), provides methods and procedures for the processing of geographical data through management information tools and also supports management through tools that make forecasts and through supporting decision-making. The system is not a mere graphic description, but a thoroughgoing process control tool. In order to achieve this, a series of management applications have been developed that use the geographical information system as a base, which they provide a complete process control system. The process is not confined to Insula's sphere but also embraces the fundamental actors of the public administration and all the bodies that operate in the area worked, and this is actualized through web technology.

In the core of the system lays the organization of geographical information levels in a database engine, which not only finds references in accordance with the principles that have been established for it, but also ensures simultaneous access and edit for users with different roles in various parts of the area covered. In accordance with the same principle, application instruments have been designed based on a web-oriented logic to enable different categories of operators (authorities, citizens, the company and the third parties) share the information and instruments. The planning, design and execution processes of the maintenance works are managed by dedicated applications. Taken from the basic map, the geographical features involved in the work are designed, and thus the general characteristics of the job and the details of each single element in the planning of the area as a whole are identified.

In the planning and design of underground works, the management and geographical system provides a working level for cooperation among the different bodies concerned, making detailed information available regarding underground service systems and the planning of the work to be done on an area basis. The execution of maintenance works is supervised by on-line collaboration among the site engineers, the job managers and the authorities.

The architecture of the system has been developed on the basis of five fundamental geographical elements: canal, bridges, embankments, buildings and pavements. All geographical elements have been subjected to detailed technical surveys. Geometrical and structural data and information sheets regarding the state of repair of the different elements that constitute their structures have been catalogued. Each level of information has been organised into univocal codes in order to create a system of relationships between distinct elements. The geographical elements are thus set out with relational links that allow correlation and complex analysis. The universe of the canals of Venice and the larger islands in the lagoon has been laid out in a network of segments and intersections following water flow principles.

URBAN GEOSPATIAL MANAGEMENT SYSTEM

The Development Model

A city is the synthesis of a great number of different activities that are constantly evolving. These activities are carried out by different people in different ways within different timeframes. This complexity is very hard to contain within a framework of classification and systematization. In order to manage everything, it is important to channel all the activities towards a common point of reference. Therefore, the staring point of is the detail of the operators' needs that would be then integrated into one single system because when an information system is imposed from the top, it is in danger of incurring several errors such as:

- Being perceived by the users as an additional burden
- Increasing the operative workload
- Neglecting some of the operative aspects that later create an obstacle
- Achieving some objectives while missing others

The point of departure for the methodological approach adopted was a rhetorical question: who is better qualified to generate correct information about the object being produced and the production process than the people who deal directly with the phases of production?

The lever required to get this information from these people is to make it convenient for them to structure and insert the data. If the problem is approached from the bottom up, with a focus on serving the operator, making his activity easier, avoiding repetitive processes (such as inserting the same data over again), making it easier to find information and instructions for dealing with this data, it is possible to achieve three important results:

- Optimize the processes
- Reduce the distance between the bureaucratic data and the reality of production
- Generate an enormous quantity of information that is constantly updated

The operators often do not have the capacity to communicate their needs and especially to offer adequate solutions. This means that in the development phase of management applications, it is not possible to create the system in one fell swoop, concentrating the effort into a single phase of analysis and development.

To remain on target, it is preferable to start with a minimal project and to develop it in cyclical phases of discussion and assessment with the operators. This is a longer and more costly process, but it allows both developers and users to understand the problem fully. During this phase it is important that directors and top-management interfere as little as possible, and that the collaboration between developers and users be encouraged.

Work Organization

While organizing a territorial information system to manage the database of a city, certain fundamental factors must be taken into consideration. First of all, it is important to organize a technological infrastructure to support the information system. It is necessary to establish a data infrastructure, that is, a body of rules, the list of information levels that are to be managed and a detailed description of each of these levels. It is necessary to measure the quantity of information that the system will be required to handle. A city with over one hundred thousand inhabitants will undoubtedly have a huge quantity of information to handle. This will make it necessary to structure an information system that will allow many operators to work simultaneously. It is fundamental however to structure the system in a centralized manner. Therefore, distributed information should be accompanying a centralized infrastructure. This means that the data must be stored in a centralized database and a large number of operators, scattered throughout the territory, must be able to access and handle the data simultaneously, without the risk of creating conflicts in the management of the data.

The Structure of the Data and the Metadata

A decisive factor for the success of a territorial information system is the proper configuration of the structure of the primary data. This is the formalization of the structure of the territorial information system. Basically, it is necessary to establish what information is important and how this information should be organized. The levels of information must be established, along with their contents and their formalization and standards that will be referred to, and the strategy for updating and historizing the data should be designated. Among other things, the form of the metadata must be selected together with the level of sustainable details for the metadata. This system works as the information system of the information system, gathering information about the data instead of gathering data. The information mentioned here may involve information about the source of the data and date at which the data is updated.

Organization of the Data

A correct technological solution for developing the territorial information system must focus on two fundamental principles:

- The centralized handling of the data
- The delocalization of the work tools

Centralized Data Management

For the information to be correct and up-to-date, it is indispensable to organize the data in a univocal, solid and compact manner. For this reason, the data (spatial or non-spatial) must be managed in a spatial database so that it can be possible to memorize, access, verify and handle geographic data with the solid tools of relational databases. Utilization of a spatial database also guarantees the typical modalities of a GIS environment. Therefore, all the properties of the DB as a relational database are extended to spatial data so that it is possible to apply a logic of normalization, lock, mirror, commit or rollback. This also provides a complete correspondence of data in all the different applications and in distinct access interfaces.

Another important feature of this technology is the possibility of keeping all the normalized tables on the database, connected to definition or coding tables which prevents typing errors as the data is input and also makes the searches and joins faster.

Figure 1.

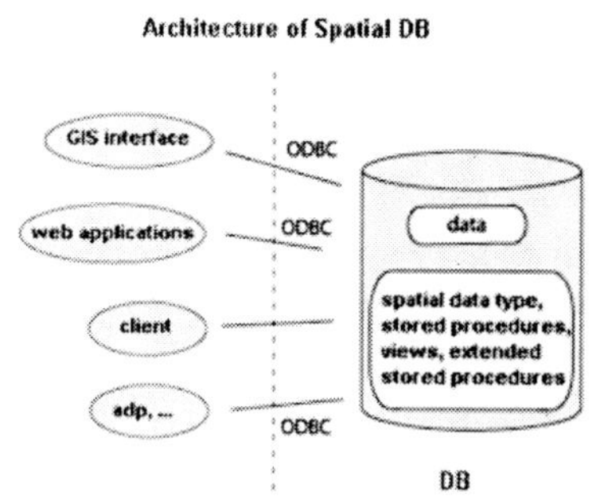

Concurrently with that, de-normalized views can be downloaded by the users so that the structure of the database can be perfectly transparent for the final users. In addition, more than one user to edit the same record at the same time with a solid and correct handling of conflicts. Thus more than one operator can work simultaneously without compromising the solidity of the information. This type of solution also allows the adoption of typical management system techniques or the activation of geographical object historicisation mechanisms.

Winstanley (2009) describes spatial databases as "systems designed specifically to include data with spatial attributes, such as geographical location, distance, and extent". An instance of spatial database architecture is represented in Figure 1.

A Spatial database provides:

- Complete integration of the attributes and spatial data
- Data protection
- A rich array of spatial functions available directly at the level of the DB
- Ease of use
- Access to DB by means of standard program interfaces (ODBC).

The Delocalization of Working Tools

There are various operators such as city government offices, municipalized companies, subservice agencies that hold different roles, points of views and requirements while operating on the same territory, and it is vital for these different agencies working all over the territory to handle the information in a coordinated manner. The solution for this situation may be found in the development of different applications that share the same database and comply with the system modality. Thereby, the univocal codes that identify the objects in the territory can also become a unit of measure shared in the dialogue between different operators. In addition, the operators are spread out over the territory; therefore it is necessary to reach them in a capillary fashion. The organization of the system presented here makes it possible to physically delocalize data management while maintaining its univocal character. There are many technical means for achieving this type of management, which enables adaptation to the different environmental contexts (both technological and human) addressed. Database mirror, direct connections, remote consoles, web application or asynchronous client serve applications are some of the different solutions that may be adopted for operation.

The Venice Geospatial Management System

The Venice Geospatial Management System was built on an architecture based on five fundamental geographic elements of the territory: canals, bridges, embankments, buildings and pavements. All the objects in the territory were subjected to detailed technical surveys. Geometrical and structural data together with reports on the state of preservation of the different constituent elements were catalogued. Each level of information was organized with univocal codes to create a system of relationships between distinct elements. The objects in the territory were thus organized by means of relational links that allowed complex correlations and analyses. The universe of canals in Venice and on the major islands of the lagoon was organized into a network of segments-intersection based on the principles of water flow.

This information system has been constituted in order to support planning. The codes that the territory was divided into are used both to plan and design the projects, and to manage the accounting. This type of organization makes it possible to manage a process to control the work that, for each element of the territory under maintenance, identifies a data sheet that describes this individual job, summarizing the technical and economic details. In order to control overall process, a series of management applications were developed that rely on a system of territorial information to provide a complete tool to oversee the process. This process should not be confined within the responsibilities of "the office" that is responsible for the maintenance of the city, but should also include entire public administration bodies and all parities operating in the territory; therefore, for the accessibility of the system by each and every party, the elements that constitute the system are based on web technology. A representation of the Urban Management System is given in Figure 2.

The Fundamental Elements

Venice has been catalogued into five fundamental elements that also serve as the basic "units of measurement" of the system: pavements, bridges, embankments, segments of canals and buildings. Every one of these elements, which also can be defined as "SMU" element, has characteristics such as typology, state of preservation, area in square meters or the presence of particular superstructures. During the management of the system, every single variation on these objects is recorded whenever necessary so that the morphological variations in the state of preservation are

Figure 2.

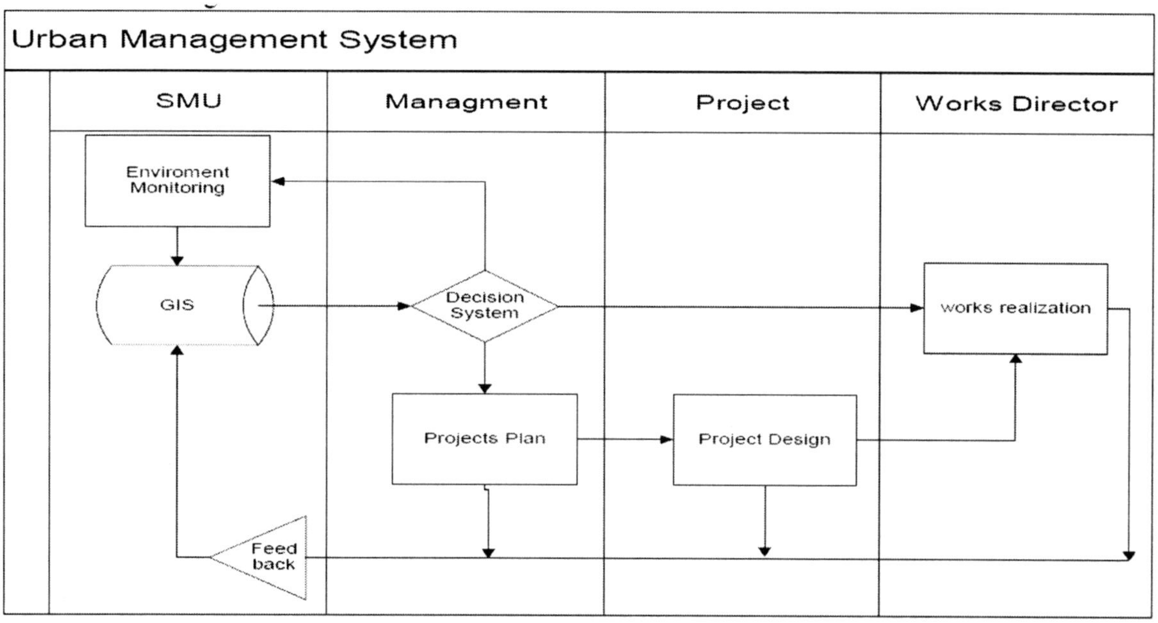

available and can be retrieved whenever needed for consultation.

Pavements: Pavements are detailed in their types, properties and other various identification properties.

Bridges: All bridges have been subjected to detailed technical survey. Geometrical and structural data have been recorded and data sheets have been compiled regarding the state of repair of their various constituent parts.

Embankments: Canal embankments, whether waterfronts, quaysides or outside walls, have been catalogued and subdivided according to criteria of structural and ownership homogeneity. A large amount of information has been obtained for each embankment element, such as the number of mooring rings, the type of material and the state of preservation.

Canal segments: The physical information about canal segments includes length, average width, area, average depth and a series of bathymetric measurements describing the beds. Their description also includes a range of embankment data regarding the structures looking on to them: ownership, material, state of repair, sewer outlets, types of construction, etc. Each canal segment is connected to the adjacent segments by means of hydraulic intersections. The canal network defined in this way can be used as input for modeling tools (hydrodynamics, traffic, inconvenience).

Buildings: Buildings are part of the structural elements viewing the Venetian canals; thence, to have precise knowledge of data regarding constructions or properties from structural as well as other points of view is of great importance.

The Elements of the Work

Urban Maintenance projects are defined within the framework of a context of "commission". A maintenance project is identified by means of several territorial information levels. The first fundamental level is therefore the commission, which identifies the geographical area the projects are related to. This object is also linked to the most important information about the project, such as the construction site manager, the fundamental dates of the project and the associated documents.

The commission object is also characterized by its status, for example the preliminary planning phase or final design.

The commission object is associated with the Territory Design Units (UTP), which are geographical objects that identify areas on which homogenous types of work are carried out. They are directly derived from the basic elements, and at the beginning inherit both their physical and alphanumerical characteristics; however, they can subsequently be modified according to a geographical point of view to fit the project better. One of the important characteristics of the Territory Design Units is the status indication that shows whether the construction site is still going through design process or active or completed.

Organization of the territory upon objects and codes provides convenience particularly in the design phase in terms of both planning of the project and setting up the accounting procedures of the projects. In the same way, a project control process is operated which can identify technical and financial data sheet for each geographical element subject to maintenance and once the project is completed, the data sheet is compiled by the site manager who thus contributes to keeping the system up-to-date.

Another aspect of the system worth to mention is the supporting elements, which are basic maps, benchmarks, and underground services.

Basic maps constitute the base for the supporting and complimentary mapping allowing generation and consultation of the levels of the work.

In terms of benchmarks, a high-precision geodetic network has been created. Global positioning system tools were used to materialise 51 network vertices, allowing the setting up planimetric and altimetric benchmarks valid for reference, re-attachment and linking to the Punta Salute mareographic system, the IGM95 (Italian Military Geographical Institute) vertices in the surrounding area and the existing IGM geometrical levelling lines.

Underground services unravel through lanes and bridges following routes that are mostly unknown. The system intends to show the routes that are known and achieve a complete route map by continuous updates as works by Insula or cooperation with the utility companies make it possible to plot the course of sections not previously surveyed.

Management Tools

The general objective of maintenance management is to preserve the functional level of the system (city, building, configuration, park, etc.) within each interval of acceptability, thereby guaranteeing desired levels of performance for each constituent unit. In order to achieve this objective, applications such as MIMUV, PUMA, MARTE and VENERE have been developed within the system.

MIMUV (MONITORAGGIO INTERVENTI MANUTENZIONE URBANA VENEZIA – MONITORING THE URBAN MAINTENANCE JOBS IN VENICE)

The objective of this application is to support the process that starts with the reception of a complaint by a citizen regarding a problem with an urban element, and ends with the completion of the intervention, systematizing all the phases of the process. Not only does this standardize the maintenance process, but it also maintains a significant degree of organization. It adapts to the typology of asset, to the context of the territory and the specific procedures of the supervision functions; the task of the management application is to provide support and facilitate the process of managing routine urban maintenance jobs (paving, street lighting, sewers, green spaces, underground utilities grids, canals, etc.) and to create a decision-making model on the basis of historic statistics regarding inspections and interventions.

Figure 3.

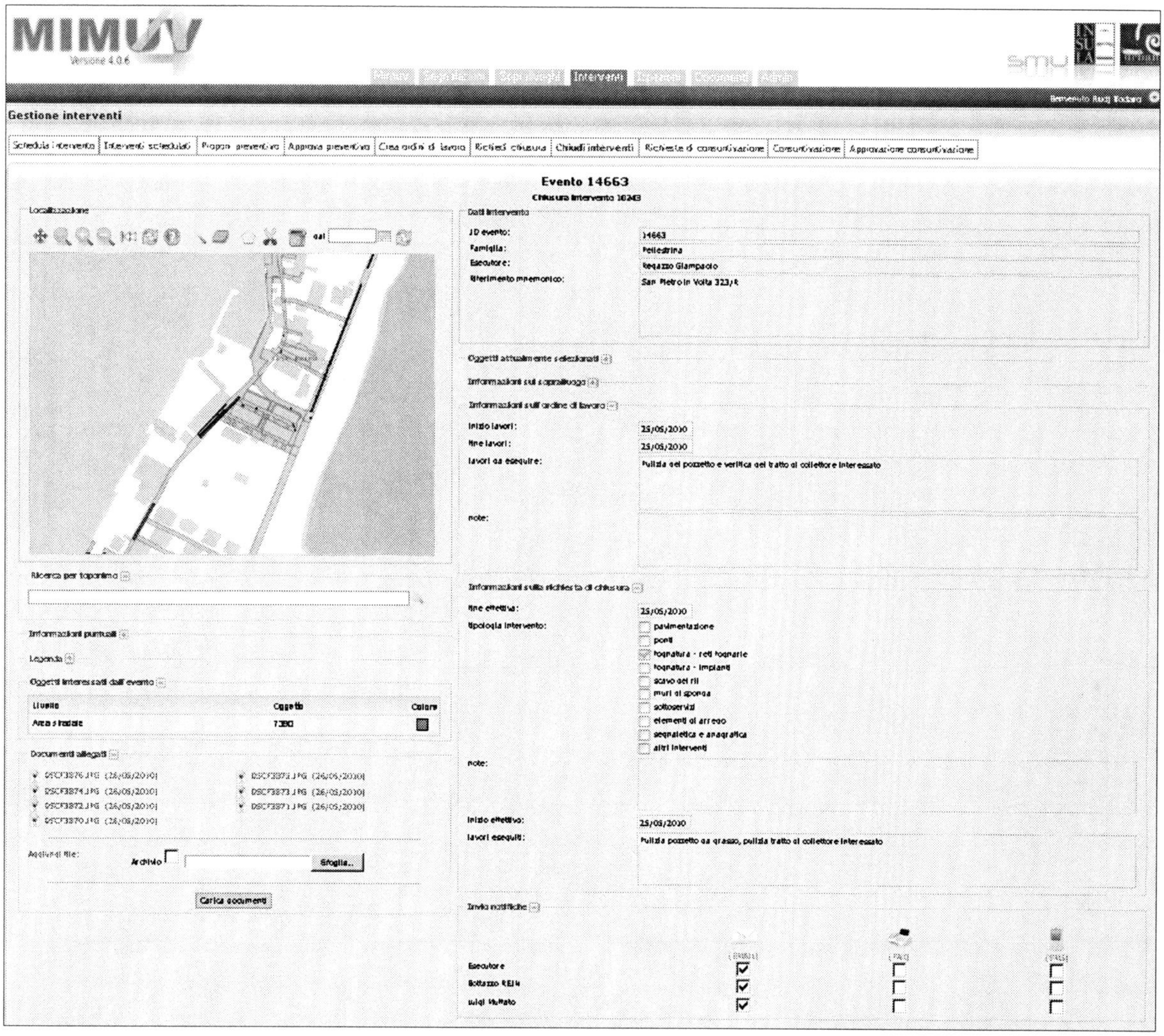

Figure 3 shows the details of an event for the maintenance of underground services on the island of Pellestrina. This is a clean-up of a cockpit and check the section of the collector concerned. In the example, inspection and reporting forms have closed. We see instead the work order information and the request for closure. It is also possible to see the photos submitted to the individual event and the options for fax, e-mail and SMS messages to other actors in this process.

In this application, the Masters such as pavements, buildings, urban furniture, electrical power grid, telephone grid, were organized into families, i.e. into groups of master data that were homogeneous in terms of maintenance logic and territorial context. For example the "Insula Venezia" family could contain bridges, paving, underground utilities grids and buildings in the context of Venice that do not require a cost estimate. "Insula Pellestrina" could contain bridges, paving and buildings in the context of Pellestrina that do not require the postponed scheduling of work orders. "Building Pellestrina" could contain the buildings in the context of Pellestrina which as their underlying cause contemplate exclusively the drafting of Work Orders. "Territorial Unit

Figure 4.

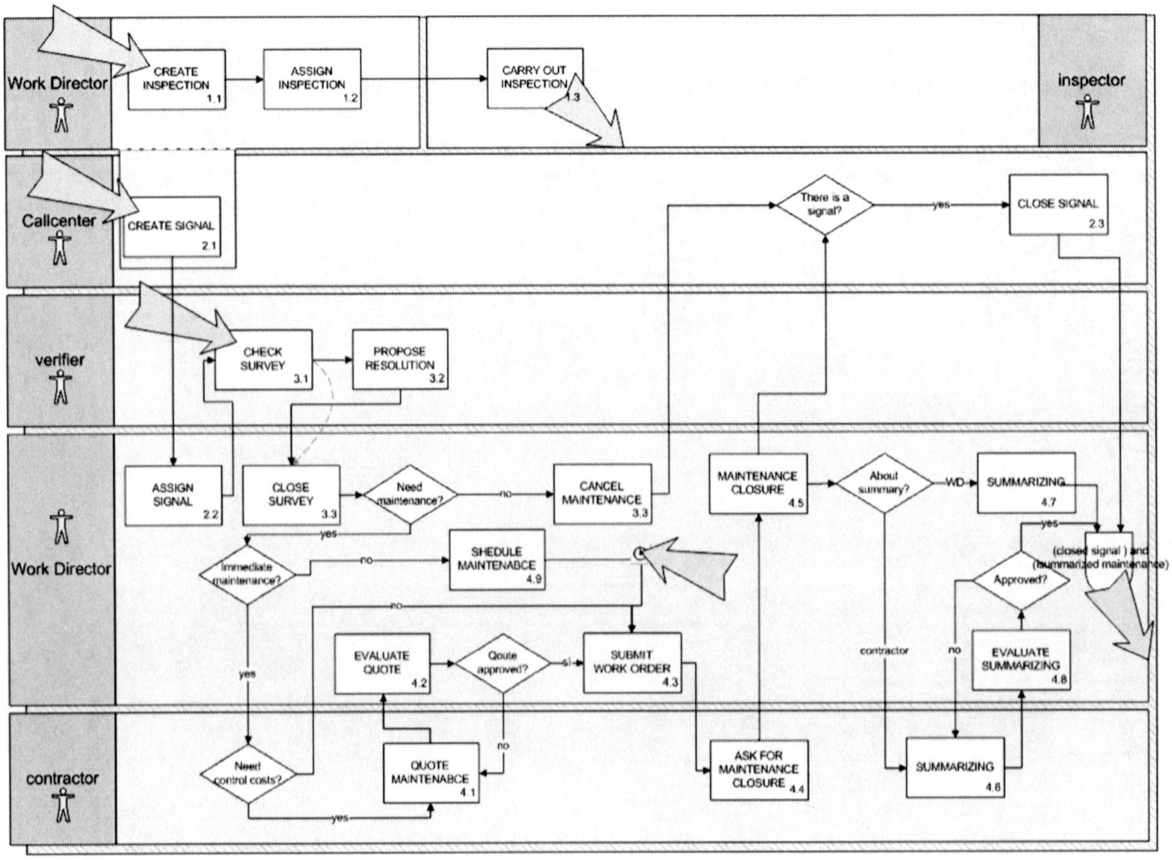

Noale" could contain paving and buildings in the context of Noale, which do not require applications for maintenance or inspections, etc. Each family could feature specific fields (visible or not); the attributions of the different fields (for example, types of intervention) could be "dedicated"; The status of each phase of the process, the printing, the flow or available functions, the aliases of the different fields could all be customized. The actions are organized into maintenance events: geo-referred portions of the master data in which a critical event, such as a malfunction, is discovered. They are contained geographically in the families, from which the actions are derived. To define them, it will be necessary, after having drawn them, to choose the appropriate family among all the ones that cover the area in question and refer to the user who is operating. The graphic object will be the intersection between the drawn area of the user and the chosen family; In case there is more than one family, and more than one is chosen, the event will be divided geographically into several parts, one for each family.

Figure 4 shows the process for maintenance of the pavement in the south of Venice.

The arrows identify the inputs and the outputs of the process.

The call center receives the signals, and passes to the work director.

The work director instructs the inspector to have a look.

Figure 5.

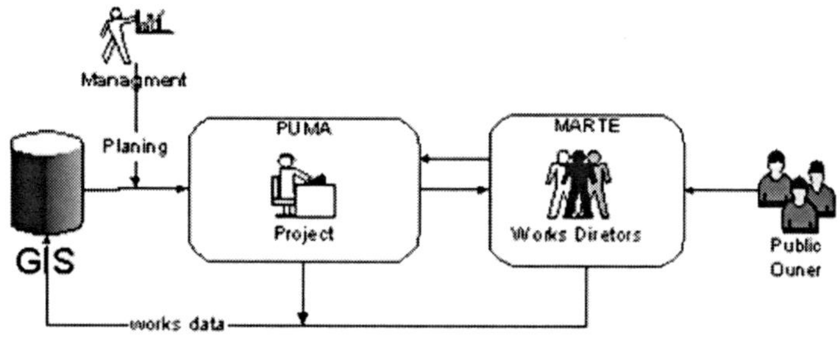

The inspector proposes a solution to the work director.
The work director decides what is needed to be done.
The company executor receives the work order, prepares a budget and is waiting for approval.

When the work director has approved the work order and the company conducted the operation, the phase of accounting will start, and then the alert will be closed.

There is also a parallel process for the inspection of the elements.

The actors in this process are:

- Applicants who discover the malfunction in the territory. In a wider sense, it covers all the citizens. The applicant can refer the problem in various ways (telephone, fax, email, etc.) to an accredited operator. He does not operate with the system.
- Inspectors are part of the technical staff that periodically monitors the situation (inspections).
- Executors are the staff that operates in the territory. They verify the complaints or discover ("verify") new malfunctions in the territory. Based on their inspection they may or may not propose a work order. They record the completion of the works they order. Executors can be construction companies or individual technicians.
- Decision managers are the ones who control the process.

PUMA and MARTE

These are two web applications that have been developed to manage large urban maintenance projects in the city of Venice and are intended to serve two different kinds of users. One of the applications (PUMA-Planning Urban Maintenance) is devoted to planning process whereas the other (MARTE-Maintenance and Restoration by Territorial Element) is for construction phase of urban maintenance projects. Figure 5 shows the process of planning and execution of the work. It can be seen in the figure how PUMA follows the part of the project and prepares the documentation for the next phase of implementation. The implementation is supported by MARTE. The information is first generated by PUMA, and then updated and exchanged between MARTE and PUMA, ie between the planner and executor. Both draw on data coming from the DB GIS and provides updates and new information.

The application for design and planning purpose, PUMA offers highly sophisticated alphanumerical and graphic tools that serve to the different typologies of territorial elements used in the works. The application is based on all geographical elements. These elements can be classified into two as basic elements, also as called SMU elements (segments and intersections, embankments, waterfronts, pavements and bridges) and other accessory elements (information layers such as, buildings, lagoon, urban furniture elements, etc.).

PUMA essentially sets four objectives:

- The design of the elements in the "project"
- The design and the definition of the UTP (Territorial Planning Units) for each project
- The publication (for consultation) of data concerning the elements of the "projects"
- The modification of the properties (not the graphic changes) of the base elements

A project can be historicised to trace its status at a certain date such as on the date of approval or rejection of a preliminary design, on the date it was stored in the achieve when an executive or final design project site was closed or after significant changes were made to key fields or areas of interest. The project itself and its geographical design units are either historicised or put away in a separate achieve where they are no longer modifiable. The attributes of the basic elements (SMU elements) may be modified by the designer after surveys and investigation campaigns conducted, and records are kept for all the changes made together with the name of the person who is responsible for the changes and the date the change is made (historicisation).

MARTE is the second application that is used to monitor and manage the physical progress of the works carried out. Access to physical description of the works and information about the different types of basic elements (segments, intersections, embankments, pavements and bridges) can be gained through Internet. MARTE enables users to take snapshots of the physical progress of the work at a specific date and to retrieve information on the state of preservation works of the bridges, embankments and pavements monitored on the same date. The data regarding each object can be updated through specific date update sheets input and each basic object may be associated to other types of information such as images, photographs or drawings. This application also opens lines of communication among project supervisors, designers and site managers and offers the company management a series of tools for supervision. The use of this application ensures constant updates regarding the status of territorial objects in the Urban Maintenance System.

In MARTE;

- Designers, project managers and site managers are all connected to one another via Internet,
- Each of these users operates on the same data and on the same territorial objects,
- All users can be connected to the system simultaneously and update the data,
- The territorial objects are updated whenever they are modified,
- Each object is also the container for all the relevant documentation,
- Each territorial object contains the history of the interventions and the monitoring processes it has undergone
- It is possible to get access to information regarding the actual progress of the works
- The use of "time" as a variable enables users to reconstruct the state of the system at any moment, and to assess the dynamics of the change (evolution of the decay, accumulation of sediment, etc.)

VENERE (Venice Network Representation)

Venere is an application developed to manage the processes for the delineation of joint interventions on underground grids and to bring the actors together in the process (city, telephone companies, electrical power company, gas company, etc.) while enabling complete management of coordination on works regarding the underground utilities grids. The use of the application meets the requirements related to coordination of works and management of data and documents. It also ensures that the state of repair of the underground utility systems is constantly updated within the Urban Maintenance System.

In VENERE:

- The team formed by the different companies can be connected via Intranet, and can participate in the process to define the intervention on the underground utilities grids;
- All the partners operate on the same territorial objects and have the same information at their disposal;
- The definition and planning of a project is guaranteed by a transparent process;
- The database is always updated with the data of completed projects:
- Each object is also the container of the related documentation and this documentation is accessible to all partners.

Publication Tools

These are applications that have been conceived for publication of the Urban Maintenance System on Internet and to provide the city's inhabitants useful information.

GISPORTAL

Venice GIS Portal is a webgis collecting the information regarding the urban elements that undergo maintenance by the Venice City Council. GISPORTAL is divided into thematic modules (Altimetry, Archelogy, Bathymetry, Maintenance, Parks Mestre, Paths, Ground Floors, Underground, Collapse Evaluation) that give citizens the opportunity to learn and use the data. Modules are as follows:

Altimetry

The module contains altimetry data compared to the average height of the ground water in all parts of the city and see the different levels of flooding. The zero reference is the average sea level measured by tide gauge at Punta della Salute in 1897. More than 4.500 monographs of the cornerstones of the town and the lagoon area are available in this module.

Archeology

The module Archaeology is a census of all archaeological findings discovered during maintenance work and renovations in Venice and its islands. In each display exact location of the findings with all monographic publications can be found.

Bathymetry

The module bathymetry shows the depth of the canals of Venice and the islands and its mathematical model. This information, along with the values of sedimentation, helps users plan more efficiently to ensure that the excavation of the rivers crossings through. It is possible to reconstruct the bathymetric sections, namely the "split " of the river with all measures related to depth.

Maintenance

The maintenance module is to map all routine and extraordinary maintenance performed by Insula since 1997.

Parks Mestre

The module will be used by the institution Parks Mestre for monitoring and planning of routine maintenance. For each service (benches, streetlights, pavement, building, tree, etc.) it is possible to view the exact location, its physical characteristics and the board with the results of inspections and intervention made so far.

Paths

The paths module allows users to identify the shortest route between two points chosen by the city. It is also possible to search for alternative routes that may be followed according to specific tidal levels chosen by user or any interruptions transitability.

Ground Floors

The module ground floor provides access to data from the census of all housing units on the ground floor of the historic center of Venice conducted by COSES (Consortium for Research and Training). In particular it gives a clearer light on the impact of high water on the ground floors of buildings.

Underground

The form shows the known routes of underground pipes for electricity, gas, water and telephone under the paving of streets and bridges. The tracks of the underground services are constantly updated, in parallel with the work of Insula and gradually the companies involved make available information on the new sections.

Collapse Evaluation

The module can monitor the state of deterioration and the structural conditions of the banks, the foundation and walls of buildings facing the canals around the historic center of Venice and provides the information of urban land on the entire system of canals and, in particular, those already involved in the Integrated Project canals.

CONCLUSION

City of Venice, which is a unique tourism destination, is under constant threat of flooding. In certain times of the year, the city experiences high water (acqua alta) that makes it difficult to stroll around streets and also causes damages to infrastructure of the city. In order to manage the infrastructure maintenance of the city, an Urban Maintenance System has been developed by Insula Spa (an urban maintenance company consisting of Venice city Council and some utility companies) tailored to the needs of the City of Venice. The information system employed combines GIS solutions with those typical of a management system in order to improve efficiency for both administrative and operative bodies.

The greatest difficulties faced in the implementation of the system in Venice are not as much related to technological or organizational issues as to human factor. Every decision-making system requires input data and these data cannot always be found through investigation campaigns but must be created step by step by those who work in the area. On the other side, the obligation to these workers is to do the job and not to record data on a keyboard. The best way to resolve this conflict is to design the system as beneficial to workers who use the system and to approach them as the first customer because any system proves useful if it is literally used. The system presented serves as a model for an application that gives priority to those who produce information as an added value to their daily operations. In such a system, data is the key concern for system users when transparency, interoperability and up-to-dateness are the points of issue. The direction undertaken in this system leads to the development of increasingly flexible geospatial management tools that enables adaptation of application to different processes without rewriting the code. The concept based on solution sharing leads to the constitution of a group of libraries and solutions that could be distributed in an open source perspective.

The experience revealed that in order to ensure the success of a project in this field, there are two things that must be taken into consideration. One is that the principal drivers for development are the operators who interact with the system frequently, and the second is that the logic of the Geographic Information Systems must be fused with the logic of management systems in order to compose an efficient geospatial information system.

REFERENCES

Abel, D. J., Yap, S. K., Ackland, R., Cameron, M. A., Smith, D. F., & Walker, G. (1992). Environmental decision support system project: An exploration of alternative architectures for geographical Information Systems. *International Journal of Geographical Information Systems*, *6*, 193–204. doi:10.1080/02693799208901905

Abudayyeha, O., Khanb, T., Yehiaa, S., & Randolph, D. (2005). The design and implementation of a maintenance information model for rural municipalities. *Advances in Engineering Software*, *36*, 540–548.

Aldosary, A. S., & Zaheer, S. A. (1996). An application mechanism for a GIS-based maintenance system: The case of KFUPM. *Computers, Environment and Urban Systems*, *20*(6), 399–412. doi:10.1016/S0198-9715(97)00005-7

Armstrong, M. P., Rushton, G., Honey, R., Dalziel, B. T., & Lolonis, P. (1991). Decision support for regionalization: A spatial decision support system for regionalizing service delivery systems. *International Journal of Computers Environment and Urban Systems*, *15*, 37–53. doi:10.1016/0198-9715(91)90044-E

Berner, L., Bogoyavlenskaya, N., Ifiushin, S. A., Kitaitseva, E., Kovalev, A., & Rochtchin, A. (1995). Integrated approach to urban facilities maintenance and alarm management. *International Journal of Computers. Environment and Urban Systems*, *19*(3), 201–206. doi:10.1016/0198-9715(95)00008-1

Boroushaki, S., & Malczewski, J. (2010). Measuring consensus for collaborative decision-making: A GIS-based approach. *Computers, Environment and Urban Systems*, *34*, 322–332. doi:10.1016/j.compenvurbsys.2010.02.006

Chang, N. B., Wei, Y. L., Tseng, C. C., & Kao, C. Y. (1997). The design of a GIS-based decision support system for chemical emergency preparedness and response in an urban environment. *International Journal of Computers. Environment and Urban Systems*, *21*, 67–94. doi:10.1016/S0198-9715(97)01009-0

Ghaemi, P., Swift, J., Sister, C., Wilson, J. P., & Wolch, J. (2009). Design and implementation of a Web-based platform to support interactive environmental planning. *International Journal of Computers. Environment and Urban Systems*, *33*, 482–491. doi:10.1016/j.compenvurbsys.2009.05.002

Hadzilacos, T., Kalles, D., Preston, N., Melbourne, P., Camarinopoulos, L., & Eimermacher, M. (2000). UtilNets: A water mains rehabilitation decision- support system. *International Journal of Computers. Environment and Urban Systems*, *24*, 215–232. doi:10.1016/S0198-9715(99)00058-7

Hamilton, A., Burns, M. C., Revilla, I., Kaklauskas, A., Alejandro, E. M., & Arayici, Y. (2005). *Final Report of IntelCities D5.4c Building Data Integration System*. Manchester, UK: University of Salford.

Hamilton, A., Wang, H., Tanyer, A. M., Arayici, Y., Zhang, X., & Song, Y. (2005). *Integration of building and urban n-dimensional data to support intelligent cities. ITcon 10* (From 3D to nD modelling) (pp. 55–67).

Huang, B. (2003). Web-based dynamic and interactive environmental visualization. *International Journal of Computers. Environment and Urban Systems, 27*, 623–636. doi:10.1016/S0198-9715(02)00063-7

Huber, G. (1990). A theory of the effects of advanced Information Technologies on organisational design, intelligence, and decision-making. *Academy of Management Review, 15*(1), 47–71.

Kim, Y. I., Pyen, M. W., & Eo, Y. D. (2000). Development of hypermap database for ITS and GIS. *International Journal of Computers Environment and Urban systems, 24*, 45-60.

Leem, Y. T., Lee, S., Han, J. H., & Lee, J. (2007). *A management of GIS DB for urban infrastructure for multi-functional administrative city.* 10th International Conference on Computers in Urban Planning and Urban Management. Retrieved March 10th, 2011, from http://www.scribd.com/doc/ 7315199/a-Management-of- Gis-Db-for-Urban- Infrastructure-for-Multifunctional -Administrative-City

Mahmoud, R., Halfawy, M. R., Pyzoha, D., & El-Hosseiny, T. (2002). *An integrated framework for GIS-based civil infrastructure management systems*. Retrieved March 11th, 2011, from http:// pedago.cegepoutaouais. qc.ca/media/0260309 /0378334/ SCGC-BON/ Documents/ GE012-Halfawy% 20et%20al.pdf.

Michele, D. S., & Daniela, L. (2011). Decision-support tools for municipal infrastructure maintenance management. *Procedia Computer Science, 3*, 36–41. doi:10.1016/j.procs.2010.12.007

Miyamoto, K., Sugiki, N., Uchida, T., & Paez, A. (1997). A GIS based land-use model dealing with building types by small unit of land in a metropolitan area. *Journal of the Eastern Asia Society for Transportation Studies, 2*, 1943–1959.

Pedersen, Å., Ø., Nyhuus, S., Blindheim, T., & Wergeland Krog, O. M. (2004). Implementation of a GIS-based management tool for conservation of biodiversity within the municipality of Oslo, Norway. *Landscape and Urban Planning, 68*, 429–438.

Pierre, S., Leclerc, G., Revaz, J., Dion, Y., & Rossi, L. (2001). Urbistics: An emerging field. In *Proceedings of World Multiconference on Systemics, Cybernetics and Informatics, (ISAS/SCI 2001)* (vol. 28, pp. 124–126), Orlando, Florida.

Pierre, S., Quintero, A., & Konaré, D. (2001). An integrated system for coordinated management of urban infrastructure: Concepts and architecture. In *Proceedings of World Multiconference on Systemics, Cybernetics and Informatics (ISAS/SCI 2001)* (vol. 28, pp. 133–137), Orlando, Florida.

Pior, M. Y., & Osman, O. (1997). Air pollution assessment applying GIS integrated system. *Journal of the Eastern Asia Society for Transportation Studies, 2*, 2071–2081.

Pior, M.-Y., & Shimizu, E. (2001). GIS-aided evaluation system for infrastructure improvements: Focusing on simple hedonic and Rosen's two-step approaches. *Computers, Environment and Urban Systems, 25*, 223–246. doi:10.1016/S0198-9715(00)00018-1

Quintero, A., & Konaré, P. S. (2005). Prototyping an intelligent decision support system for improving urban infrastructures management. *European Journal of Operational Research, 162*, 654–672. doi:10.1016/j.ejor.2003.10.019

Rubenstein-Montano, B. (2000). A survey of knowledge-based Information Systems for urban planning: Moving towards knowledge management. *International Journal of Computers Environment and Urban systems, 24*, 155-172.

Sanchez, T. W. (1993). *Measuring the impact of highways and roads on residential property values with GIS: An empirical analysis of Atlanta.* In R. E. Klosternman, & S. P. French (Eds.), 3rd International Conference on Computers in Urban Planning and Management (vol. 2).

Siderelis, K. C. (1991). Land resource Information Systems. In Maguire, D., Goodchild, J., & Rihind, D. W. (Eds.), *Geographical Information System: Principles and applications* (*Vol. 2*). New York, NY: Longman.

Wang, H., Song, Y., Hamilton, A., & Curwell, S. (2007). Urban information integration for advanced e-planning in Europe. *Government Information Quarterly, 24*, 736–754. doi:10.1016/j.giq.2007.04.002

Winstanley, A. C. (2009). Spatial databases. *International encyclopedia of human geography* (pp. 345-347).

Yeh, A. G., & Chow, M. H. (1996). An integrated GIS and location-allocation approach to public facilities planning Đ an example of open space planning. *International Journal of Computers. Environment and Urban Systems, 20*, 339–350. doi:10.1016/S0198-9715(97)00010-0

ENDNOTE

[1] INSPIRE- Infrastructure for Spatial Information in Europe: The directive aims to provide an infrastructure for spatial information established and operated by the Member States of the European Union in order to support Community environmental policies, and policies or activities which may have effects on the environment. Retrieved March, 20 2011 from http://inspire.jrc.ec.europa.eu/index.cfm

Compilation of References

Abel, D. J., Yap, S. K., Ackland, R., Cameron, M. A., Smith, D. F., & Walker, G. (1992). Environmental decision support system project: An exploration of alternative architectures for geographical Information Systems. *International Journal of Geographical Information Systems*, *6*, 193–204. doi:10.1080/02693799208901905

Abrahamson, M. (2004). *Global cities*. New York, NY: Oxford University Press.

Abudayyeha, O., Khanb, T., Yehiaa, S., & Randolph, D. (2005). The design and implementation of a maintenance information model for rural municipalities. *Advances in Engineering Software*, *36*, 540–548.

ACI-Europe. (2007). *Airport traffic statistics*. Retrieved from http://www.aci-europe.org

Acs, Z. J., Glaeser, E., Litan, R., Fleming, L., Goetz, S. J., Kerr, W., & Strange, W. (2008). *Entrepreneurship and urban success: Toward a policy consensus*. Kansas City, MO: Ewing Marion Kauffman Foundation. Retrieved from http://ssrn.com/ abstract=1092493

Adult, S. (2005). Some important distinctions in place branding. *Place Branding*, *1*(2), 116–121. doi:10.1057/palgrave.pb.5990011

Agrawal, R., & Ramakrishnan, S. (1994). Fast algorithms for mining association rules. In J. B. Bocca, M. Jarke, & C. Zaniolo (Eds.), *Proceedings of the 20th International Conference on Very Large Data Bases (VLDB), Santiago de Chile, Chile* (pp. 487-499). Waltham, MA: Morgan Kaufmann Publishers.

Ahern, D. K. (2007). Challenges and opportunities of e-health research. *American Journal of Preventive Medicine*, *32*(5), 75–82. doi:10.1016/j.amepre.2007.01.016

Air France. (2010). *Tgv high-speed train*. Retrieved July 26, 2010, from http://corporate.airfrance.com/ en/ alliances/competition/ tgv-high-speed-train/

Akçelik, R. (1988). The highway capacity manual delay formula for signalized intersections. *ITE Journal*, *58*(3), 23–27.

Aldosary, A. S., & Zaheer, S. A. (1996). An application mechanism for a GIS-based maintenance system: The case of KFUPM. *Computers, Environment and Urban Systems*, *20*(6), 399–412. doi:10.1016/S0198-9715(97)00005-7

Aliadière, L. (2008). Fast track to sustainable mobility. *Railway Gazette International*, *164*(3), 141–143.

Alkan, M., & Bulut, G. (2010). GIS and remote-sensing-based urban-information system design and development: A case study for Kozlu, Zonguldak. *Scientific Research and Essays*, *5*(19), 2889–2899.

Almeida, M. (2009). *Ubiquitous computing and natural interfaces for environmental information*. Lisbon, Portugal: Faculdade de Ciências e Tecnologia da Universidade Nova de Lisboa.

Amin, A., & Roberts, J. (2008). Knowing in action: Beyond communities of practice. *Research Policy*, *37*(2), 353–369. doi:10.1016/j.respol.2007.11.003

Anderson, N. (2008). Light rail for freight. *Logistics & Transport Focus*, *10*(8), 42–45.

Anderson, S. (2010). Improving data center efficiency. *Energy Engineering*, *107*(5), 42–63.

Andersson, D. E., & Andersson, Å. E. (2008). Infrastructural change and secular economic development. *Technological Forecasting and Social Change*, *75*(6), 799–816. doi:10.1016/j.techfore.2007.08.003

Annual Survey of Health Information Exchange. (2009). *Migrating toward meaningful use: The state of health information exchange. A report based on the results of the eHealth Initiative's 2009 sixth annual survey of health information exchange.* Washington, DC: eHealth Initiative.

Arduin, J. P., & Ni, J. (2005). 40 years of high-speed railways - French TGV network development. *Japan Railway & Transport Review, 40*, 22–28.

Armstrong, M. P., Rushton, G., Honey, R., Dalziel, B. T., & Lolonis, P. (1991). Decision support for regionalization: A spatial decision support system for regionalizing service delivery systems. *International Journal of Computers Environment and Urban Systems, 15*, 37–53. doi:10.1016/0198-9715(91)90044-E

Asheim, B. (1996). Industrial districts as 'learning regions': A condition for prosperity. *European Planning Studies, 4*(1), 379–400. doi:10.1080/09654319608720354

Asheim, B., Coenen, L., & Vang, J. (2007). Face-to-face, buzz, and knowledge bases: Sociospatial implications for learning, innovation, and innovation policy. *Environment and Planning. C, Government & Policy, 25*(5), 655–670. doi:10.1068/c0648

Asheim, B., Boschma, R., & Cooke, P. (2008). *Constructing regional advantage: Platform policies based on related variety and differentiated knowledge bases.* Unpublished manuscript, Utrecht, The Netherlands.

Ashworth, J., & Voogd, H. (1994). Place promotion: The use of publicity and marketing to sell towns and regions. *Place Branding, 1*(2), 39–52.

ATAG. (2008). *The economic and social benefits of air transport.* Retrieved from http://www.atag.org/ our-publications /latest/1/23-the-economic-and- social-benefits-of-air-transport.html

Aurigi, A., & De Cindio, F. (Eds.). (2008). *Augmented urban spaces.* Hampshire, UK: Ashgate.

Avraham, E., & Ketter, E. (2008). *Media strategies for marketing places in crisis - Improving the image of cities, countries and tourist destinations.* Burlington, MA: Elsevier.

Bagchia, M., & White, P. R. (2005). The potential of public transport smart card data. *Transport Policy, 12*, 464–474. doi:10.1016/j.tranpol.2005.06.008

Baldassi, V. (2000). *The Venice card: Our city at our fingers.* In Major cities of Europe: ITusers group. Retrieved from http://www.majorcities.org/ pics/medien/1_1102426321 /Baldassi_eins.pdf

Ball, M. (2010). *Exploring the Role of GIS in the Enterprise.* Retrieved June 24, 2010, from http://vector1media.com/dialog/ interviews/7320-interview-exploring -the-role-of-gis-in-the-enterprise.pdf

Baloglu, S., & Pekcan, Y. A. (2006). The website design and internet site marketing practices of upscale and luxury hotels in Turkey. *Tourism Management, 27*(1), 171–176. doi:10.1016/j.tourman.2004.07.003

Banister, D. (2008). The sustainable mobility paradigm. *Transport Policy, 15*, 73–80. doi:10.1016/j.tranpol.2007.10.005

Banister, D., & Berechman, Y. (2001). Transport investment and the promotion of economic growth. *Journal of Transport Geography, 9*(3), 209–218. doi:10.1016/S0966-6923(01)00013-8

Banister, D., Stead, D., & Hickman, R. (2008). Looking over the horizon: Visioning and backcasting. In Himanen, V., Lee-Gosselin, M., & Perrels, A. (Eds.), *Building blocks for sustainable transport* (pp. 25–54). Helsinki, Sweden: VATT and Emerald.

Baradaran, S., & Ramjerdi, F. (2001). Performance of accessibility measures in Europe. *Journal of Transportation and Statistics, 4*(2-3), 31–38.

Bathelt, H., Malmberg, A., & Maskell, P. (2004). Clusters and knowledge: Local buzz, global pipelines and the process of knowledge creation. *Progress in Human Geography, 28*(1), 31–56. doi:10.1191/0309132504ph469oa

Batty, M., & Densham, P. J. (1996). *Decision support, GIS, and urban planning.* London, UK: Center for Advanced Spatial Analysis, University College London.

Batty, M. (2001). Contradictions and conceptions of the digital city. *Environment and Planning. B, Planning & Design, 28*(4), 479–480. doi:10.1068/b2804ed

Baum, S., Kendall, E., Muenchberger, H., Gudes, O., & Yigitcanlar, T. (2010). Geographical Information Systems: An effective planning and decision-making platform for community health coalitions in Australia. *Health Information Management Journal, 39*(3), 28–33.

Baur, C. (2008). An analysis of factors underlying e-health disparities. *Cambridge Quarterly of Healthcare Ethics*, *17*(4), 417–428. doi:10.1017/S0963180108080547

BBSR. (2011). *Metropolitan areas in Europe*. Bonn, Germany: Federal Institute for Research on Building, Urban Affairs and Spatial Development.

Beath, J. (2002). UK industrial policy: Old tunes on new instruments? *Oxford Review of Economic Policy*, *18*(2), 221–239. doi:10.1093/oxrep/18.2.221

Becattini, G. (1990). In F. Pyke, G. Becattini, & W. Sengenberger (Eds.), *The Marshallian industrial districts as a socio-economic notion*. Geneva, Switzerland: International Institute of Labor Studies, IIL Publisher.

Begg, I. (2000). *Cities and competitiveness*. London, UK: South Bank University.

Begg, I. (2002). *Urban competitiveness*. Bristol, UK: Policy Press.

Begg, I. (1999). Cities and competitiveness. *Urban Studies (Edinburgh, Scotland)*, *36*(5), 795–809. doi:10.1080/0042098993222

Bell, R. (2008). Five ways for communities to shape their broadband futures. *Economic Development Journal*, *7*(3), 13–19.

Belson, K. (2005, June 6). Technology: Fast phones in Japan- But no pot of gold. Retrieved from http://query.nytimes.com/ gst/fullpage.html?res= 9D04E7DA-1538F935A 35755C0A9639C8B63

Ben-Akiva, M., & Lerman, S. R. (1979). Disaggregate travel and mobility choice models and measures of accessibility. In Hensher, D. A., & Stopher, P. R. (Eds.), *Behavioural travel modelling* (pp. 654–679). London, UK: Croom Helm.

Beninger, J. (1986). *The control revolution: Technological and economic origins of the information society*. Cambridge, MA: Harvard University Press.

Benneworth, P. (2006). *The role of university spin-off firms in strengthening regional innovation systems in weaker places*. Paper presented at the 6th European Urban & Regional Studies Conference, Roskilde, Denmark.

Benneworth, P., Coenen, L., & Truffer, B. (2010). *Back to the future? The geographical dilemma of contemporary transition studies*. Paper presented at 2010 AAG Annual Meeting, Washington, DC.

Berg, L., & Braun, E. (1999). Urban competitiveness, marketing and the need for organizing capacity. *Urban Studies (Edinburgh, Scotland)*, *36*(5), 987–999. doi:10.1080/0042098993312

Berg, L. V. D., & Braun, E. (1999). Urban competitiveness, marketing and the need for organising capacity. *Urban Studies (Edinburgh, Scotland)*, *36*(5-6), 987–999. doi:10.1080/0042098993312

Berg, L. (1987). Urban system in a dynamic society. Avebury, UK: Aldershot.

Bergek, A., Jacobsson, S., Carlsson, B., Lindmark, S., & Rickne, A. (2008). Analyzing the functional dynamics of technological innovation systems: A scheme of analysis. *Research Policy*, *37*(3), 407–429.

Berger, C., Möslein, K., Piller, F., & Reichwald, R. (2005). Cooperation between manufacturers, retailers, and customers for user co-design: Learning from exploratory research. *European Management Review*, *2*(1), 70–87. doi:10.1057/palgrave.emr.1500030

Bernard, P., Charafeddine, R., Frohlich, K. L., Daniel, M., Kestens, Y., & Potvin, L. (2007). Health inequalities and place: A theoretical conception of neighbourhood. *Social Science & Medicine*, *65*(9), 1839–1852. doi:10.1016/j.socscimed.2007.05.037

Berner, L., Bogoyavlenskaya, N., Ifiushin, S. A., Kitaitseva, E., Kovalev, A., & Rochtchin, A. (1995). Integrated approach to urban facilities maintenance and alarm management. *International Journal of Computers. Environment and Urban Systems*, *19*(3), 201–206. doi:10.1016/0198-9715(95)00008-1

Berry, J. K., Delgado, J. A., & Khosla, R. (2008). *Precision farming and conservation advances agricultural sustainability*. Fort Collins, CO: USDA Soil Plant Nutrient Research Unit (SPNR). Retrieved from http://www.ars.usda.gov/ research/publications/ publications.htm?seq_no_115=223229

Bertolini, L., & le Clercq, F. (2003). Urban development without more mobility by car? Lessons from Amsterdam: A multimodal urban region. *Environment & Planning A*, *35*(4), 575–589. doi:10.1068/a3592

Bertolini, L., le Clercq, F., & Kapoen, L. (2005). Sustainable accessibility: A conceptual framework to integrate transport and land use plan-making. Two test-applications in the Netherlands and a reflection on the way forward. *Transport Policy*, *12*(3), 207–220. doi:10.1016/j.tranpol.2005.01.006

Bertolini, L. (2008). Station areas as nodes and places in urban networks: An analytical tool and alternative development strategies. In Bruinsma, F., Pels, E., Priemus, H., Rietveld, P., & van Wee, B. (Eds.), *Railway development: Impacts on urban dynamics* (pp. 35–57). Heidelberg, Germany: Physica-Verlag.

Beyer, K. M. M., & Rushton, G. (2009). Mapping cancer for community engagement. *Preventing Chronic Disease*, *6*(1).

Binder, T., Brandt, E., & Gregory, J. (2008). Design participation(-s): A creative commons for ongoing change. *CoDesign*, *4*(2), 79–83. doi:10.1080/15710880802114458

Bishop, S. (2010). Building programmatic capacity at the grassroots level: The reactions of local nonprofit organization to public participation geographic information system. *Nonprofit and Voluntary Sector Quarterly*, *39*(6), 991–1013. doi:10.1177/0899764009334984

Blichfeldt, B. S. (2005). Unmanageable place brands? *Place Branding*, *1*(4), 388–401. doi:10.1057/palgrave.pb.5990036

Blum, U., Haynes, K. E., & Karlsson, C. (1997). The regional and urban effects of high-speed trains. *The Annals of Regional Science*, *31*(1), 1–20. doi:10.1007/s001680050036

Böcker, J., Lind, J., & Zirkler, B. (2001). Using a multi-agent approach to optimise the train coupling and sharing system. *European Journal of Operational Research*, *131*(2), 242–252. doi:10.1016/S0377-2217(00)00124-7

Boisier, S. (1994). *Los escenarios del desarrollo descentralizados en America Latina (Serie Investigación, documento 94/09)*. Santiago, Chile: ILPES.

Bonnafous, A. (2002). The regional impact of the TGV. *Classics in Transport Analysis*, *3*, 480–490.

Boos, S., & Mueller, H. (2009). Evaluation of spatial Information Technology applications for mega city management. SDI Convergence. Research, Emerging Trends, and Critical Assessment. In van Loenen, B., Besemer, J. W. J., & Zevenbergen, J. A. (Eds.), *Nederlandse Commissie voor Geodesie (Netherlands Geodetic Commission), 48.*

Boroushaki, S., & Malczewski, J. (2010). Measuring consensus for collaborative decision-making: A GIS-based approach. *Computers, Environment and Urban Systems*, *34*, 322–332. doi:10.1016/j.compenvurbsys.2010.02.006

Boschma, R. (2004). Competitiveness of regions from an evolutionary perspective. *Regional Studies*, *38*(9), 1001–1014. doi:10.1080/0034340042000292601

Boyd, J. (1995). *The essence of winning and losing* [Presentation]. United States Air Force. Retrieved from http://www.danford.net/ boyd/essence.htm

Bradford, C. I. Jr. (1994). The new paradigm of systemic competitiveness: Why it matters, what it means and implications for policy. In Bradford, C. (Ed.), *The new paradigm of systemic competitiveness: Toward more integrated policies in Latin America*. Paris, France: OECD Development Centre Documents.

Braunerhjelm, P., & Feldman, M. P. (2006). *Cluster genesis: Technology-based industrial development*. New York, NY: Oxford University Press.

Bristow, G. (2005). Everyone's a'winner': Problematising the discourse of regional competitiveness. *Journal of Economic Geography*, *5*(3), 285. doi:10.1093/jeg/lbh063

Brocker, J., & Rietveld, P. (2009). Infrastructure and regional development. In Capello, R., & Nijkamp, P. (Eds.), *Handbook of regional growth and development theories*. Cheltenham, UK: Edward Elgar.

Brons, M., Givoni, M., & Rietveld, P. (2009). Access to railway stations and its potential in increasing rail use. *Transportation Research Part A, Policy and Practice*, *43*(2), 136–149. doi:10.1016/j.tra.2008.08.002

Bruinsma, F., Nijkamp, P., & Rietveld, P. (1992). Regional economic transformation and social overhead investments. *Tijdschrift voor Economische en Sociale Geografie*, *83*(1), 3–12. doi:10.1111/j.1467-9663.1992.tb00604.x

Bruinsma, F., & Rietveld, P. (1998). The accessibility of European cities: Theoretical framework and comparison of approaches. *Environment & Planning A*, *30*(3), 499–521. doi:10.1068/a300499

Brunello, L., Bunker, J., & Ferreira, L. (2008). *Continuous railway systems: An innovative approach to improve high speed rail sustainability*. Paper presented at the 10th International Conference on the Application of Advanced Technologies in Transportation.

Bruns, A. (2007, September). Tech Magnet. *Site Selection*, *52*(5), 744–759.

Burckhart, K., & Blair, C. (2009). *Urban intermodality: Potentials for connecting the cities' public transport*. Paper presented at the XV Urban Transport Conference.

Burmeister, A., & Colletis-Wahl, K. (1998). Proximity in networks. The role of transport infrastructure. In Reggiani, A. (Ed.), *Accessibility, trade and locational behaviour* (pp. 201–218). Aldershot, UK: Ashgate.

BusinessWeek. (2006, June 12). *Servers as high as an elephant's eye*. Retrieved from http://www.businessweek.com/ magazine/content/ 06_24/b3988087

Bustillos, B., Urista, V., Rentería, G., Vega, G., Vazquez, E., & Delgadillos, C. (2008). A model to measure the degree of competitiveness of medium-sized cities in the Chihuahua region of Mexico: *Annual Statistics Report 2005*.

Butler, C. J. (1999). Cities in competition: Equity issues. *Urban Studies (Edinburgh, Scotland)*, *36*(5-6), 865–891. doi:10.1080/0042098993259

Button, K. J. (2010). Economic aspects of regional airport development. In Postorino, M. N. (Ed.), *Developments of regional airports* (pp. 11–27). Ashurst, UK: WIT Press. doi:10.2495/978-1-84564-143-6/01

Buyck, C. (2008). *Planes versus trains*. Retrieved July 29, 2010, from http://atwonline.com/ it-distribution/article/ planes-versus-trains-0309

Byrne, A. (2000, August 28). Management by Web. *Business Week*, 84-96.

Cai, L., Card, J. A., & Cole, S. T. (2004). Content delivery performance of World Wide Web sites of US tour operators focusing on destinations in China. *Tourism Management*, *25*(2), 219–227. doi:10.1016/S0261-5177(03)00095-5

Cairncross, F. (2000). Inside the machine: A survey of e-management. *The Economist*, *11*, 8-10.

Camagni, R. (Ed.). (1991). *Innovation networks: Spatial perspectives*. London, UK: Bellhaven.

Camagni, R. (2002). On the concept of territorial competitiveness: Sound or misleading? *Urban Studies (Edinburgh, Scotland)*, *39*(13), 2395. doi:10.1080/0042098022000027022

Campbell, J. (1991). *Map use and analysis*. Dubuque, IA: Wm. C. Brown Publishers.

Campbell, F., & Deakin, M. (2005). Cities as leading examples of digitally-inclusive knowledge societies: The e-citizenship course, representative users, pedagogy and engagement matrix. In Osborne, M., & Wilson, B. (Eds.), *Making knowledge work. Stirling, Scotland*. UK: Stirling University.

Campos, J., & de Rus, G. (2009). Some stylized facts about high-speed rail: A review of hsr experiences around the world. *Transport Policy*, *16*(1), 19–28. doi:10.1016/j.tranpol.2009.02.008

Cantos, P., Gumbau-Albert, M., & Maudos, J. (2005). Transport infrastructures, spillover effects and regional growth: Evidence of the spanish case. *Transport Reviews*, *25*(1), 25–50. doi:10.1080/014416410001676852

Capello, R., & Rietveld, P. (1998). The concept of network synergy in economic theory: Policy implications. In Button, K., Nijkamp, P., & Priemus, H. (Eds.), *Transport networks in Europe* (pp. 57–83). Cheltenham, UK: Edward Elgar.

Carlin, W., Haskel, J., & Seabright, P. (2001). Understanding the essential fact about capitalism: Markets, competition and creative destruction. *National Institute Economic Review*, *175*(1), 67–84. doi:10.1177/002795010117500107

Carvalho, L., van Tuijl, E., & van Haaren, J. (2009). *Developing locations in the knowledge economy: Arabianranta, Helsinki*. Rotterdam, The Netherlands: Eramus University Rotterdam.

Carvalho, L., Mingardo, G., & Van Haaren, J. (2010). *Green urban transport policies and cleantech innovations. Evidence from Curitiba, Göteborg and Hamburg*. Paper presented at 2010 AAG Annual Meeting, Washington, DC.

Cascetta, E. (2001). *Transportation systems engineering: Theory and methods*. Dordrecht, The Netherlands: Kluwer Academic Publishers.

Cascetta, E., & Pagliara, F. (2008). Integrated railways-based policies: The regional metro system (RMS) project of Naples and Campania. *Transport Policy*, *15*(2), 81–93. doi:10.1016/j.tranpol.2007.11.001

Castells, M. (2003). *The Internet galaxy. Reflections on the Internet, bussiness and society*. Oxford, UK: Oxford University Press.

Castells, M. (1989). *The informational city: Information Technology, economic restructuring and the urban-regional process*. London, UK: Basil Blackwell.

Castells, M. (1996). *The information age: Economy, society, and culture. The rise of the network society* (*Vol. 1*). London, UK: Blackwell.

Castells, M. (1991). *Informational city: Information Technology, urban restructuring and the urban regional process*. Oxford, UK: Blackwell Publishers.

Castells, M. (1996). *The rise of the network society*. Malden, MA: Wiley-Blackwell.

Castells, M. (1998). *The rise of the network society* (*The information age: Economy, society, and culture,* vol. 1). West Sussex, UK: Wiley-Blackwell Publisher.

Cellini, R., & Soci, A. (2002). Pop competitiveness. *Quarterly Review. Indexes - Banca Nazionale del Lavoro*, *55*, 71 101.

Cendrowski, S. (2010). *20 most profitable companies*. Retrieved from http://money.cnn.com

Central Northern Adelaide Health Service. (2008). *Central Northern Adelaide Health Service: Social health atlas 2008*. Retrieved November 24, 2008, from http://www.publichealth.gov.au /interactive_mapping/ sa_sha_cnahs_2005/ atlas.html

Chambers, R. (2006). Participatory mapping and geographic information systems: Whose map? Who is empowered and who is disempowered? Who gains and who loses? *The Electronic Journal on Information Systems in Developing Countries, 25*.

Chang, N. B., Wei, Y. L., Tseng, C. C., & Kao, C. Y. (1997). The design of a GIS-based decision support system for chemical emergency preparedness and response in an urban environment. *International Journal of Computers. Environment and Urban Systems*, *21*, 67–94. doi:10.1016/S0198-9715(97)01009-0

Chapleau, R., Trépanier, M., & Chu, K. K. (2008). The ultimate survey for transit planning: Complete information with smart card data and GIS. In *8th International Conference on Survey Methods in Transport: Harmonisation and Data Comparability,* Annecy, France [Proceedings] (pp. 1-25).

Cheshire, P. (1999). Cities in competition: Articulating the gains from integration. *Urban Studies (Edinburgh, Scotland)*, *36*(5-6), 843–864. doi:10.1080/0042098993240

Cheshire, P. (2009). Spatial policies, planning and urban competitiveness. In Karlsson, C., Johansson, B., & Stongh, R. (Eds.), *The particular case of London*. Cheltenham, UK: Edward Elgar.

Cheung, W. M., & Huang, W. (2002). An investigation of commercial usage of the World Wide Web: A picture from Singapore. *International Journal of Information Management*, *22*(5), 337–388. doi:10.1016/S0268-4012(02)00026-9

Chorianopoulos, I., Pagonis, T., Koukoulas, S., & Drymoniti, S. (2010). Planning, competitiveness and sprawl in the Mediterranean city: The case of Athens. *Cities (London, England)*, *27*(4), 249–259. doi:10.1016/j.cities.2009.12.011

Chorus, C. G., Molin, E. J. E., & Van Wee, B. (2006). Use and effects of Advanced Traveller Information Services (ATIS): A review of the literature. *Transport Reviews*, *26*(2), 127–149. doi:10.1080/01441640500333677

Chrisman, N. (1992). *Exploring geographic Information Systems*. New York, NY: John Wiley & Sons.

ComScore. (2009). *2009 US search engine rankings.* Retrieved from http://www.comscore.com/ Press_Events/ Press_Releases/ 2009/3/US_Search_Engine _Ranking

Conway, M. D., Rizzuto, C. D., & Weiss, L. M. (2008). *A better way to speed the adoption of vaccines*. Retrieved November 20, 2008, from http://www.mckinseyquarterly.com /home.aspx

Cooke, P. (2001). Regional innovation systems, clusters, and the knowledge economy. *Industrial and Corporate Change, 10*(4), 945–974. doi:10.1093/icc/10.4.945

Cooke, P., Uranga, M., & Extebarria, G. (1998). Regional innovation systems: An evolutionary perspective. *Environment & Planning A, 30*(9), 1563–1584. doi:10.1068/a301563

Couto, A., & Graham, D. J. (2008). The impact of high-speed technology on railway demand. *Transportation Planning and Analysis, 35*, 111–128.

Cox, K. R. (1998). Spaces of dependence, spaces of engagement and the politics of scale, or: looking for local politics. *Political Geography, 17*(1). doi:10.1016/S0962-6298(97)00048-6

Crockett, J., & Hounsell, N. (2005). Role of the travel factor convenience in rail travel and a framework for its assessment. *Transport Reviews, 25*(5), 535–555. doi:10.1080/01441640500064389

Croll, P. R., & Croll, J. (2007). Investigating risk exposure in e-health systems. *International Journal of Medical Informatics, 76*(5-6), 460–465. doi:10.1016/j.ijmedinf.2006.09.013

Crouch, G. I., & Ritchie, J. R. B. (1999). Tourism, competitiveness, and societal prosperity. *Journal of Business Research, 44*(3), 137–152. doi:10.1016/S0148-2963(97)00196-3

Cuadrado, J. R. (2008). Towards new European peripheries. In Stongh, R. (Ed.), *Innovation, agglomeration and regional competition*. Cheltenham, UK: Edward Elgar.

Curwell, S., Deakin, M., Cooper, I., Paskaleva-Shapira, K., Ravetz, J., & Babicki, D. (2005). Citizens expectations of information cities: Implications for urban planning and design. *Building Research and Information, 22*(1), 55–66. doi:10.1080/0961321042000329422

Daily Mail Reporter. (2009). *Taxi of tomorrow: The space-age pods with no driver, no fumes... Just a 'start' button*. Retrieved November 26, 2009, from http://www.dailymail.co.uk/ sciencetech/article-1205970/ Taxi-future-Space-age-pods

Davey, K. (1996). The structure and functions of urban government. *Urban Management: The challenge of growth*. Avebury, UK: Aldershot.

De Bruijn, H., & Veeneman, W. (2009). Decision-making for light rail. *Transportation Research Part A, Policy and Practice, 43*(4), 349–359. doi:10.1016/j.tra.2008.11.003

de Neufville, R., & Odoni, A. (2003). *Airport systems: Planning, design and management*. New York, NY: McGraw-Hill.

Deakin, M., & Allwinkle, S. (2006). The IntelCities community of practice: The e-learning platform, knowledge management system and digital library for semantically-interoperable e-governance services. *International Journal of Knowledge. Culture and Change Management, 6*(3), 155–162.

Deakin, M., & Allwinkle, S. (2007). E-Topia, SUD and ICTs: The post-human nature, embedded intelligence, cyborg-self and agency of digitally-inclusive regeneration platforms. *International Journal of the Humanities, 5*(2), 199–208.

Deakin, M., & Allwinkle, S. (2007). Urban regeneration and sustainable communities: The role of networks, innovation and creativity in building successful partnerships. *Journal of Urban Technology, 14*(1), 77–91. doi:10.1080/10630730701260118

Deakin, M., Van Isacker, K., & Wong, A. (2004). *Review of the IntelCities knowledge capture requirements using a S.W.O.T. analysis*. Edinburgh, Scotland: Napier University.

Deakin, M. (2009). The IntelCities community of practice: The eGov services model for socially inclusive and participatory urban regeneration programs. In Karacapilidis, N. (Ed.), *Web-based learning solutions for communities of practice: Developing virtual environments for social and pedagogical advancement*. Hershey, PA: Information Science Reference. doi:10.4018/978-1-60566-711-9.ch018

Deakin, M. (2010a). A review of city portals: The transformation of service provision under the democratization of the fourth phase. In Reddick, C. (Ed.), *Politics, democracy and e-government: Participation and service delivery*. Hershey, PA: IGI Global. doi:10.4018/978-1-61520-933-0.ch013

Deakin, M. (2010b). SCRAN: The SmartCities (inter) regional academic network supporting the development of a trans-national comparator for the standardisation of egovernment services. In Reddick, C. (Ed.), *Comparative e-government: An examination of e-government across countries*. Berlin, Germany: Springer Press.

Deakin, M. (2007). E-Topia, SUD and ICTs: Taking the digitally-inclusive urban regeneration thesis full circle. *Journal of Urban Technology, 14*(3), 131-139.

Debrezion, G., Pels, E., & Rietveld, P. (2009). Modelling the joint access mode and railway station choice. *Transportation Research Part E, Logistics and Transportation Review, 45*(1), 270–283. doi:10.1016/j.tre.2008.07.001

Derrible, S., & Kennedy, C. (2010). Characterizing metro networks: State, form, and structure. *Transportation, 37*(2), 275–297. doi:10.1007/s11116-009-9227-7

Dewan, S. (2007, March 15). *Google is reviving hopes for ex-furniture makers*. Retrieved from http://www.nytimes.com/ 2007/03/15/us/ 15google.html?_r= 2&0ref=slogin

Dias, L. C. (2003). *Redes: Emergência e organização*. Retrieved from http://www.cfh.ufsc.br/~imprimat /artigos/leila-dias.htm

Díaz, P., Parapar, J., & Touriño, J. (2009). *Web-GIS based system for the management of objections to a comprehensive municipal land use plan*. Research Study of the Department of Agroforestry of the University of Santiago de Compostela, 2009.

Diaz-Luque, P. (2009). Official tourism Web sites and city marketing. In Gascó-Hernández, M., & Torres-Coronas, T. (Eds.), *Information communication technologies and city marketing: Digital opportunities for cities around the world* (pp. 152–183). Hershey, PA: Information Science Publishing. doi:10.4018/978-1-60566-134-6.ch008

Dinnie, K. (2004). Place branding: Overview of an emerging literature. *Place Branding, 1*(1), 106–110. doi:10.1057/palgrave.pb.5990010

Dirks, S., Keeling, M., & Dencik, J. (2009). *How smart is your city?* [Report]. Somers, NY: IBM Institute for Business Value.

Djurdjanovic, D., Liu, J., Marko, K. M., & Ni, J. (2010). Immune systems inspired approach to anomaly detection, fault localization, and diagnosis in automotive engines. In Schumann, J. M., & Liu, Y. (Eds.), *Applications of Neural Networks in High Assurance Systems (Studies in Computational Intelligence 268)* (pp. 141–163). Berlin: Springer. doi:10.1007/978-3-642-10690-3_7

Domanski, R. (1979). Accessibility, efficiency and spatial organization. *Environment & Planning A, 11*, 1189–1206. doi:10.1068/a111189

Ducruet, C. (2007). Incheon, vitrine de la Corée du Sud et masque de Séoul. *M@appemonde, 85*(1).

Duhl, L. J. (1996). An ecohistory of health: The role of 'healthy cities'. *American Journal of Health Promotion, 10*(4), 258–261.

Duhl, L. J., & Sanchez, A. K. (1999). *Healthy cities and the city planning process. A background documentation on links between health and urban planning*. Copenhagen, Denmark: WHO, Regional Office for Europe.

Dur, F., Yigitcanlar, T., & Bunker, J. (2009, March 26). *A decision support system for sustainable urban development: the integrated land use and transportation indexing model*. Paper presented at the Postgraduate Infrastructure Conference, Brisbane, Australia, Queensland University of Technology.

E.C. (2001). *White paper European transport policy for 2010: Time to decide*. Brussels, Belgium: Energy and Transport DG.

E.C. (2005). *Trans-European transport network: TEN-T priority axes and projects 2005*. Brussels, Belgium: Energy and Transport DG.

E.C. (2008). *TEN-T trans-European transport network: Implementation of the priority projects progress report*. Brussels, Belgium: Energy and Transport DG.

ECORYS. (2010). *Competitiveness of regions: What works and what doesn't work?* Retrieved May 16, 2010, from http://www.ecorys.com/ competitiveness/resources/ competitiveness_remi.ppt#1

Eguchi, M., Fredholm, S., Liu, S., Ponce de León Baridó, P., & Ye, J. (2007). *Policy issues in implementing smart cards in urban public transit systems*. Cambridge, MA: Massachusetts Institute of Technology, Technology and Policy Program.

Elgamal, A., Conte, J. P., Masri, S., Fraser, M., Fountain, T., & Gupta, A. … El Zarki, M. (2003). Health monitoring framework for bridges and civil infrastructure. In F. K. Chang (Ed.), *Proceedings of 4th International Workshop on Structural Health Monitoring*, Stanford University, Stanford, CA (pp. 123-130). Lancaster, PA: DEStech Publications.

Ellis, D., Oldridge, R., & Vasconcelos, A. (2004). Community and Virtual Community. *Annual Review of Information Science & Technology*, *38*, 146–186.

ENAC. (2008). *Runway incursions – Avoidance and airport management. Circular APT-30*. Airport Series.

Erke, A., Sagberg, F., & Hagman, R. (2007). Effects of route guidance variable message signs (VMS) on driver behaviour. *Transportation Research Part F: Traffic Psychology and Behaviour*, *10*(6), 447–457. doi:10.1016/j.trf.2007.03.003

ESPON. (2006). *Territory matters for competitiveness and cohesion*. Luxembourg: Ministry of Interior and Spatial Development.

Eurocontrol. (2007). A place to stand: Airports in the European air network. *Eurocontrol Trends in Air Traffic* (vol. 3).

Eurocontrol. (2008). *Navigation application and NAVAID infrastructure strategy for the ECAC area up to 2020*. Retrieved May 2008 from http://www.ecacnav.com/downloads /NAV%20Application%20+%20 NAVAID%20Infrastructure%20 Strategy%2015MAY08%20 Agreed%20at%20SCG-8.pdf

European Commission. (2009). *A sustainable future for transport: Towards an integrated, technology-led and user-friendly system*. Luxembourg: Publications Office of the European Union.

European Commission. (2003). *A study on the factors of regional competitiveness*. Brussels, Belgium: Directorate-General Regional Policy.

European Commission. (2001, September 12). *European transport policy for 2010: Time to decide* [COM(2001) 370]. Bruxelles, EU: Commission of the European Communities.

European Commission. (2005). Community guidelines on financing of airports and start-up aid to airlines departing from regional airports. *Official Journal of the European Union, C 312*(1-C), 312-314.

European Commission. (2008, December 16). *Action plan for the deployment of intelligent transport systems in Europe* [COM(2008) 886]. Bruxelles, EU: Commission of the European Communities.

European Union. (2004). Trans-European transpot network (TEN-T). Priority axes and projects. Retrieved January 14, 2011, from http://ec.europa.eu/ten/ transport/maps/doc/ axes/pp00.pdf

Evans, G. (2009). Creative cities, creative spaces and urban policy. *Urban Studies (Edinburgh, Scotland)*, *46*(5-6). doi:10.1177/0042098009103853

Federal Aviation Administration. (1983). Airport Capacity and Delay. *Advisory Circular*, *150*, 5060–5065.

Federal Transit Administration. (2000). *Benefit assessments of advanced public transportation system technologies*. Washington, DC: Office of Research, Demonstration and Innovation, Office of Mobility Innovation Advanced Public Transportation Systems Division.

Feldman, M., & Audretsch, D. (1999). Innovation in cities: Science-based diversity, specialization and localized competition. *European Economic Review*, *43*(2), 409–429. doi:10.1016/S0014-2921(98)00047-6

Fernald, J. G., & Ramnath, S. (2004). The Acceleration in the US total factor productivity after 1995: The role of Information Technology. *Economic Perspectives*, *28*(1), 52–68.

Fernandez-Cavia, J., & Huertas-Roig, A. (2009). City brands and their communication through Web sites: Identification of problems and proposals for improvement. In Gascó-Hernández, M., & Torres-Coronas, T. (Eds.), *Information communication technologies and city marketing: Digital opportunities for cities around the world* (pp. 26–49). Hershey, PA: Information Science Publishing. doi:10.4018/978-1-60566-134-6.ch002

Feuerle, T., Techt, T., & Altenscheidt, M. (2010, March). *GBAS - Pilots and airlines perspective*. Paper presented at the ICAO GBAS Workshop, Paris, France.

Firmino, R. J., Duarte, F., & Moreira, T. (2008). Pervasive technologies and urban planning in the augmented city. *Journal of Urban Technology*, *15*(2). doi:10.1080/10630730802401983

Firmino, R., Duarte, F., & Ultramari, C. (Eds.). (2010). *ICTs for mobile and ubiquitous urban infrastructures: Surveillance, locative media and global networks*. Hershey, PA: IGI Global. doi:10.4018/978-1-60960-051-8

Fleishman, D., Schweiger, C., Lott, D., & Pierlott, G. (1998). *Multipurpose transit payment media* (TCRP Report 32, pp. 1-132). Washington, DC: National Academy Press.

Flora, C. B., & Flora, J. L. (1993). Entrepreneurial social infrastructure: A necessary ingredient. *The Annals of the American Academy of Political and Social Science*, *529*(1), 48–58. doi:10.1177/0002716293529001005

Florek, M., Insch, A., & Gnoth, J. (2006). City council websites as a means of place brand identity communication. *Place Branding*, *2*(4), 276–296. doi:10.1057/palgrave.pb.6000036

Florida, R. (2002). *The rise of the creative class*. New York, NY: Basic Books.

Florida, R. (2005). *Cities and the creative class*. New York, NY: Routledge.

Florida, R. (2005). *The flight of the creative class: The new global competition for talent*. New York, NY: HarperBusiness.

Florida, R. (2008). *Who's your city?*New York, NY: Basic Books.

Florida, R. L. (2002). *The rise of the creative class: And how it's transforming work, leisure, community and everyday life*. New York, NY: Basic Civitas Books.

Flynn, B. C. (1996). Healthy cities: Toward worldwide health promotion. *Annual Review of Public Health*, *17*(1), 299–309. doi:10.1146/annurev.pu.17.050196.001503

Forgy, C. (1982). Rete: A fast algorithm for the many pattern/many object pattern match problem. *Artificial Intelligence*, *19*(1), 17–37. doi:10.1016/0004-3702(82)90020-0

Fotheringham, A. S. (1981). Spatial structure and distance-decay parameters. *Annals of the Association of American Geographers. Association of American Geographers*, *71*(3), 425–436.

Fotheringham, A. S. (1983). Some theoretical aspects of destination choice and their relevance to production-constrained gravity models. *Environment & Planning A*, (15): 1121–1132. doi:10.1068/a151121

Freilich, R. H. (1998). The land-use implications of transit-oriented development: Controlling the demand side of transportation congestion and urban sprawl. *Urban Lawyer*, *30*(3), 547–572.

Freire, J. R. (2005). Geo-branding-Are we talking nonsense? A theoretical reflection on brands applied to places. *Place Branding*, *1*(4), 347–362. doi:10.1057/palgrave.pb.5990033

Fröidh, O., & Nelldal, B.-L. (2008). Regional high-speed trains on the svealand line: Evaluation of effects. In Bruinsma, F., Pels, E., Rietveld, P., Priemus, H., & van Wee, B. (Eds.), *Railway development: Impacts on urban dynamics* (pp. 295–314). Heidelberg, Germany: Physica-Verlag.

Fryer, C. E. J. (1997). *A history of slipping and slip carriages*. Oxford, UK: The Oakwood Press.

Fuller, B. (2009, October 30). *Iredell one of five counties to form NC data center corridor*. Retrieved from http://www2.mooresville tribune.com

Gardiner, B., Martin, R., & Tyler, P. (2004). Competitiveness, productivity and economic growth across the European regions. *Regional Studies*, *38*(9), 1045–1067. doi:10.1080/0034340042000292638

Garmendia, M., Urena, J. M., Ribalaygua, C., Leal, J., & Coronado, J. M. (2008). Urban residential development in isolated small cities that are partially integrated in metropolitan areas by high speed train. *European Urban and Regional Studies*, *15*(3), 249–264. doi:10.1177/0969776408090415

Garmendia, M., Urena, J. M., Rivas, A., Coronado, J. M., Menendez, M., Gallego, J. I., & Romero, V. (2009). *High speed rail, a new mode of suburban metropolitan transport*. Paper presented at the XV Urban Transport Conference.

Gatrell, A. C. (1991). Concepts of space and geographical data. In Maguire, D. J., Goodchild, M. F., & Rhind, D. W. (Eds.), *Geographical Information Systems* (pp. 119–146). New York, NY: Longman Scientific & Technical.

George, P. (1965). *Cidades, redes urbanas, região. Conferências no Brasil*. Rio de Janeiro, Brasil: Instituto Brasileiro de Geografia FIBGE.

Getis, A. (1989). A spatial causal model of economic interdependency among neighboring communities. *Environment & Planning A*, (21): 115–120. doi:10.1068/a210115

Geurs, K. T., van Wee, B., & Rietveld, P. (2006). Accessibility appraisal of integrated land-use-transport strategies: Methodology and case study for the netherlands randstad area. *Environment and Planning. B, Planning & Design*, *33*(5), 639–660. doi:10.1068/b31099

GfK Sverige AB. (2009). *GfK Roper and Anholt-City brands index*. Retrieved May 3, 2010, from http://www.mynewsdesk.com/ se/view/pressrelease/ gfk-roper-and-anholt-city- brands-index-287066

Ghaemi, P., Swift, J., Sister, C., Wilson, J. P., & Wolch, J. (2009). Design and implementation of a Web-based platform to support interactive environmental planning. *International Journal of Computers. Environment and Urban Systems*, *33*, 482–491. doi:10.1016/j.compenvurbsys.2009.05.002

Ghere, R. K., & Rismiller, L. S. (2001). Information Technology's potential to improve urban neighborhoods: Some citizen planning dilemmas'. *Journal of Urban Technology*, *8*(2). doi:10.1080/106307301316904781

Givoni, M. (2006). Development and impact of the modern high-speed train: A review. *Transport Reviews*, *26*(5), 593–611. doi:10.1080/01441640600589319

Givoni, M., & Rietveld, P. (2007). The access journey to the railway station and its role in passengers' satisfaction with rail travel. *Transport Policy*, *14*(5), 357–365. doi:10.1016/j.tranpol.2007.04.004

Glaeser, E. L. (2000). The new economics of urban and regional growth. In Clark, G. L., Feldman, M. P., & Gertler, M. S. (Eds.), *The Oxford handbook of economic geography*. Oxford, UK: Oxford University Press.

Gold, E. R. (2006). Intellectual architecture as place brand. *Place Branding*, *2*(3), 220–228. doi:10.1057/palgrave.pb.5990059

Golledge, R. G. (1997). *Dynamics and ITS: Behavioural responses to information available from ATIS*. Paper presented at the Conference Pre-Print IATBR '97, Austin, Texas.

Gomez, L., Haeseroets, S., Kuispers, B., & Vaisman, A. (2009). Spatial aggregation: Data model and implementation. *Information Systems*, *34*(6), 551–576. doi:10.1016/j.is.2009.03.002

Google. (n.d.). *Homepage*. Retrieved from www.google.com

Gordon, I. (1999). Internationalisation and urban competition. *Urban Studies (Edinburgh, Scotland)*, *36*(5-6), 1001–1016. doi:10.1080/0042098993321

Gordon, C. (2010). *Density and LRT: The case of Canberra, Australia*. Paper presented at the 89th Transportation Research Board Annual Meeting.

Gould, M. (1996). *Modeling third wave (Virtual) accessibility*. Retrieved from http://www.ncgia.ucsb.edu/ conf/ BALTIMORE/authors /gould/paper.html

Gouveira, C., Fonseca, A., Camara, A., & Ferreira, F. (2004). Promoting the use of environmental data collected by concerned citizens through information and communication technologies. *Journal of Environmental Management*, *71*(2), 135–154. doi:10.1016/j.jenvman.2004.01.009

Graham, S. (2009). *Disrupted cities*. London, UK: Routledge.

Graham, S., & Marvin, S. (1996). *Telecommunications and the city: Electronic spaces, urban places*. London, UK: Routledge. doi:10.4324/9780203430453

Graham, S. (2000). Constructing premium network spaces: Reflections on infrastructure networks and con-temporary urban development. *International Journal of Urban and Regional Research*, *24*(1), 183–200. doi:10.1111/1468-2427.00242

Graham, S., & Marvin, S. (1996). *Telecommunications and the city*. London, UK: Routledge. doi:10.4324/9780203430453

Greenair. (2008). *Air France plans to enter high-speed rail venture with Veolia on routes to London and Amsterdam*. Retrieved July 29, 2010, from http://www.greenaironline.com/ news.php?viewStory=253

Greenbiz Staff. (2009, July 14). *IBM and Cisco to help Amsterdam become a 'smart city'*. Retrieved November 30, 2010, from http://www.greenbiz.com/ news/2009/07/14/ibm-and-cisco-help-amsterdam- become-smart-city

Grodach, C. (2009). Urban branding: An analysis of city homepage imagery. *Journal of Architectural and Planning Research*, *26*(3), 181–197.

Grundy, J., & Boudreau, J. A. (2008). Living with culture: Creative citizenship practices in Toronto. *Citizenship Studies*, *12*(4). doi:10.1080/13621020802184226

Gudes, O., Kendall, E., Yigitcanlar, T., Pathak, V., & Baum, S. (2010). Rethinking health planning: A framework for organising information to underpin collaborative health planning. *Health Information Management Journal*, *39*(2), 18–29.

Gudes, O., Yigitcanlar, T., & Pathak, V. (2009). *A community health support system for the planning of healthy cities*. Paper presented at the Postgraduate Infrastructure Conference, Brisbane, Australia, Queensland University of Technology.

Guihéry, L. (2002). The future of railway transport in europe. In Black, W. R., & Nijkamp, P. (Eds.), *Social change and sustaianble transport* (pp. 241–247). Bloomington, IN: Indiana University Press.

Gutiérrez, J. (2001). Location, economic potential and daily accessibility: An analysis of the accessibility impact of the high-speed line Madrid-Barcelona-French border. *Journal of Transport Geography*, *9*(4), 229–242. doi:10.1016/S0966-6923(01)00017-5

Gutiérrez, J., Condeco-Melhorado, A., & Martín, J. C. (2010). Using accessibility indicators and gis to assess spatial spillovers of transport infrastructure investment. *Journal of Transport Geography*, *18*(1), 141–152. doi:10.1016/j.jtrangeo.2008.12.003

Gutiérrez, J., Monzón, A., & Piñero, J. M. (1998). Accessibility, network efficiency, and transport infrastructure planning. *Environment & Planning A*, *30*(8), 1337–1350. doi:10.1068/a301337

H. M. Treasury/DTI. (2001). *Productivity in the UK: 3 – The regional dimension*. London, UK: H. M. Treasury.

Ha, H. Y., & Perks, H. (2005). Effects of consumer perceptions of brand experience on the Web: Brand familiarity, satisfaction and brand trust. *Journal of Consumer Behaviour*, *4*(6), 438–452. doi:10.1002/cb.29

Haal, P., & Hay, D. (1980). *Growth centers in the European urban system*. London, UK: Heinemann Educational Publishers.

Hadzilacos, T., Kalles, D., Preston, N., Melbourne, P., Camarinopoulos, L., & Eimermacher, M. (2000). UtilNets: A water mains rehabilitation decision- support system. *International Journal of Computers. Environment and Urban Systems*, *24*, 215–232. doi:10.1016/S0198-9715(99)00058-7

Hall, P. (1995). *Towards a general urban theory: Cities in competition*. Melbourne, Australia: Longman.

Hall, P. (1999). The european high-speed train and urban development: Experiences in 14 European regions. *Progress in Human Geography*, *23*(4), 670–671.

Hamel, P., Boudreau, J. A., Keil, R., & Jouve, B. (2006). Arrested metropolitanism: Limits and contradictions of municipal governance reform in Los Angeles, Montreal and Toronto. In Heinelt, H., & Kübler, D. (Eds.), *Metropolitan governance in the 21st century: Governing capacity, democracy and the dynamics of place*. Milton Park, UK: Routledge.

Hamilton, A., Burns, M. C., Revilla, I., Kaklauskas, A., Alejandro, E. M., & Arayici, Y. (2005). *Final Report of IntelCities D5.4c Building Data Integration System*. Manchester, UK: University of Salford.

Hamilton, A., Wang, H., Tanyer, A. M., Arayici, Y., Zhang, X., & Song, Y. (2005). *Integration of building and urban n-dimensional data to support intelligent cities. ITcon 10* (From 3D to nD modelling) (pp. 55–67).

Han, T., & Yang, B. S. (2006). Development of e-maintenance system integrating advanced techniques. *Computers in Industry*, *57*(6), 569–580. doi:10.1016/j.compind.2006.02.009

Handy, S., & Niemeier, D. A. (1997). Measuring accessibility: An exploration of issues and alternatives. *Environment & Planning A*, *29*(7), 1175–1194. doi:10.1068/a291175

Hankinson, G. (2001). Location branding: A study of the branding practices of 12 English cities. *Brand Management*, *9*(2), 127–142. doi:10.1057/palgrave.bm.2540060

Hankinson, G. (2004). Relational network brands: Towards a conceptual model of place brands. *Journal of Vacation Marketing*, *10*(2), 109–121. doi:10.1177/135676670401000202

Hansen, W. G. (1959). How accessibility shapes land-use. *Journal of the American Institute of Planners*, *25*, 73–76. doi:10.1080/01944365908978307

Hanson, S. (1999). Reconceptualizing accessibility. In Janelle, D. G., & Hodge, D. C. (Eds.), *Information, Place, and Cyberspace* (pp. 267–278). Berlin, Germany: Springer-Verlag.

Harloe, M. (2001). Social justice and the city. *International Journal of Urban and Regional Research*, *25*(4), 799–797. doi:10.1111/1468-2427.00350

Harman, R. (2006). *High speed trains and the development and regeneration of cities*. Retrieved January 12, 2010, from http://www.greengauge21.net/

Harrison, J. (2007). From competitive regions to competitive city-regions: A new orthodoxy, but some old mistakes. *Journal of Economic Geography*, *7*(3), 311. doi:10.1093/jeg/lbm005

Haynes, K. E. (1997). Labor markets and regional transportation improvements: The case of high-speed trains - an introduction and review. *The Annals of Regional Science*, *31*(1), 57–76. doi:10.1007/s001680050039

Haywood, R. (2009). *Railways, urban development and town planning in Britain: 1948-2008*. Aldershot, UK: Ashgate.

Healthy Cities Illawarra. (2010). *Working with the community for a healthier, safer, greener and more caring Illawarra*. Wollongong, Australia: Illawarra city council. Retrieved on October 10, 2010, from http://www.healthy-citiesill.org.au /contactus.htm

Heintze, T., & Bretschneider, S. (2000). Information Technology and restructuring in public organizations: Does adoption of Information Technology affect organizational structures, communications, and decision making? *Journal of Public Administration: Research and Theory*, *10*(4), 801–830.

Helft, M. (2010, March 22). *Hoping for gift from Google? Go jump in the lake*. Retrieved from http://www.nytimes.com/2010 /03/22/technology/ 22stunts.html

(1993). Introduction: Changing forms of state power in South East Asia. InHewison, K., Rodan, G., & Robison, R. (Eds.), *Southeast Asia in the 1990s: Authoritarianism, democracy and capitalism* (pp. 2–8). Sydney, Australia: Allen and Unwin.

Hickman, M. D., & Wilson, N. H. M. (1995). Passenger travel time and path choice implications of real-time transit information. *Transportation Research*, *3*(4), 211–226. doi:10.1016/0968-090X(95)00007-6

Ho, K. (2001). Hong Kong – Moving into the 21st century. In Gar-On Yeh, A., Hills, P. R., & Ka-Wing Ng, S. (Eds.), *Modern transport in Hong Kong for the 21st century* (p. 283). Hong Kong: Centre of Urban Planning and Environmental Management, University of Hong Kong.

Hoffman-Wellenhof, B., Lichtenegger, H., & Collins, J. (1997). *Global positioning system: Theory and practice*. New York, NY: Springer-Verlag.

Hoogma, R., Kemp, R., Schot, J., & Truffer, B. (2002). *Experimenting for sustainable transport. The approach of strategic niche management*. London: Spon Press.

Huang, B. (2003). Web-based dynamic and interactive environmental visualization. *International Journal of Computers. Environment and Urban Systems*, *27*, 623–636. doi:10.1016/S0198-9715(02)00063-7

Huber, G. (1990). A theory of the effects of advanced Information Technologies on organisational design, intelligence, and decision-making. *Academy of Management Review*, *15*(1), 47–71.

Huerta-Carvajal, M. I., & Luna-Reyes, L. F. (2009). City boosterism through Internet marketing: An institutional perspective. In Gascó-Hernández, M., & Torres-Coronas, T. (Eds.), *Information communication technologies and city marketing: Digital opportunities for cities around the world* (pp. 130–151). Hershey, PA: Information Science Publishing. doi:10.4018/978-1-60566-134-6.ch007

IBM. (2006). *Santa Barbara sheriff's department wields innovation to fight fraud and inefficiency*. Retrieved November 30, 2010, from ftp://ftp.software.ibm.com/ software/solutions/pdfs/ ODB-0141-00A_ITM.pdf

IBM. (2008). *Unlock opportunities for innovation while addressing security risks in air travel*. Retrieved November 30, 2010, from IBM http://www.ibm.com

IBM. (2009). *New strides in mobility signal opportunity calling*. Retrieved November 30, 2010, from http://www-304.ibm.com/ businesscenter/cpe/download0/ 175785/ FV0905_Mobility.pdf

IBM. (2010a, October 10). *IBM chosen by EnergyAustralia for smart grid, smart city initiative*. Retrieved November 30, 2010, from http://www.tmcnet.com/ usubmit/2010/10/08/ 5057234.htm

IBM. (2010b). *IBM smarter public safety solutions*. Retrieved November 30, 2010, from http://www.ibm.com

IBM. (2010c). *Optimize your software with IBM Cognos 8 business intelligence*. Retrieved November 30, 2010, from http://www.ibm.com/ cognos

ICAO. (2004). *Advanced surface movement guidance and control systems (A-SMGCS) manual (DOC 9830)*. Montreal, Canada: ICAO.

ICAO. (2005a). *Comparison for SSR Mode S system* (annex 10, amendment 73-77, ed. 4). Montreal, Canada: ICAO.

ICAO. (2005b). *Communications systems & Surveillance radar and collision avoidance systems* (annex 10, amendment 77, vol. 3-4). Montreal, Canada: ICAO.

IFEZ. (2010). *Business hub of Asia–Incheon free economic zone*. Incheon, South Korea: IFEZ Public Relations Office.

Inn, K. (2004). Plan for city identity establishment and city marketing the case of Kimpo city. *Dela*, *21*, 233–240.

In-Stat. (2006). *RFID tags and chips: Opportunities in the second generation*. Retrieved November 30, 2010, from http://www.instat.com/ abstract.asp?id=3& SKU=IN0502115WT

Iseki, H., Yoh, A. C., & Taylor, B. D. (2006). *Survey on status of knowledge and interest of smartcard fare collection systems among US transit agencies* (pp. 1–43). Los Angeles, CA: California PATH Research Report, University of California, Los Angeles.

Ishida, T. (2000). Understanding digital cities. *Digital Cities,* 7-17.

Israel, B. A., Schulz, A. J., Parker, E., & Becker, A. B. (2001). Community-based participatory research: Policy recommendations for promoting a partnership approach in health research. *Education for Health*, *14*(2), 182–197. doi:10.1080/13576280110051055

ISTAT. (2010). *Principali indicatori demografici della popolazione residente*. Retrieved March 26, 2010, from http://demo.istat.it/

ITDP. (2007). *Bus rapid transit planning guide*. New York, NY: Institute for Transportation & Development Policy.

Jackson, S. (2008). The city from thirty thousand feet: Embodiment, creativity, and the use of geographic information systems as urban planning tools. *Technology and Culture*, *49*(2), 325–346. doi:10.1353/tech.0.0039

Jakubauskas, G. (2006). Improvement of urban passenger transport ticketing systems by deploying intelligent transport systems. *Transport*, *21*(4), 252–259.

Jardak, C., Rerkrai, K., Kovacevic, A., Riihijarvi, J., & Mahonen, P. (2009). *Email from the vineyard.* Paper presented at 5th International Conference on Testbeds and Research Infrastructures for the Development of Networks & Communities and Workshops (TridentCom), Washington, DC.

Jeon, N. J., Leem, C. S., Kim, M. H., & Shin, H. G. (2007). A taxonomy of ubiquitous computing applications. *Wireless Personal Communications*, *43*(4), 1229–1239. doi:10.1007/s11277-007-9297-9

Jessop, B. (2002). Liberalism, neoliberalism and urban governance: A state-theoretical perspective. *Antipode*, *34*(3). doi:10.1111/1467-8330.00250

Jiang, B., Claramunt, C., & Batty, M. (1999). Geometric accessibility and geographic information: Extending desktop GIS to space syntax. *Computers, Environment and Urban Systems*, *23*, 127–146. doi:10.1016/S0198-9715(99)00017-4

Johansson, B., Karlson, C., & Stough, R. (Eds.). (2006). *The emerging digital economy: Entrepreneurship, clusters, and policy*. Berlin, Germany: Springer-Verlag.

Johnson, C. (2001). A survey of current research on online communities of practice. *The Internet and Higher Education*, *4*(1), 45–60. doi:10.1016/S1096-7516(01)00047-1

Johnson, J. (2002). A conceptual model for enhancing community competitiveness in the new economy. *Urban Affairs Review*, *37*(6), 763–779. doi:10.1177/107874037006001

Joly, O. (1999). Geographical position: *State of French art of spatial accessibility indicators* [Working paper]. France: SPESD (Study Program on European Spatial Development) working group 1.1. Retrieved from www.nordregio.se.spespn/files/1.1annex5

Josefsson, U. (2005). Coping with illness online: The case of patients online communities. *The Information Society*, *21*(2), 143–153. doi:10.1080/01972240590925357

Kang, J. M., Shekhar, S., Wennen, C., & Novak, P. J. (2008). Discovering flow anomalies: A SWEET approach. In *Proceedings of the 8th International Conference on Data Mining (ICDM),* Pisa, Italy (pp. 851-856).

Kanter, M. (1995). *World class: Thriving locally in the global economy*. New York, NY: Simon and Schuster.

Karaki, J. N., & Kamal, A. E. (2006). A taxonomy of routing techniques in wireless sensor networks. In Ilyas, M., & Mahgoub, I. (Eds.), *Handbook of Sensor Network Protocols: Compact wireless and wired sensing systems* (pp. 1–24). Boca Raton, FL: CRC Press.

Kasturirangan, K., & Rao, P. (1997). *Urban planning and management: Space technology options-Computers in urban planning and urban management*. Bombay, India: Narosa.

Katsikas, S., Lopez, J., & Pernul, G. (2008). The challenge for security and privacy services in distributed health settings. *Studies in Health Technology and Informatics*, *134*, 113–125.

Kauchakje, S., Penna Neto, M. C., Frey, K., & Duarte, F. (2006). Redes socio-técnicas y participación ciudadana: Propuestas conceptuales y analíticas para el uso de las TICs. *REDES. Revista Hispana para el Análisis de Redes Sociales*, *11*, 1–26.

Kavaratzis, M. (2004). From city marketing to city branding: Towards a theoretical framework for developing city brands. *Place Branding*, *1*(1), 58–73. doi:10.1057/palgrave.pb.5990005

Kavaratzis, M., & Ashworth, G. J. (2005). City branding: An effective assertion of identity or a transitory marketing trick? *Tijdschrift voor Economische en Sociale Geografie*, *96*(5), 506–514. doi:10.1111/j.1467-9663.2005.00482.x

Keil, R. (2002). Commonsense neoliberalism: Progressive conservative urbanism in Toronto, Canada. *Antipode*, *34*(3). doi:10.1111/1467-8330.00255

Kelly, T., Gray, V., & Minges, M. (2003). *Broadband Korea: Internet case study*. Seoul, South Korea: International Telecommunication Union.

Kemp, R., Schot, J., & Hoogma, R. (1998). Regime shifts to sustainability through processes of niche formation: The approach of strategic niche management. *Technology Analysis and Strategic Management*, *10*(2), 175–198. doi:10.1080/09537329808524310

Kenney, M., & Patton, D. (2006). The co-evolution of technologies and institutions: Silicon Valley as the iconic high-technology cluster. In Braunerhjelm, P., & Feldman, M. P. (Eds.), *Cluster genesis: Technology-based industrial development*. New York, NY: Oxford University Press.

Kerr, G. (2006). From destination brand to location brand. *Brand Management*, *13*(4-5), 276–283. doi:10.1057/palgrave.bm.2540271

Kerr, G., & Johnson, S. (2005). A review of a brand management strategy for a small town: Lessons learnt! *Place Branding*, *1*(4), 373–387. doi:10.1057/palgrave.pb.5990035

Kim, T. J. (2008). Planning for knowledge cities in ubiquitous technology spaces: Opportunities and challenges. In Yigitcanlar, T., Velibeyoglu, K., & Baum, S. (Eds.), *Creative urban regions: Harnessing urban technologies to support knowledge city initiatives*. Hershey, PA: IGI Global. doi:10.4018/978-1-59904-838-3.ch013

Kim, Y. I., Pyen, M. W., & Eo, Y. D. (2000). Development of hypermap database for ITS and GIS. *International Journal of Computers Environment and Urban systems, 24*, 45-60.

King, J., & Kraemer, K. (1995). *Information infrastructure national policy*. Irvine, CA: Center for research on Information Technology and organizations, University of California (UCI).

Kipfer, S., & Keil, R. (2002). Toronto Inc.? Planning the competitive city in the new Toronto. *Antipode: A Radical. The Journal of Geography*, *34*(2), 227–264.

Kitson, M., Martin, R., & Tyler, P. (2004). Regional competitiveness: An elusive yet key concept? *Regional Studies*, *38*(9), 991–999. doi:10.1080/0034340042000320816

Kluge, E. H. (2007). Secure e-health: Managing risks to patient health data. *International Journal of Medical Informatics*, *76*(5-6), 402–406. doi:10.1016/j.ijmedinf.2006.09.003

Knapp, W., Kunzmann, K. R., & Schmitt, P. (2004). A cooperative spatial future for Rheinruhr. *European Planning Studies*, *12*(3), 323–349. doi:10.1080/0965431042000195047

Kohavi, R., Henne, R. M., & Sommerfield, D. (2007). Practical guide to controlled experiments on the Web: Listen to your customers not to the HIPPO. In *Proceedings of the 13th ACM SIGKDD international conference on knowledge discovery and data mining* (pp. 959-967). New York: ACM.

Kolbe, R. H., & Burnett, M. S. (1991). Content-analysis research: An examination of applications with directives for improving research reliability and objectivity. *The Journal of Consumer Research*, *18*(2), 243–250. doi:10.1086/209256

Kolko, J. (1999, November). *The death of cities? The Death of distance? Evidence from the geography of commercial Internet usage*. Paper presented at 27th Annual Telecommunications Policy Research Conference 1999, Alexandria, VA.

Komninos, N. (2002). *Intelligent cities: Innovation, knowledge systems, and digital spaces*. London, UK: Spon Press.

Komninos, N. (2006). The architecture of intelligent cities. *Intelligent Environments*, *6*, 53–61.

Kotler, P., Haider, D., & Rein, I. (1993). *Marketing places*. New York, NY: The Free Press.

Kotler, P. (2000). *Marketing management: The millennium edition*. Upper Saddle River, NJ: Prentice-Hall.

Krugman, P. (1979). Increasing returns, monopolistic competition, and international trade. *Journal of International Economics*, *9*(4), 469–479. doi:10.1016/0022-1996(79)90017-5

Krugman, P. (1994). Competitiveness: A dangerous obsession. *Foreign Affairs (Council on Foreign Relations)*, *73*(2), 28. doi:10.2307/20045917

Krugman, P. (1997). *Pop internationalism*. Cambridge, MA: MIT Press.

Kuby, M., Barranda, A., & Upchurch, C. (2004). Factors influencing light-rail station boardings in the United States. *Transportation Research Part A, Policy and Practice*, *38*(3), 223–247. doi:10.1016/j.tra.2003.10.006

Kurkjian, C., & Glass, J. (2007, February). Meeting the needs of 24/7 data centers. *ASHRAE Journal*, *49*(2), 24–35.

Kwan, M. P., & Weber, J. (2008). Scale and accessibility: Implications for the analysis of land use-travel interaction. *Applied Geography (Sevenoaks, England)*, *28*(2), 110–123. doi:10.1016/j.apgeog.2007.07.002

Kwok, A. (2008). *City Hall and GIS: Government GIS, local government, technology*. Retrieved October 2010, from http://www.futuregov.asia/ articles/2008/oct/10/city-hall-and-gis/

Laguerre, M. S. (2005). *The digital city. The American metropolis and Information Technology*. New York, NY: Palgrave MacMillan.

Laird, J. J., Nellthorp, J., & Mackie, P. J. (2005). Network effects and total economic impact in transport appraisal. *Transport Policy*, *12*(6), 537–544. doi:10.1016/j.tranpol.2005.07.003

Landry, C. (2000). *The creative city: A toolkit for urban innovators*. London, UK: Earthscan.

Lane, B. W. (2008). Significant characteristics of the urban rail renaissance in the United States: A discriminant analysis. *Transportation Research Part A, Policy and Practice*, *42*(2), 279–295. doi:10.1016/j.tra.2007.10.001

Larner, W. (2000). Neo-liberalism: Policy, ideology, governmentality. *Studies in Political Economy*, *63*, 5–26.

Latour, B. (1993). *Aramis, ou l'amour des techniques*. Paris, France: Editions de la Découverte.

Lee, W., Tseng, S., & Tsai, S. (2009). A knowledge based real-time travel time prediction system for urban network. *Expert Systems with Applications*, *36*(3), 4239–4247. doi:10.1016/j.eswa.2008.03.018

Lee, S. H., Yigitcanlar, T., Han, J. H., & Leem, Y. T. (2008). Ubiquitous urban infrastructure: Infrastructure planning and development in Korea. *Innovation: Management. Policy & Practice*, *10*(2-3), 282–292. doi:10.5172/impp.453.10.2-3.282

Leem, Y. T., Lee, S., Han, J. H., & Lee, J. (2007). *A management of GIS DB for urban infrastructure for multi-functional administrative city*. 10th International Conference on Computers in Urban Planning and Urban Management. Retrieved March 10th, 2011, from http://www.scribd.com/doc/ 7315199/a-Management-of-Gis-Db-for-Urban- Infrastructure-for-Multifunctional -Administrative-City

Lefebvre, H. (1991). *The production of space*. Cambridge, MA: Blackwell.

Lengyel, I. (2003). The pyramid model: Enhancing regional competitiveness in Hungary. *Acta Oeconomica*, *54*(3), 323–242. doi:10.1556/AOecon.54.2004.3.3

Lengyel, I., & Lucovics, M. (2006). An attempt for the measuring of regional competitiveness in Hungary. In *Proceedings of the 46th Congress of Regional Science*, Volos, Greece.

Leonardi, G. (1978). Optimum facility location by accessibility maximizing. *Environment & Planning A*, *10*, 1287–1305. doi:10.1068/a101287

Lever, W. F., & Turok, I. (1999). Competitive cities: Introduction to the review. *Urban Studies (Edinburgh, Scotland)*, *36*(5-6), 791–793. doi:10.1080/0042098993213

Lewis, N. M., & Donald, B. (2010). A new rubric for Canada's creative cities. *Urban Studies (Edinburgh, Scotland)*, *47*(1). doi:10.1177/0042098009346867

Liao, C., To, P. L., & Shih, M. L. (2006). Website practices: A comparison between the top 1000 companies in the US and Taiwan. *International Journal of Information Management*, *26*(3), 196–211. doi:10.1016/j.ijinfomgt.2006.02.007

Lindsay, G. (2010, February 1). *Cisco's big bet on New Songdo: Creating cities from scratch*. Retrieved January 17, 2011, from http://www.fastcompany.com/ magazine/142/ the-new-new-urbanism.html

Linnamaa, R. (1999). Competitive advantage and leadership of urban regions in information society, Suomen Kuntaliitto, Helsinki, Finland, Acta-sarja 106.

Linneker, B., & Spence, N. (1996). Road transport infrastructure and regional economic development: The regional development effects of the m25 London Orbital Motorway. *Journal of Transport Geography*, *4*(2), 77–92. doi:10.1016/0966-6923(96)00001-4

Linstone, H., & Turoff, M. (2002). The Delphi method. In M. Turoff, & H. Linstone (Eds.), *Techniques and applications*. Retreived November 26, 2009 from http://www.is.njit.edu/ pubs/delphibook/

Litman, T. (2007). Evaluating rail transit benefits: A comment. *Transport Policy*, *14*(1), 94–97. doi:10.1016/j.tranpol.2006.09.003

Logan, J., & Molotch, H. (1987). *Urban fortunes: The political economy of place*. Berkeley, CA: University of California Press.

Lombardi, P., & Curwell, S. (2005). INTELCITY Scenarios for the city of the future. In Miller, D., & Patassini, D. (Eds.), *Beyond benefit cost analysis*. Aldershot, UK: Ashgate.

Lombardi, P., Cooper, I., Paskaleva, K., & Deakin, M. (2009). The challenge of designing user-centric e-services: European dimensions. In Riddeck, C. (Ed.), *Research strategies for e-government service adoption*. Hershey, PA: Idea Group Publishing.

López, E., Gutiérrez, J., & Gómez, G. (2008). Measuring regional cohesion effects of large-scale transport infrastructure investments: An accessibility approach. *European Planning Studies*, *16*(2), 277–301. doi:10.1080/09654310701814629

López, E., Monzon, A., Ortega, E., & Mancebo, S. (2008). *Strategic assessment of transport infrastructure plans on European integration: Application for the Spanish strategic transport and infrastructure plan 2005-2020*. Paper presented at the 87th Transportation Research Board Annual Meeting.

López, E., Monzón, A., Ortega, E., & Mancebo, S. (2009). Assessment of cross-border spillover effects of national transport infrastructure plans: An accessibility approach. *Transport Reviews: A Transnational Transdisciplinary Journal, 29*(4), 515 - 536.

Lourenco, J., Alves, E., & Mendes, J. F. G. (1998). *Evaluating urban quality of life in Portugal*. Braga, Portugal: University of Minho.

Lyon, D. (2002). Everyday surveillance: Personal data and social classifications. *Information Communication and Society*, *5*(2), 242–257. doi:10.1080/13691180210130806

Lyon, D. (2007). *Surveillance studies*. Cambridge, UK: Polity Press.

Machado, L. O. (1992). Sociedade urbana, inovação tecnológica e nova geopolítica. Boletim de Geografia Teorética, AGETEO. *Rio Claro*, *22*(43), 398–403.

Maeng, D., & Nedovic-Budic, Z. (2010). Relationship between ICT and urban form in knowledge-based development: Empirical analysis of Washington, DC metro region. *International Journal of Knowledge-Based Development*, *1*(1-2), 97–117. doi:10.1504/IJKBD.2010.032588

Maggi, R., Masser, I., & Nijkamp, P. (1992). Missing networks in european transport and communications. *Transport Reviews: A Transnational Transdisciplinary Journal, 12*(4), 311 - 321.

Mahmoud, R., Halfawy, M. R., Pyzoha, D., & El-Hosseiny, T. (2002). *An integrated framework for GIS-based civil infrastructure management systems*. Retrieved March 11th, 2011, from http://pedago.cegepoutaouais. qc.ca/ media/0260309 /0378334/ SCGC-BON/ Documents/ GE012-Halfawy% 20et%20al.pdf.

Mahon, R., & Keil, R. (2006). *The political economy of scale: An introduction* (manuscript).

Mahoney, R. P. (1991). GIS and utilities. In Maguire, D. J., Goodchild, M. F., & Rhind, D. W. (Eds.), *Geographical Information Systems* (pp. 101–114). New York, NY: Longman Scientific & Technical.

Malecki, E. (2002). Hard and soft networks of urban competitiveness. *Urban Studies (Edinburgh, Scotland)*, *29*(5/6), 929–946. doi:10.1080/00420980220128381

Malecki, E. J. (2007). Cities and regions competing in the global economy: Knowledge and local development policies. *Environment and Planning. C, Government & Policy*, *25*(5), 638–654. doi:10.1068/c0645

Malecki, E. J. (2010). Everywhere? The geography of knowledge. *Journal of Regional Science*, *50*(1), 493–513. doi:10.1111/j.1467-9787.2009.00640.x

Malmberg, A., & Maskell, P. (2006). Localized learning revisited. *Growth and Change*, *37*(1), 1–18. doi:10.1111/j.1468-2257.2006.00302.x

Markusen, A. (2003). Fuzzy concepts, scanty evidence, policy distance: The case for rigour and policy relevance in critical regional studies. *Regional Studies*, *37*(6), 701–717. doi:10.1080/0034340032000108796

Marshall, A. (1920). *Principles of economics* (revised edition). London, UK: Macmillan.

Marshall, S. (2001). The challenge of sustainable transport. In Layard, A., Davoudi, S., & Batty, S. (Eds.), *Planning for a Sustainable Future* (pp. 131–147). London, UK: Spon.

Martin, F. (1997). Justifying a high-speed rail project: Social value vs. Regional growth. *The Annals of Regional Science*, *31*(2), 155–174. doi:10.1007/s001680050043

Martin, R., & Sunley, P. (2003). Deconstructing clusters: Chaotic concept or policy panacea? *Journal of Economic Geography*, *3*(1), 5. doi:10.1093/jeg/3.1.5

Martín, J. C., Gutiérrez, J., & Román, C. (2004). Data envelopment analysis (dea) index to measure the accessibility impacts of new infrastructure investments: The case of the high-speed train corridor Madrid-Barcelona-French border. *Regional Studies*, *38*(6), 697–712. doi:10.1080/003434042000240987

Martín, J. C., & Reggiani, A. (2007). Recent methodological developments to measure spatial interaction: Synthetic accessibility indices applied to high-speed train investments. *Transport Reviews*, *27*(5), 551–571. doi:10.1080/01441640701322610

Maskell, P., Eskelinen, H., Hannibalsson, I., Malmberg, A., & Vatne, E. (1998). *Competitiveness, locational learnings and regional development: Specialisation and prosperity in small open economies*. London, UK: Routledge. doi:10.4324/9780203276921

Maskell, P., & Malmberg, A. (1999). The competitiveness of firms and regions: 'Ubiquitification' and the importance of localized learning. *European Urban and Regional Studies*, *6*(1), 9–25. doi:10.1177/096977649900600102

Maskell, P., & Malmberg, A. (2007). Myopia, knowledge development and cluster evolution. *Journal of Economic Geography*, *7*(5), 603–618. doi:10.1093/jeg/lbm020

Mayer, M. (2006, May 17). MS&E 472 course: Entrepreneurial thought leaders seminar series. *ETL Seminar Series*. Stanford, CA: Stanford University.

Mayes, K. (2008). *Smart cards, tokens, security and applications*. New York, NY: Springer. doi:10.1007/978-0-387-72198-9

Maynard, M., & Tian, Y. (2004). Between global and glocal content analysis of the Chinese Web sites of the 100 top global brands. *Public Relations Review*, *30*(3), 285–291. doi:10.1016/j.pubrev.2004.04.003

McCall, M. (1988). The implications of Eastern African rural social structure for local-level development: The case for participatory development based on indigenous knowledge systems. *Regional Development Dialogue*, *9*(2), 41–69.

McCall, M. (2003). Seeking good governance in participatory-GIS: A review of processes and governance dimensions in applying GIS to participatory spatial planning. *Habitat International*, *27*(4), 549–573. doi:10.1016/S0197-3975(03)00005-5

McCann, E. (2007). Inequality and politics in the creative city-region: Questions of livability and strategy. *International Journal of Urban and Regional Research*, *31*(1). doi:10.1111/j.1468-2427.2007.00713.x

Menerault, P. (1998). Processus de territorialisation des réseaux: Analyse de la grande vitesse ferroviaire à l'échelle régionale. *Networks and communication studies*, *12*(1-3), 161-184.

Menzies, T., & Hu, Y. (2003). Data mining for very busy people. *Computer*, *36*(11), 22–29. doi:10.1109/MC.2003.1244531

Meyer, D., & Miller, E. J. (2001). *Urban transportation planning* (2nd ed.). New York, NY: McGraw-Hill.

Michele, D. S., & Daniela, L. (2011). Decision-support tools for municipal infrastructure maintenance management. *Procedia Computer Science*, *3*, 36–41. doi:10.1016/j.procs.2010.12.007

Migliorini, F. (2007). *Un corridoio tutto da inventare: L'alta velocità per far crescere città e distretti*. Venezia, Italy: Marsilio.

Miller, H. J., & Wu, Y. (2005). Gis software for measuring space-time accessibility in transportation planning and analysis. *GeoInformatica*, *4*(2), 141–159. doi:10.1023/A:1009820006075

Miller, R. (2007a, April 4). *Google announces $600M SC project. Data Center Knowledge*. Retrieved from http://www.datacenter knowledge.com/ archives/2007/04/04/google-announces- 600m-so-car-project/

Miller, R. (2007b, June 19). *Google confirms data center in Council Bluffs, Iowa. Data Center Knowledge*. Retrieved from http://www.datacenter knowledge.com/archives/2007/06/19/ google-confirms-data-center -in-council-buffs-iowa/

Miller, R. (2008, March 27). *Google Data Center FAQ. Data Center Knowledge*. Retrieved from http://www.datacenterknowledge.com/archives/2008/03/27/google-data-center-faq/

Miller, R. (2010, March 17). *Google won't claim SC incentives. Data Center Knowledge*. Retrieved from http://www.datacenter knowledge.com/archives /2010/03/17/google-wont-claim-s-c-incentives/

Miller, R. (October 9, 2009). *When incentives don't work. Data Center Knowledge*. Retrieved from http://www.datacenter knowledge.com/archives /2009/10/09/when-incentives-dont-work-out/

Minkler, M. (2000). Using participatory action research to build healthy communities. *Public Health Reports*, *115*(2-3), 191–197. doi:10.1093/phr/115.2.191

Minkler, M., & Wallerstein, N. (2003). Introduction to community based participatory research. In M. Minkler, & N. Wallerstein, N (Eds.), *Community based participatory research for health* (pp. 3-26). San Francisco, CA: Jossey-Bass.

Mishra, B., & Rolland, E. (2009). A strategic framework for city marketing: The SSRM approach. In Gascó-Hernández, M., & Torres-Coronas, T. (Eds.), *Information communication technologies and city marketing: Digital opportunities for cities around the world* (pp. 50–63). Hershey, PA: Information Science Publishing. doi:10.4018/978-1-60566-134-6.ch003

Mitchell, W. J. (2005). *Placing words symbols, space, and the city*. Cambridge, MA: The MIT Press.

Mitchell, W. J., Borroni-Bird, C. E., & Burns, L. D. (2010). *Reinventing the automobile: Personal urban mobility for the 21st century*. Cambridge, MA: The Mit Press.

Mitchell, W. (1995). *City of bits: Space, place, and the Infobahn*. Cambridge, MA: MIT Press.

Mitchell, W. (2000). *E-Topia: Urban life, Jim-But not as you know it*. Cambridge, MA: MIT Press.

Miyamoto, K., Sugiki, N., Uchida, T., & Paez, A. (1997). A GIS based land-use model dealing with building types by small unit of land in a metropolitan area. *Journal of the Eastern Asia Society for Transportation Studies*, *2*, 1943–1959.

Mooney, G., & Fohtung, N. G. (2008). Issues in the measurement of social determinants of health. *Health Information Management Journal*, *37*(3), 26–30.

Moss, M. L., & Townsend, A. M. (1997, November). *The role of the real city in cyberspace: Understanding regional variations in Internet accessibility and utilisation*. Unpublished manuscript, Taub Urban Research Centre, New York University, NY.

Moutinho, J. (2006). From technological determinism to regional development: Lessons learned from the Portuguese digital cities program (1998-2006). In Cunningham, P., & Cunningham, M. (Eds.), *Exploiting the knowledge economy: Issues, applications, case studies*. Amsterdam, The Netherlands: IOS Press.

Müller, B., Flores, L., Agrebi, M., & Chandon, J. L. (2008). The branding impact of brand websites: Do newsletters and consumer magazines have a moderating role? *Journal of Advertising Research*, *48*(3), 465–472. doi:10.2501/S0021849908080471

Muñiz-Martínez, N., & Cervantes-Blanco, M. (2009). Identity and marketing of cities. In Gascó-Hernández, M., & Torres-Coronas, T. (Eds.), *Information communication technologies and city marketing: Digital opportunities for cities around the world* (pp. 1–25). Hershey, PA: Information Science Publishing. doi:10.4018/978-1-60566-134-6.ch001

Munnell, A. H. (1992). Policy watch infrastructure investment and economic growth. *The Journal of Economic Perspectives*, *6*(4), 189.

Murakami Wood, D., & Firmino, R. (2010). Empowerment or repression? Opening up questions of identification and surveillance in Brazil through a case of identity fraud. *Identity in the Information Society*, *2*(3), 297–317. doi:10.1007/s12394-010-0038-y

Murray, A., & Wu, X. (2003). Accessibility tradeoffs in public transit planning. *Journal of Geographical Systems*, *5*(1), 93–107. doi:10.1007/s101090300105

Nelson, A. C., & Niles, J. (1999). *Essentials for transit-oriented development planning: Analysis of non-work activity patterns and a method for predicting success*. Paper presented at the 7th Transportation Research Board Conference on the Application of Transportation Planning Methods.

Neo, I. T. (2006). Offshore Insight. *Market Report Series*, *4*(8), 4–14.

Nijkamp, P., & Van Geenhuizen, M. (1997). European transport: Challenges and opportunities for future research and policies. *Journal of Transport Geography*, *5*(1), 4–11. doi:10.1016/S0966-6923(96)00041-5

Nikolaus, F., Keinz, P., & Schreier, M. (2008). Complementing mass customization toolkits with user communities: How peer input improves customer self-design. *Journal of Product Innovation Management*, *25*(6), 546–559. doi:10.1111/j.1540-5885.2008.00321.x

Niu, P. (2010). *MEMSIC adds structural health monitoring capabilities to wireless sensor platform*. Retrieved from http://investor.memsic.com/ releasedetail.cfm?ReleaseID=462923

Nonaka, I., & Takeuchi, H. (1995). *The knowledge-creating company*. Oxford, UK: Oxford University Press.

Noronha, V. T., & Goodchild, M. F. (2005). Modeling inter-regional interaction: Implications for functional regions. *Annals of the Association of American Geographers. Association of American Geographers*, *82*(1), 86–102. doi:10.1111/j.1467-8306.1992.tb01899.x

Northridge, M. E., Sclar, E. D., & Biswas, P. (2003). Sorting out the connections between the built environment and health: A conceptual framework for navigating pathways and planning healthy cities. *Journal of Urban Health*, *80*(4), 556–568. doi:10.1093/jurban/jtg064

Ntu-Petry, R., & Cruz, D. (1999). *Sistemas automatizados de arrecadação tarifária para o transporte urbano*. Retrieved from http://www.ntu.org.br/ frame_publicacoes.htm

OECD. (2006). *Competitive cities in the global economy*. Paris, France: OECD.

Ogden, K. (Ed.). (2003). World Congress on intelligent transport systems. *Transportation Research Part C, Emerging Technologies*, *11*(5), 329–342. doi:10.1016/S0968-090X(03)00028-7

Okazaki, S., & Škapa, R. (2009). Understanding brand website positioning in the new EU member States: The case of the Czech Republic. In Bandyopadhyay, S. (Ed.), *Contemporary Research in e-Branding* (pp. 193–215). Hershey, PA: Information Science Publishing.

OpenCity Projects Inc. (2007). *Town of Markham* (case by W. Gold, & J. Rutgers). Retrieved on November 29, 2010, from http://www.opencityprojects.com

OpenStreetMap. (2011). *Street map of North-Eastern Italy*. Retrieved January 14, 2011, from http://www.openstreetmap.org/ ?lat=45.89&lon=12.25& zoom=8&layers=M

Orr, J. (1996). *Talking about machines: An ethnography of a modern job*. New York, NY: IRL Press.

Ortuzar, J. de D., & Willumsen, L. G. (2001). *Modelling transport* (3rd ed.). Hoboken, NJ: John Wiley and Sons.

Overman, H. (2010). GIS a job: What use geographical Information Systems in spatial economics. *Journal of Regional Science*, *50*(1), 165–180. doi:10.1111/j.1467-9787.2009.00644.x

Page, M., Kelly, C., May, A., Jones, P., & Forrester, J. (2009). Enhancing appraisal methods to support sustainable transport and land use policies. *European Journal of Transport and Infrastructure Research*, *9*(3), 296–313.

Pandia. (2007, July 2) Google: One million servers and counting. *Pandia Search Engine News*. Retrieved from http://www.pandia.com

Parkerson, B., & Saunders, J. (2005). City branding: Can goods and services branding models be used to brand cities? *Place Branding*, *1*(3), 242–264. doi:10.1057/palgrave.pb.5990026

Patriksson, M. (1994). *The traffic assignment problem: Models and methods*. Utrecht, The Netherlands: VSP.

Paulsson, B. (1992). Urban applications of satellite remote sensing and GIS analysis, World Bank/ UNCHS/ UNDP. *Urban Management Program, DP*(9), 1-40.

Pazour, J. A., Meller, R. D., & Pohl, L. M. (2010). A model to design a national high-speed rail network for freight distribution. *Transportation Research Part A, Policy and Practice*, *44*(3), 119–135. doi:10.1016/j.tra.2009.11.006

Peck, J. (1998). Geographies of governance: TECs and the neoliberalisation of local interests. *Space and Polity*, *2*(1). doi:10.1080/13562579808721768

Peck, J. (2007). Banal urbanism: Cities and the creativity fix. *Monu*, *7*, 35–47.

Peck, J., & Tickell, A. (2002). Neoliberalizing space. *Antipode*, *34*(3). doi:10.1111/1467-8330.00247

Peck, J., Theodore, N., & Brenner, N. (2009). Neoliberal urbanism: Models, moments, mutations. *SAIS Review*, *29*(1).

Pedersen, Å., Ø., Nyhuus, S., Blindheim, T., & Wergeland Krog, O. M. (2004). Implementation of a GIS-based management tool for conservation of biodiversity within the municipality of Oslo, Norway. *Landscape and Urban Planning*, *68*, 429–438.

Pelletier, M. P., Trépanier, M., & Morency, C. (2009). Smart card data in public transit planning: A review. *Cirrelt*, *46*, 1–19.

Pengfei, N., & Qinghu, H. (2006). *Comparative research on the global urban competitiveness*. Beijing, China: Chinese Academy for Social Sciences.

Pereira, A. M., & Roca-Sagalés, O. (2003). Spillover effects of public capital formation: Evidence from the Spanish regions. *Journal of Urban Economics*, *53*(2), 238–256. doi:10.1016/S0094-1190(02)00517-X

Perlmutt, D. (2001). Bank of America executive to retire after 23 years. *Charlotte Observer*. Retrieved January 31, 2001, from http://www.charlotte.com/ mld/charlotte/archives/

Perry, M., & Bodkin, C. (2000). Content analysis of Fortune 100 company Web sites. *Corporate Communications: An International Journal*, *5*(2), 87–96. doi:10.1108/13563280010338331

Philo, C., & Kearns, G. (1993). *Culture, history, capital: A critical introduction to the selling of places*. Oxford, UK: Pergamon Press.

Piatetsky-Shapiro, G. (1991). Discovery, analysis, and presentation of strong rules. In Piatetsky-Shapiro, G., & Frawley, J. (Eds.), *Knowledge Discovery in Databases* (pp. 229–248). Menlo Park, CA: AAAI Press.

Pierre, S., Leclerc, G., Revaz, J., Dion, Y., & Rossi, L. (2001). Urbistics: An emerging field. In *Proceedings of World Multiconference on Systemics, Cybernetics and Informatics, (ISAS/SCI 2001)* (vol. 28, pp. 124–126), Orlando, Florida.

Pierre, S., Quintero, A., & Konaré, D. (2001). An integrated system for coordinated management of urban infrastructure: Concepts and architecture. In *Proceedings of World Multiconference on Systemics, Cybernetics and Informatics (ISAS/SCI 2001)* (vol. 28, pp. 133–137), Orlando, Florida.

Pior, M. Y., & Osman, O. (1997). Air pollution assessment applying GIS integrated system. *Journal of the Eastern Asia Society for Transportation Studies*, *2*, 2071–2081.

Pior, M.-Y., & Shimizu, E. (2001). GIS-aided evaluation system for infrastructure improvements: Focusing on simple hedonic and Rosen's two-step approaches. *Computers, Environment and Urban Systems*, *25*, 223–246. doi:10.1016/S0198-9715(00)00018-1

Plassard, F. (1992). L'impact territorial des liaisons à grande vitesse. In Derycke, P. H. (Ed.), *Espace et dynamiques territoriales* (pp. 243–262). Paris, France: Economica.

Plotkin, D. (1997). Carrying freight on high-speed rail lines. *Journal of Transportation Engineering*, *123*(3), 199–201. doi:10.1061/(ASCE)0733-947X(1997)123:3(199)

Polanyi, K. (2001). *1957). Great transformation: The political and economic origins of our time*. Boston, MA: Beacon Press.

Poltras, C. (2009). Downtowns, past and present. *Urban History Review. Revue d'Histoire Urbaine*, *37*(2), 3.

Porter, M., & Kramer, M. (2006, December). Strategy and society: The link between competitive advantage and corporate social responsibility. *Harvard Business Review*, 2–15.

Porter, M. E. (1995). The competitive advantage of the inner city. *Harvard Business Review*, *74*(6), 61–78.

Porter, M. (1990). *The comparative advantage of nations*. New York, NY: Free Press.

Porter, M. E. (2000). Location, competition, and economic development: Local clusters in a global economy. *Economic Development Quarterly*, *14*(1), 15–34. doi:10.1177/089124240001400105

Postorino, M. N., Barrile, V., & Cotroneo, F. (2006). Surface movement ground control by means of a GPS-GIS system. *Journal of Air Traffic Management*, *12*(6), 375–381. doi:10.1016/j.jairtraman.2006.09.003

Postorino, M. N. (Ed.). (2010). *Developments of regional airports: Theoretical Analyses and Case Studies* (pp. 29-53). Boston, MA: WIT Press. RITA. (2010-2014). *ITS Strategic Research Plan 2010-2014*. US Department of Transportation, Research and Innovative Technology Administration. Retrieved from www.its.dot.gov/strat_plan

Prats-Planagumà, L., & Camprubí, R. (2009). E-tourism image: The relevance of networking for Web sites destination marketing. In Gascó-Hernández, M., & Torres-Coronas, T. (Eds.), *Information communication technologies and city marketing: Digital opportunities for cities around the world* (pp. 184–202). Hershey, PA: Information Science Publishing. doi:10.4018/978-1-60566-134-6.ch009

PRC Ministry of Railways. (2010). *Planning of the National Railway Network* (in Chinese). Retrieved July 13, 2010, from http://www.china-mor.gov.cn/

Priemus, H., & Konings, R. (2001). Light rail in urban regions: What Dutch policymakers could learn from experiences in France, Germany And Japan. *Journal of Transport Geography*, *9*(3), 187–198. doi:10.1016/S0966-6923(01)00008-4

Priemus, H., Nijkamp, P., & Banister, D. (2001). Mobility and spatial dynamics: An uneasy relationship. *Journal of Transport Geography*, *9*(3), 167–171. doi:10.1016/S0966-6923(01)00007-2

Pryor, S., & Grossbart, S. (2007). Creating meaning on main street: Towards a model of place branding. *Place Branding and Public Diplomacy*, *3*(4), 291–304. doi:10.1057/palgrave.pb.6000080

Quintero, A., & Konaré, P. S. (2005). Prototyping an intelligent decision support system for improving urban infrastructures management. *European Journal of Operational Research*, *162*, 654–672. doi:10.1016/j.ejor.2003.10.019

Rakodi, C. (2003). Politics and performance: The implications of emerging governance arrangements for urban management approaches and Information Systems. *Habitat International*, *27*(4), 523–547. doi:10.1016/S0197-3975(03)00004-3

Rambaldi, G., Kyem, P., McCall, M., & Weiner, D. (2006). Participatory spatial information management and communication in developing countries. *The Electronic Journal on Information Systems in Developing Countries*, *25*, 2006.

Rankl, W., & Effing, W. (2002). *Smart cards handbook*. West Sussex, UK: Wiley.

Reader, J. (2004). *Cities: A swift examination of the pleasures and problems of city life*. New York, NY: Grove Press.

Reggiani, A. (1998). Accessibility, trade and locational behaviour: An introduction. In Reggiani, A. (Ed.), *Accessibility, trade and locational behaviour* (pp. 1–14). Aldershot, UK: Ashgate.

Regional Studies. (2004). Special issue on regional competitiveness. *Regional Studies, 38/39*.

Regione del Veneto. (2001). *Il sistema ferroviario metropolitano regionale dell'area centrale veneta*. Este, Italy: Grafica Atestina.

Regione del Veneto. (2004). *Piano regionale dei trasporti del veneto: Quaderno di sintesi*. Venezia, Italy: Segreteria Regionale Infrastrutture e Mobilitá.

Resor, R., Blaze, J., & Morlok, E. (2004). Short-haul rail intermodal: Can it compete with trucks? *Transportation Research Record*, *1873*, 45–52. doi:10.3141/1873-06

Reusser, D. E., Loukopoulos, P., Stauffacher, M., & Scholz, R. W. (2008). Classifying railway stations for sustainable transitions - Balancing node and place functions. *Journal of Transport Geography*, *16*(3), 191–202. doi:10.1016/j.jtrangeo.2007.05.004

Reuters. (2007). *U.S. blames airline, FAA for deadly Florida crash*. Retrieved November 30, 2010, from http://www.reuters.com/ article/idUSN30280 19120070530

Rezende, D. (2005). *Planejamento de Informações Públicas Municipais: Guia para Planejar Sistemas de Informação, Informática e Governo Eletrônico nas Prefeituras e Cidades* (1st ed.). São Paulo, Brazil: Atlas.

Rice, J. A., & Spencer, B. F., Jr. (2009). *Flexible smart sensor framework for autonomous full-scale structural health monitoring* (NSEL Report No. 18). Urbana-Champain, IL: University of Illinois at Urbana-Champaign. Retrieved from http://hdl.handle.net/ 2142/13635

Ridley, J., & Jones, L. (2002). User and public involvement in health services: A literature Review. Edinburgh, Scotland: Partners in Change, SHS Trust.

Rietveld, P., & Bruinsma, F. (1998). *Is transport infrastructure effective?*Berlin, Germany: Springer-Verlag.

Ritsema van Eck, J., & Jong, T. (1999). Accessibility analysis and spatial competition effect in context of GIS-supported service location planning. *Journal of Computers. Environment and Urban System*, *23*(2), 75–89. doi:10.1016/S0198-9715(99)00016-2

Robinson, J. (2003). Future subjunctive: Backcasting as social learning. *Futures*, *35*(8), 839–856. doi:10.1016/S0016-3287(03)00039-9

Robinson, J. (2005). *Eurocult21 Integrated Report. Helsinki: Book on Demand, Lasipalatsi Media Centre* (p. 114). Retrieved from http://eurocul21.org

Rogerson, R. J. (1999). Quality of life and city competitiveness. *Urban Studies (Edinburgh, Scotland)*, *36*(5-6), 969–985. doi:10.1080/0042098993303

Rotmans, J. (2005). *Societal innovation: Between dream and reality lies complexity*. Rotterdam, The Netherlands: Erasmus University Rotterdam.

Rowe, G., & Frewer, L. J. (2009). Public participation methods: A framework for evaluation. *Science, Technology & Human Values*, *25*(1), 3–29. doi:10.1177/016224390002500101

Rowe, G., & Wright, G. (1999). The delphi technique as a forecasting tool: Issues and analysis. *International Journal of Forecasting*, *15*(4), 353–375. doi:10.1016/S0169-2070(99)00018-7

Roy, R. (2003). European versus national-level evaluation: The case of the PBKAL high-speed rail project. In Pearman, A., Mackie, P., & Nellthorp, J. (Eds.), *Transport projects, programmes and policies: Evaluation needs and capabilities* (pp. 125–138). Aldershot, UK: Ashgate.

Rubenstein-Montano, B. (2000). A survey of knowledge-based Information Systems for urban planning: Moving towards knowledge management. *International Journal of Computers Environment and Urban systems, 24*, 155-172.

Russo, A. P. (1999). Venice: Coping with culture vultures. *The UNESCO Courier*. *Ano*, *57*, 42–43.

Sanchez, T. W. (1993). *Measuring the impact of highways and roads on residential property values with GIS: An empirical analysis of Atlanta*. In R. E. Klosternman, & S. P. French (Eds.), 3rd International Conference on Computers in Urban Planning and Management (vol. 2).

Sands, G., & Reese, L. A. (2008). Cultivating the creative class: And what about Nanaimo? *Economic Development Quarterly*, *22*(1). doi:10.1177/0891242407309822

Santos, M. (1985). *Espaço e Método*. São Paulo, Brasil: Nobel.

Santos, M. (1996). *A natureza do espaço, técnica e tempo, razão e emoção*. São Paulo, Brasil: Editora Hucitec.

Sasaki, K., Ohashi, T., & Ando, A. (1997). High-speed rail transit impact on regional systems: Does the shinkansen contribute to dispersion? *The Annals of Regional Science*, *31*(1), 77–98. doi:10.1007/s001680050040

World Health Organisation (WHO). (1999). Health: Creating healthy cities in the 21st century. In Satterthwaite, D. (Ed.), *The earthscan reader in sustainable cities* (pp. 137–172). London, UK: Earthscan.

Saxenian, A. L. (1990). Regional networks and the resurgence of Silicon Valley. *California Management Review*, *33*(1), 89–112.

Saxenian, A. L. (1996). *Regional advantage: Culture and competition in Silicon Valley and Route 128*. Cambridge, MA: Harvard University Press.

Saxenian, A. L. (2007). *The new Argonauts: Regional advantage in a global economy*. Cambridge, MA: Harvard University Press.

Schneider, J. B. (1993). Selecting and evaluating intermodal stations for intercity high-speed ground transportation. *Transportation Quarterly*, *47*(2), 221–245.

Schulz, A., & Northridge, M. E. (2004). Social determinants of health: Implications for environmental health promotion. *Health Education & Behavior*, *31*(4), 455–471. doi:10.1177/1090198104265598

Schwab, K. (2010). *The Global Competitiveness Report 2009 – 10*. Geneva, Switzerland: World Economic Forum.

Scott, A. J., & Storper, M. (Eds.). (1992). *Pathways to industrialization and regional development*. Oxford, UK: Routledge.

Scott, A. (1999). The cultural economy: Geography and the creative field. *Media Culture & Society*, *21*, 807–817. doi:10.1177/016344399021006006

Segel, A. I. (2005). *New Songdo City* (Harvard Business School case no. 9-206-019). Boston, MA: Harvard Business School Publishing.

Selwyn, N. (2002). E-stablishing an inclusive society: Technology, social inclusion and UK government policy. *Journal of Social Policy*, *21*(1), 1–20.

Sheffi, Y. (1985). *Urban transportation networks*. Englewood Cliffs, NJ: Prentice Hall.

Shim, J. P., Warkentin, M., Courtney, J. F., Power, D. J., Sharda, R., & Carlsson, C. (2002). Past, present, and future of decision support technology. *Decision Support Systems*, *33*(2), 111–126. doi:10.1016/S0167-9236(01)00139-7

Shin, D. H. (2009). Ubiquitous city: Urban technologies, urban infrastructure and urban informatics. *Journal of Information Science*, *35*(5), 515. doi:10.1177/0165551509100832

Short, J. R., Benton, L. M., Luce, W. B., & Walton, J. (1993). Reconstructing the image of an industrial city. *Annals of the Association of American Geographers. Association of American Geographers*, *83*(2), 207–224. doi:10.1111/j.1467-8306.1993.tb01932.x

Siderelis, K. C. (1991). Land resource Information Systems. In Maguire, D., Goodchild, J., & Rihind, D. W. (Eds.), *Geographical Information System: Principles and applications* (*Vol. 2*). New York, NY: Longman.

Siemens, GlobeScan, & MRC McLean Hazel. (2007). *Megacity challenges: A stakeholder perspective*. Munich, Germany: Siemens.

Siemens. (2010). *ICT for city management: Using information and communications technology to enable, engage, and empower city stakeholders*. South Africa: SIS Global Center of Competence for City Management, sponsored by Siemens AG.

Sim, L., Ong, S., Adeesh, A., Parsa, A., & Keivani, R. (2003). Singapore's competitiveness as a global city: Development strategy, institutions and business environment. *Cities (London, England)*, *20*(2), 115–127. doi:10.1016/S0264-2751(02)00119-1

Simeon, R. (1999). Evaluating domestic and international Web site strategies. *Internet Research: Electronic Networking Applications and Policy*, *9*(4), 297–308. doi:10.1108/10662249910286842

Simeon, R. (2001). Evaluating the branding potential of Web sites across borders. *Marketing Intelligence & Planning*, *19*(6), 418–424. doi:10.1108/02634500110405432

Simmons, G. J. (2007). I-Branding: Developing the Internet as a branding tool. *Marketing Intelligence & Planning*, *25*(6), 544–562. doi:10.1108/02634500710819932

Sivakumaran, K., Lu, X. L., & Hanson, M. (2010). *The use of passenger transit infrastructure for goods movement: A bay area economic feasibility study*. Paper presented at the 89th Transportation Research Board Annual Meeting.

Slobodian, B. (2000). *Spatial analysis*. Retrieved 9/4, 2000 from http://www.mala.bc.ca/~slobodian /concepts/GC_20_8.htm

Sohn, J., Kim, T. J., & Hewings, G. (2004). Intra-metropolitan agglomeration, Information Technology, and polycentric urban development. In Cappelo, R., & Nijkamp, P. (Eds.), *Contributions to economic analysis*. Amsterdam, The Netherlands: Elsevier.

Spatial Vision. (2010). *Spatially enabling government: 'Healthtracks' wins 2010 WA Spatial Excellence Award*. Retrieved November 30, 2010, from http://www.spatialvision.com.au /index.php/latest-news -spatially-enabling-government.html

Spiekermann, K., & Wegener, M. (1994). The shrinking continent - New time-space maps of Europe. *Environment and Planning. B, Planning & Design*, *21*(6), 653–673. doi:10.1068/b210653

Starner, R. (2006). Energizing the Internet. *Site Selection*, *51*(5), 674.

Stern, E., Gudes, O., & Svoray, T. (2009). Web-based traditional public participation comprehensive planning: A comparative study. *Environment and Planning*, *36*(6), 1067–1085. doi:10.1068/b34113

Stoker, G. (1998). *Public-private partnerships and urban governance*. Basingstoke, UK: MacMillan.

Storper, M., & Venables, A. J. (2004). Buzz: Face-to-face contact and the urban economy. *Journal of Economic Geography*, *4*(4), 351–370.

Straatemeier, T. (2008). How to plan for regional accessibility. *Transport Policy*, *15*(2), 127–137. doi:10.1016/j.tranpol.2007.10.002

Straight Up Search. (2008, April 29). *Google builds data center in small town, North Carolina*. Retrieved from http://www.straightup search.com/archives /2008/04/ google_builds_d.html

Studies, U. (1999). Special issue on competitive cities. *Urban Studies (Edinburgh, Scotland)*, *36*(5-6).

Sui, D. Z. (1998). GIS based urban modeling: Practices, problems and prospects. *International Journal of Geographical Information Science*, *12*(7), 651–671. doi:10.1080/136588198241581

Sweedberg, C. (2010, November 18). Driscoll's monitors its berry shipments in real time. *RFID Journal*. Retrieved from http://www.rfidjournal.com/ article/view/8011/

Swyngedouw, E., Moulaert, F., & Rodriguez, A. (2002). Neoliberal urbanization in Europe: Large-scale urban development projects and the new urban policy. *Antipode*, *34*(3). doi:10.1111/1467-8330.00254

Symonds, M. (2000). The next revolution: A survey of government and the Internet. *The Economist*. Retrieved from http://www.economist.com/ surveys/ showsurvey.cfm?issue= 20000624

Tapia, E. M., Intille, S. S., & Larson, K. (2004). Activity recognition in the home using simple and ubiquitous sensors. In A. Ferscha & F. Mattern (Eds.), *Pervasive computing: Proceedings of the Second International Conference on Pervasive Computing* (pp. 158-175). Berlin: Springer.

Tapiador, F. J., Burckhart, K., & Marti-Henneberg, J. (2009). Characterizing European high speed train stations using intermodal time and entropy metrics. *Transportation Research Part A, Policy and Practice*, *43*(2), 197–208. doi:10.1016/j.tra.2008.10.001

TCRP. (2007). *Bus Rapid Transit Practitioner's Guide (TCRP Rep. 18)*. Washington, DC: Transportation Research Board.

Teague, P. (2010, July 30). *Dacentec makes official announcement of new data center*. Retrieved from http:// www.newstopic.net/ printer_friendly/ 8933471

Teleplay. (2002). *Telepayment system for multimodal transport services using portable phones*. Telepay Conference, Lyon, France. Retrieved from www.ertico.com/ download/ telepay_documents/ 2_TELEPAY.pdf

Thompson, E. R. (2003). A grounded approach to identifying national competitive advantage: A preliminary exploration. *Environment & Planning A*, *35*, 631–657. doi:10.1068/a35110

Thrift, N. (2000). Performing cultures in the new economy. *Annals of the Association of American Geographers. Association of American Geographers*, *90*(4), 674–692. doi:10.1111/0004-5608.00217

Torchin, F., Grilly, D., Combes, S., Hasiak, S., & Menerault, P. (2008). *High speed rail for regional transport: Case studies in European countries*. Paper presented at the European Transport Conference.

Torres, J., Izquierdo, A., & Sierra, J. M. (2007). Advances in network smart cards authentication. *Computer Networks*, *51*, 2249–2261. doi:10.1016/j.comnet.2007.01.010

Transmodel. (2001). *Transmodel: The reference data model for public transport*. Retrieved from http://www.transmodel.org/ en/cadre1.html

Transportation Research Board. (2000). TRB publications index. *Highway Capacity Manual*, *193*, 22–23.

TRB. (2008). *AVL systems for bus transit: Update (TCRP synthesis 73)*. Washington, DC: Transportation Research Board.

Trippl, M., Tödtling, F., & Lengauer, L. (2009). Knowledge sourcing beyond buzz and pipelines: Evidence from the Vienna software sector. *Economic Geography*, *85*(4), 443–462. doi:10.1111/j.1944-8287.2009.01047.x

Troin, J. F. (1995). *Rail et aménagement du territoire, des héritages aux nouveaux défis*. Aix-en-Provence: Edisud.

Truffer, B., Metzner, A., & Hoogma, R. (2002). The coupling of viewing and doing: Strategic niche management and the electrification of individual transport. *Greener Management International*, *37*, 111–124.

Tschirhart, M. (2008). Evaluation of brand use on city government websites: A search for brand type, repetition, and consistency. *Journal of Nonprofit & Public Sector Marketing*, *19*(1), 35–53. doi:10.1300/J054v19n01_02

Turban, E. (1993). *Decision support and expert systems: Management support systems*. Upper Saddle River, NJ: Prentice Hall PTR.

Turok, I. (2004). Cities, regions and competitiveness. *Regional Studies*, *38*(9), 1069–1083. doi:10.1080/0034340042000292647

Turok, I., & Bailey, N. (2004). Twin track cities? Competitiveness and cohesion in Glasgow and Edinburgh. *Progress in Planning*, *62*(3), 135–204. doi:10.1016/j.progress.2004.04.001

Urban, F. (2002). Small town, big Web site? Cities and their representation on the Internet. *Cities (London, England)*, *19*(1), 49–59. doi:10.1016/S0264-2751(01)00045-2

Urena, J. M., Menerault, P., & Garmendia, M. (2009). The high-speed rail challenge for big intermediate cities: A national, regional and local perspective. *Cities (London, England)*, *26*(5), 266–279. doi:10.1016/j.cities.2009.07.001

Van den Berg, L., & Braun, E. (1999). Urban competitiveness, marketing and the need for organising capacity. *Urban Studies (Edinburgh, Scotland)*, *36*(5-6), 987. doi:10.1080/0042098993312

van Exel, J., Rienstra, S., Gommers, M., Pearman, A., & Tsamboulas, D. (2002). EU involvement in TEN development: Network effects and European value added. *Transport Policy*, *9*(4), 299–311. doi:10.1016/S0967-070X(02)00016-1

van Ham, J., & Muilerman, G. (2002). *Opportunities for time-sensitive intermodal rail freight*. Paper presented at the International congress on Freight Transport Automation and Multimodality.

Van Winden, W., Carvalho, L., Van Tuijl, E., Van Haaren, J., & Van den Berg, L. (2011). *Creating Knowledge Locations in Cities*. Rotterdam, The Netherlands: Erasmus University Rotterdam, Euricur.

Van Winden, W., Van den Berg, L., & Pol, P. (2007). European cities in the knowledge economy: Towards a typology. *Urban Studies (Edinburgh, Scotland)*, *44*(3), 525–549. doi:10.1080/00420980601131886

Van Winden, W., & Carvalho, L. (2009). Exploration and exploitation networks in space: The case of the shipbuilding cluster of Turku, Finland. In Mahnken, G., & Matthiesen, U. (Eds.), *Das Wissen der Städte*. Wiesbaden, Germany: VS Verlag. doi:10.1007/978-3-531-91648-4_25

Van Winden, W. (2006). Globalisation and urban competitiveness: Challenges for different types of urban regions. In OECD (Ed.), *Competitive cities in the global economy*. Paris, France: OECD.

Vickerman, R. (1997). High-speed rail in Europe: Experience and issues for future development. *The Annals of Regional Science*, *31*(1), 21–38. doi:10.1007/s001680050037

Vickerman, R. (2008). Transit investment and economic development. *Research in Transportation Economics*, *23*(1), 107–115. doi:10.1016/j.retrec.2008.10.007

Vickerman, R., Spiekermann, K., & Wegener, M. (1999). Accessibility and economic development in europe. *Regional Studies*, *33*(1), 1–15. doi:10.1080/00343409950118878

Virtsonis, N., & Haridge-March, S. (2008). Website elements in B2B online communications: A case from the UK print industry. *Marketing Intelligence & Planning*, *26*(7), 699–718. doi:10.1108/02634500810916672

Vitiello, R., & Willcocks, M. (2006). The difference is in the detail: Its potential as a place branding tool and impact upon perceptions and responses. *Place Branding*, *2*(3), 248–262. doi:10.1057/palgrave.pb.5990061

Vizard, M. (2010, June 8). *IBM helps build smarter city in Florida*. Retrieved November 30, 2010, from http://www.itbusinessedge.com/ cm/blogs/vizard/ibm-helps-build-smarter-city-in-florida/

Voskuhl, D. E. F. (1995). Interlinking the region with its centre: The example of the Karlsruhe region in Germany. *Journal of Transport Geography*, *3*(4), 281–285. doi:10.1016/0966-6923(95)00026-7

Vuchic, R. V., & Casello, J. M. (2002). An evaluation of maglev technology and its comparison with high speed rail. *Transportation Quarterly*, *56*(2), 33–49.

Wade, R. (1990). *Governing the market: Economic theory and the role of government in East Asian industrialization*. Princeton, NJ: Princeton University Press.

Waitt, G. (1999). Playing games with Sydney: Marketing Sydney for the 2000 Olympics. *Urban Studies (Edinburgh, Scotland)*, *36*(7), 1055–1077. doi:10.1080/0042098993097

Wan, C. S. (2002). The Web sites of international tourist hotels and tour wholesalers in Taiwan. *Tourism Management*, *23*(2), 155–160. doi:10.1016/S0261-5177(01)00048-6

Wang, H., Song, Y., Hamilton, A., & Curwell, S. (2007). Urban information integration for advanced e-planning in Europe. *Government Information Quarterly*, *24*, 736–754. doi:10.1016/j.giq.2007.04.002

Wardman, M., & Tyler, J. (2000). Rail network accessibility and the demand for inter-urban rail travel. *Transport Reviews*, *20*(1), 3–24. doi:10.1080/014416400295310

Warf, B. (2001). Segueways into cyberspace: Multiple geographies of the digital divide. *Environment and Planning. B, Planning & Design*, *28*(1), 3–19. doi:10.1068/b2691

Webster, D., & Muller, L. (2000). *Urban competitiveness assessment in developing country urban regions: The road forward.* Paper presented at the Urban Group, INFUD, World Bank, Washington, DC.

Wegener, M., & Fürst, F. (1999). *Land-use transport interaction: State of the art*. Dortmund, Germany: Institut für Raumplanung.

Weisbrod, G. (2008). Models to predict the economic development impact of transportation projects: Historical experience and new applications. *The Annals of Regional Science*, *42*(3), 519–543. doi:10.1007/s00168-007-0184-9

Weisbrod, G., & Treyz, F. (1998). Productivity and accessibility: Bridging project specific and macroeconomic analyses of transportation investments. *Journal of Transportation and Statistics*, *1*(3), 65–79.

Weiser, M. (1991). The computer for the 21st century. *Scientific American*, *265*(3), 94–104. doi:10.1038/scientificamerican0991-94

Wells, A. T. (2000). *Airport planning & management* (4th ed.). New York, NY: McGraw-Hill.

Wenger, E. (1998). *Communities of practice: Learning, meaning, and identity*. Cambridge, MA: Cambridge University Press.

Wenger, E. (2000). Communities of practice and social learning systems. *Organization*, *7*(2), 225–246. doi:10.1177/135050840072002

Wickramasinghe, N. S., Fadlalla, A. M. A., Geisler, E., & Schaffer, J. L. (2005). A framework for assessing e-health preparedness. *International Journal of Electronic Healthcare*, *1*(3), 316–334. doi:10.1504/IJEH.2005.006478

Williams, G. (1999). Metropolitan governance and strategic planning: A review of experience in Manchester, Melbourne and Toronto. *Progress in Planning*, *52*(1), 1–100. doi:10.1016/S0305-9006(99)90003-X

Willigers, J., & Floor, H. (2007). Accessibility indicators for location choices of offices: An application to the intraregional distributive effects of high-speed rail in the Netherlands. *Environment & Planning A*, *39*(9), 2086–2098. doi:10.1068/a3913

Wilson, D., & Atkeson, C. (2005). Simultaneous tracking & activity recognition (STAR) using many anonymous binary sensors. In *Pervasive computing: Proceedings of the Third International Conference on Pervasive Computing* (pp. 62-79). Berlin: Springer.

Wind, J., & Rangaswamy, A. (2001). Customerization: The next revolution in mass customization. *Journal of Interactive Marketing*, *15*(1), 13–32. doi:10.1002/1520-6653(200124)15:1<13::AID-DIR1001>3.0.CO;2-#

Winstanley, A. C. (2009). Spatial databases. *International encyclopedia of human geography* (pp. 345-347).

Wireman, T. (1990). *World class maintenance management*. New York, NY: Industrial Press.

World Health Organisation (WHO). (2008). Framework and standards for country health Information Systems (2nd ed.). *Health metrics network*. Geneva, Switzerland: Health Metrics Network.

Worldtravels.com. (2010). *South Korea map*. Retrieved August 12, 2010, from http://www.wordtravels.com/Travelguide/Countries/ South+Korea/Map

Wyatt, D., Philipose, M., & Choudhury, T. (2005). Unsupervised activity recognition using automatically mined common sense. *American Association for Artificial Intelligence*, *1*, 21–27.

Xi, Y., Zhuang, Y., Huang, W., She, C., & Zhang, Z. (2007). The quality assessment and content analysis of corporate Web sites in China: An empirical study. *International Journal of Information Technology & Decision Making*, *6*(2), 389–405. doi:10.1142/S0219622007002472

Xu, W., Bodik, P., & Patterson, D. (2004). *A flexible architecture for statistical learning and data mining from system log streams*. Paper presented at the Workshop on Temporal Data Mining: Algorithms, Theory and Applications at the 4th IEEE International Conference on Data Mining (ICDM '04), Brighton, UK.

Yang, C. W., & Sung, Y. C. (2010). Constructing a mixed-logit model with market positioning to analyze the effects of new mode introduction. *Journal of Transport Geography*, *18*, 175–182. doi:10.1016/j.jtrangeo.2009.01.005

Yeh, A. G., & Chow, M. H. (1996). An integrated GIS and location-allocation approach to public facilities planning Ð an example of open space planning. *International Journal of Computers. Environment and Urban Systems*, *20*, 339–350. doi:10.1016/S0198-9715(97)00010-0

Yigitcanlar, T., Velibeyoglu, K., & Baum, S. (2008). *Knowledge-based urban development: Planning and applications in the information era*. Hershey, PA: Information Science Reference. doi:10.4018/978-1-59904-720-1

Zanfei, A. (2000). Transnational firms and the changing organization of innovative activities. *Cambridge Journal of Economics*, *24*(5), 515–542. doi:10.1093/cje/24.5.515

Zhang, L., & Zhao, S. X. (2009). City branding and the Olympic effect: A case study of Beijing. *Cities (London, England)*, *26*(5), 245–254. doi:10.1016/j.cities.2009.05.002

About the Contributors

Melih Bulu has worked at various sections of the private sector as both a professional and an entrepreneur. While doing these, he has also worked on competitiveness related projects for cities during the last decade as an Academician and a Consultant. He has visited various cities to examine their competitiveness, in terms of their infrastructures and for consultancy purposes. Since 2004, Dr. Bulu has been the General Coordinator of International Competitiveness Research Institute (URAK), an NGO working on economic competitiveness of cities and countries. He is leading various competitiveness related projects on cities in URAK. He especially likes working on the improvement of cities in order to make them competitive. Dr. Bulu teaches strategy related courses at Istanbul Sehir University. His main interest areas are city competitiveness, regional development, game theory, cluster theory, and organization theory. He has various publications in academic and popular media.

* * *

Lara Brunello, an Architect and Researcher, was born in Vicenza (Italy) in 1973. During her school years, she has moved several times from the north to the south and from Eastern to Western Italy, not to exclude a few years abroad, such as in Wisconsin (USA). She received her degree in Architecture in Venice (Italy), where she had lived for 15 years before moving to Brisbane (Australia) to complete a PhD course of studies in transport research at the Queensland University of Technology. Now settled in Pordenone (Italy), her interests range from architecture to infrastructure and planning, with vivid attention to accessibility issues linked to the transportation and land use system.

Ahmet Bulut received the Bachelor of Engineering degree from Bilkent University, Ankara, Turkey, and the PhD degree from the University of California, Santa Barbara. He worked as a Senior Research Scientist on the GotoMeeting product line platform improvements at Citrix Systems, and as a Senior Researcher at Like.com on A/B framework development and research. His current research interests include platforms enabling infrastructure as a service, online business optimization using A/B testing frameworks, applications of wireless sensor networks, and informed policy making in smart cities using an Information and Communication Technologies. He is a member of the IEEE.

Jonathan Bunker is a Senior Lecturer in Civil Engineering with the School of Urban Development at Queensland University of Technology (QUT) and researches in traffic engineering, transport planning, airports, highway engineering and infrastructure asset management, and sustainable development practices. Jonathan has a Bachelor of Engineering (Civil) with Honours 1 and a Doctor of Philosophy

(QUT). Prior to joining faculty at QUT he practiced as a Consulting Transport Engineer in Australia and the United States of America. Jonathan is presently the National President of the Australian Institute of Traffic Planning and Management, a Fellow of AITPM, and a Member of the Institution of Engineers Australia.

Ayça Cangel completed her undergraduate education at University of Bahçeşehir, obtaining Bachelor Degrees in Industrial Engineering and Computer Engineering. Since 2004, she has participated in several projects on competitiveness at micro, meso and macro level. She currently holds the position of Projects Coordinator of International Competitiveness Research Institute (URAK), a Non-Governmental organization aiming to foster competitiveness of cities and countries. Since 2006, Ayça Cangel has also worked on "City Competitiveness Index of Turkey", a yearly publication of URAK that evaluates the competitiveness levels of Turkey's cities. She has several publications both in academic and non-academic media. She is particularly interested in cluster theory, regional development, city competitiveness, and optimization.

Luís Carvalho is a Researcher at the European Institute for Comparative Urban Research, specialized in the fields of urban management, regional development, and economic geography. During the last few years, he has been looking at the dynamics between global knowledge networks and localized innovation. He currently participates in two large international comparative studies on the regional trajectories to the knowledge economy and development of new urban knowledge quarters. Luís is the Coordinator of the module "Managing the Urban Economy" in the MSc "Management of the European Metropolitan Region" at Erasmus University Rotterdam. He recently published a book on the role of "manufacturing in the new urban economy" and he has done comparative urban research in several European cities, Latin America, and East Asia.

Mark Deakin is a Professor in the School of Engineering and Built Environment, Edinburgh Napier University. Mark's recent research has been inter-disciplinary, cutting across academic, scientific and technical boundaries within the built environment. Much of his research has explored the relationship between ICTs, information society, and the knowledge economy. This research has been project-based, focusing on the networked communities, social informatics and economic instruments of two scientific and technical developments: those known as IntelCities and SmartCities. Over the past five years, Mark has acted as Principle Investigator for the following projects funded under the European Commission's IST programme: the IntelCity Roadmap, IntelCities, and SmartCities. He is currently Director of the SCRAN (SmartCities (inter) Regional Academic Network) and works with his counterparts in the Universities of Amsterdam and Turin to exploit the knowledge-transfer potential of the scientific and technical applications developed by this organisation.

Fábio Duarte is an Urban Planner, and has coordinated the consultant team for the urban motility plan of Curitiba. Duarte has a PhD in Communications. A Professor at the Post-Graduation Program on Urban Management at the Pontifical Catholic University, Curitiba, Brazil, Duarte has been a Visiting Professor at the Polytechnic University of Cataluña, Barcelona, and received the Faculty Enrichment Award, from the Canadian government. Duarte has co-edited the book ICTs for mobile and ubiquitous urban infrastructures: Surveillance, locative media and global networks, IGI, 2010.

Sandro Fabbro, Urban and Regional Planner, was awarded a PhD in territorial public policies at the University IUAV of Venice (Italy) in 1993. Professor of Urban and Regional Planning at the Faculty of Engineering of the University of Udine (Italy), he is an Author of more than 130 publications, several of them on international books and journals. Currently he is President of the National Commission for Infrastructural Policies of the INU, Italian National Town Planning Institute, and Advisor of the Italian Ministry of Infrastructure and Transport.

Renzo Ferrara, born in 1949, graduated in Architecture in Venice, Italy. Collaborating with Rete Ferroviaria Italiana until 2010, Renzo has worked for the upgrading of several lines in the Venetian railway compartment. In particular, he contributed to the layout of the strategic plan for the node of Mestre (Italy) and supervised the construction of the new high speed rail line between Mestre and Padua (Italy). Now working as Independent Consultant, Renzo participates in studies and projects regarding road and rail infrastructures, ports and bridges, as well as co-authoring journal articles and supervising students at the University IUAV of Venice.

Ori Gudes holds a MA in planning and GIS from Ben-Gurion University of the Negev, Israel. He has a diverse knowledge of GIS and over eight years of work experience in private consulting, government, and academia in Israel. He is currently a Research Fellow and GIS Specialist at Griffith University and a PhD candidate at the School of Urban Development, Queensland University of Technology Brisbane, Australia. The main foci of his research are developing Web-based decision support systems for urban planning and healthy communities' promotion. These two broad research areas are clustered around several themes: Web-based decision support systems, urban development, public participation, and health planning. Mr. Gudes received an award in 2007 for his MA work from the Israeli Planning Association.

Diana Sarita Hamburger is currently Professor at Federal University of ABC. She teaches undergraduate courses in Science and Technology, Environmental Engineering, and Urban Engineering. She has a PhD in Engineering with emphasis in Spatial Information from the University of São Paulo (2001), a Masters in Remote Sensing at the National Institute for Space Research (1993), a Degree in Geography from the University of São Paulo (in 1986), and a Degree in Public Administration from Fundação Getulio Vargas - SP (in 1985). Dr. Hamburger has experience in Geography with emphasis in Remote Sensing and GIS (Geographical Information System). Her special interests areas are field-based projects in urban network, geographic Information Systems, and distance and spatial analysis.

Jung Hoon Han possesses extensive experience in all aspects of the urban planning process including initiating, designing, managing, completing, and publishing research in both the scholarly and consulting environments. Dr Han has designed and conducted many research projects from a wide range of both discipline-specific and multi-disciplinary perspectives, including the areas of housing market, property economics, urban planning and design, health and built environment, and spatially integrated social science research methods. Over the past 15 years, Dr Han developed a comprehensive knowledge of environmental and of contemporary urban planning issues including housing affordability (particularly for income support recipients and older people), urban consolidation (inner city renewal) vs. decentralisation (Greenfield development), social polarization, and sustainable development in the area of urban planning (in an Australian context). Some of Dr Han's specific works focus on environmental and planning issues, including planning for the improvements of appropriateness, and affordability of housing (mix in type and tenure) in an attempt to produce more equitable cities through the reduction of locational disadvantage. Dr Han currently sits on an editorial board as Member of the *International Journal of City, Culture and Society* published by Elsevier (ISSN: 1877-9166) and has been serving for a Guest Editor of the *International Journal of Knowledge-based Development* (ISSN: 2040-4476).

Elizabeth Kendall completed her PhD in 1997 on the topic of adjustment following traumatic injury, for which she won the Dean's Commendation for Outstanding PhD Thesis in 1998 (UQ). In 1996, she was the first MAIC-funded Research Fellow, focusing on disability and rehabilitation. Later, she became an Associate Director of CONROD. In this role, she has designed and directed complex community-based evaluations and randomised controlled trials of major health reform projects over the last decade, including the Ambulatory Care demonstration projects, the national brain injury working party, the Queensland Self-Management Alliance, the Chronic Disease Place-based Initiative, the Logan-Beaudesert Health Coalition, the Qld Health Self-Management Working Party, the Coordinated Care Trials, the Sharing Healthcare Demonstration project, the Community Rehabilitation Workforce Reform Project, the Chronic Disease Smart State Working Party, the Spinal Injury Response Project on Integrated Services, and the Youngcare Alternative Service Model trial. She is currently managing the implementation and evaluation of the Multicultural Natural Helpers project in Logan. Elizabeth runs a collaborative research program with several significant partners in addition to MAIC, including Queensland Health, General Practice Queensland, Wesley Mission, and Qld Police. She manages a research team of 20 researchers, all focusing on the role of communities in well-being.

Júlia Maria Lourenço, Assistant Professor at the Department of Civil Engineering of the University of Minho, teaches MSc in Urban Engineering and MSc in Civil Engineering. She holds a PhD degree in Civil Engineering (IST/UTL, 1997). Relevant academic roles of Lourenço include Vice-Dean, and Dean of Civil Engineering Studies, making her responsible for the curricula of 1000 students. She is author of over 70 publications. She published a book on Management of Urban Growth that won the 2003 Gerd Albers Award from the International Society of City and Regional Planners. She has been involved in several international research projects at home and abroad in the last 10 years.

Franco Migliorini, an architect and town planner, was born in Udine (Italy) in 1945. Graduating in Venice (Italy) in 1970, Franco has 40 years of professional practice experience in the public sector as well as a private consultant. Active in EU transnational territorial cooperation projects, he has been researching and teaching at the University IUAV of Venice. Franco is Member of INU, Italian National Town Planning Institute; Member of ECTP, European Council of Spatial Planners; permanent Member in the organizing party of European T&C planning Biennials; and an Editor and Co-editor of books and essays on Italian and European planning practice, with interests in urban and regional planning, transport planning, and landscape.

Erkan Özdemir is currently an Assistant Professor at Uludag University in Bursa, Turkey. Dr Özdemir received his PhD from the Uludag University in 2005. His current research interests include e-marketing, online consumer behaviors, high-tech marketing, and marketing ethics. His previously published works include two chapters named Ethics in e-Marketing, and Gender and e-Marketing. These chapters were published by IGI Global. Dr. Özdemir has also published two books and many articles, nationally.

Virendra Pathak has a PhD in Civil Engineering. He has 20 years of experience in academia and is currently associated with the Queensland University of Technology, Australia. He is involved in teaching spatial science related subjects and his research interests are in the area of applications of geospatial technologies for urban planning and development.

Maria Nadia Postorino got a Civil Engineering degree in Transportation at the Mediterranea University of Reggio Calabria with full marks. In 1995, she got her PhD in Transport Engineering at the University of Rome "La Sapienza". From 1994 to 2002, she was an Assistant Professor in Transport Engineering, and since 2002, she has been a Professor in Transport Engineering at the Engineering Faculty of the "Mediterranea" University of Reggio Calabria, where she teaches Transport Systems Theory. Since 1991, she has been involved with many research projects, financed by both national and local public bodies and companies, and with various scientific committees. She has cooperated and still cooperates for many refereed activities, for both national and international reviews and scientific meetings. From March 2008, she has been a component of the "Conseil scientifique de l'école nationale des ponts et chaussées" in Paris, France. Her research interests are mainly focused on: air transport system models; transport demand estimation and SP techniques; performances and impacts on the transportation system; and public transport system analysis and simulation. She has published more than 70 papers in journals and books of national and international interest.

Khan Rubayet Rahaman is a Postdoctoral Researcher in the Department of Civil Engineering at University of Minho, Portugal. Dr. Khan graduated in Urban and Rural Planning from Khulna University, Bangladesh, in 1998. He has completed his MSc in Spatial Planning from the Royal Institute of Technology, Stockholm, in 2002, and his PhD in Urban Engineering from the University of Tokyo in 2006. He has been working as an Expert Team Member on advisory boards of international projects and as an International Consultant as an Urban Development Specialist.

Jacqueline Bueno Sousa is a seasoned Researcher and Editor who has been covering business and economic issues for more than 20 years. A former Writer with The Wall Street Journal, she also was Editor of the Business Monday section of the Miami Herald and Editor-in-Chief of the Daily Business Review. During her career, she has reported on issues ranging from real estate development trends to the economic impact of business regulations. Ms. Sousa is a Research Fellow at Florida International University's Metropolitan Center, an urban think tank that focuses on South Florida's growth needs. At FIU, she is completing graduate work in public administration. Ms. Sousa obtained a Bachelor's of Science in Journalism from the University of Florida, where she graduated with honors.

Gustavo Taniguchi is a Transportation Engineer with a Master degree in Urban Management from the Pontifical Catholic University, Curitiba, Brazil. Taniguchi has been working with Information Technology, designing Intelligent Transport Systems (ITS) for urban environment, developing innovative projects, and providing solutions to cities' traffic and planning problems. His projects include Metro and Bus Rapid Transit (BRT) systems in South America, in which he plays a key role by providing new approaches combining technology and planning tools. Since 2000, he has been specialized in fare collection systems based on smart cards, designing systems for Curitiba and other cities in Brazil and South America.

Rudj Maria Todaro was born in Venice on the 5th of March in 1969. He graduated with a degree in Electronics Engineering from Padua University. He is currently responsible for the Urban Maintenance System of Insula S.p.A., and he is in charge of Design, Direction and Coordination of the Geographic Information System connected with urban maintenance works in Venice. He has built a management system based on geographic objects in order to support the urban maintenance works. These activities also include an environmental monitoring system and an Information Technology system. He worked on the Enterprise Resource Planning (ERP) software applications. He worked as Product Manager on a Web-based application developed for industries related to foreign trade. Before that, he was in Coca-Cola Bevande, Italia, where he was responsible for the activities involving the management and maintenance of Information Systems in Nord-est.

Evren Tok is a collaborative PhD candidate at the School of Public Policy and Administration and Institute of Political Economy at Carleton University, Ottawa/Canada. He obtained his MA degree from the Institute of Political Economy at Carleton University. He is currently working as a Sessional Instructor in the Department of Political Science and Department of Social Work at Carleton University; teaching courses on International Development, Political Economy, Public Opinion, and Politics of Developing Countries. His recent publications include: "G-20 and the new rules of global governance" *International Economic Issues*, Ministry of Foreign Affairs, Turkey (November, 2010); "Rethinking G-20 as a new global actor: Paradoxes, problems and prospects" Center for Research on Globalization and Democratic Governance (GLODEM) Working Paper (October 2010); *A la Turka Neoliberalism: Gecekondus as Turkey's New State Spaces*, Lambert Academic Publishing, Cologne (2009); "Anatolian cities and new spirit of Turkish capitalism" *Turkish Policy Quarterly*, 7(4).

Index

F

G

H

I

K

L

M

N

O

P

Q

R

S

T

U

V

W

CPSIA information can be obtained at www.ICGtesting.com
Printed in the USA
BVOW020924251011
R4277300001B/R42773PG273893BVX3B/2/P

$ 180.00 CI 373 15-11-2011